Multimedia Environmental Models

Multimedia Environmental Models

The Fugacity Approach

Third Edition

J. Mark Parnis and Donald Mackay

CRC Press
Taylor & Francis Group
Boca Raton London New York

CRC Press is an imprint of the
Taylor & Francis Group, an **informa** business

Errata sheet (2022) is available for download at www.routledge.com/9780367507589. The corrected errors will be incorporated into all future printings of the book.

Third edition published 2021
by CRC Press
6000 Broken Sound Parkway NW, Suite 300
Boca Raton, FL 33487-2742

and by CRC Press
2 Park Square, Milton Park, Abingdon, Oxon, OX14 4RN

First edition published by CRC Press 1991
Second Edition published by CRC Press 2001

CRC Press is an imprint of Taylor & Francis Group, LLC

Library of Congress Cataloging-in-Publication Data
Names: Mackay, Donald, 1936- author.
Title: Multimedia environmental models : the fugacity approach / Donald
Mackay, Mark Parnis.
Description: Third edition. | Boca Raton, FL : CRC Press, 2020. | Includes
bibliographical references and index.
Identifiers: LCCN 2020010637 | ISBN 9780367407827 (hardback) |
ISBN 9780367809829 (ebook)
Subjects: LCSH: Organic compounds—Environmental aspects—Mathematical
models. | Cross-media pollution—Mathematical models.
Classification: LCC TD196.O73 .M33 2020 | DDC 628.5/2015118—dc23
LC record available at https://lccn.loc.gov/2020010637

ISBN: 978-0-367-40782-7 (hbk)
ISBN: 978-0-367-50758-9 (pbk)
ISBN: 978-0-367-80982-9 (ebk)

Typeset in Times
by codeMantra

Contents

Preface

This book is about the behavior of organic chemicals in our multimedia environment or biosphere of air, water, soil, and sediments, and the diversity of biota that reside in these media. It is a response to the concern that we have unwisely contaminated our environment with a large number of chemicals in the mistaken belief that the environment's enormous capacity to dilute and degrade will reduce concentrations to negligible levels. We now know that the environment has only a finite capacity to dilute and degrade. Certain chemicals have persisted and accumulated to levels that have caused adverse effects on wildlife and even humans. Some chemicals have the potential to migrate throughout ecosystems, reaching unexpected destinations in unacceptably high concentrations. We need to understand these processes, not only qualitatively in the form of general assertions about fate and transport but also quantitatively in terms of kg and ng L^{-1}.

Chemicals are subject to the laws of nature, which dictate chemical partitioning and rates of transport and transformation. Most fundamentally, chemicals are subject to the law of conservation of mass, i.e., a mass balance exists for the chemical that is a powerful constraint on quantities, concentrations, and fluxes. By coupling the mass-balance principle with expressions based on our understanding of the laws of nature, we can formulate a quantitative accounting of chemical inputs and outputs. This book is concerned with developing and applying these expressions in the form of mathematical statements or "models" of chemical fate. These accounts or models are invaluable summaries of chemical behavior and can form the basis of credible remedial and proactive strategies.

The models are formulated using the concept of fugacity, a criterion of equilibrium that has proven to be a very convenient and elegant method of calculating multimedia equilibrium partitioning. It has been widely and successfully used in chemical processing calculations. In this book, we exploit fugacity as a convenient and elegant method of explaining and deducing the environmental fate of chemicals. Since the publication of the first edition of this book in 1991 and the second edition in 2001, there has been increased acceptance of the benefits of using fugacity to formulate models and interpret environmental fate. Multimedia fugacity models are now routinely used for evaluating chemicals before and after production. Mathematical simulations of chemical fate are now more accurate, comprehensive, and reliable. They have been demonstrated to be consistent with monitoring data and have gained greater credibility as decision-support tools. No doubt this trend will continue, especially as young environmental scientists and engineers take over the reins of environmental science and continue to develop new fugacity models.

This book has been written to support those teaching graduate-level courses in multimedia modeling, and for practitioners of the environmental science of chemical fate in government, industry, and the private consulting sector. The simpler concepts are entirely appropriate for undergraduate courses. In this third edition, we emphasize the need to understand the principles on which models are based by providing step-by-step explanations of the modeling process. Numerous examples are provided and explained in full detail.

Creating and maintaining a healthful and clean environment is best guaranteed by building up a quantitative understanding of chemical fate in our total multimedia environment, how chemicals are transported and transformed, and where and to what extent they may accumulate. It is hoped that this book is one step toward this goal and will be useful to all those who value the environment and seek its more enlightened stewardship.

As authors we have sought to explain the fundamentals of fugacity modeling for both professional readers and especially for students as part of their learning experiences in environmental sciences. Accordingly, much effort has been devoted to presenting worked examples of model applications and problems suitable for training and testing purposes as a foundation for using available fugacity models from the CEMC web site and from other sources. Since multi-media fugacity modeling is an evolving science we intend to make available new examples and models as well as emerging information of data sources. Readers are encouraged to consult the CEMC website at www.trentu.ca/cemc/resources-and-models for new models, updated data sources and errata.

J. Mark Parnis
Donald Mackay

Acknowledgments

It is a pleasure to acknowledge the contribution of many colleagues. Much of the credit for the approaches devised in this book is due to the pioneering work by George Baughman, who saw most clearly the evolution of multimedia environmental modeling as a coherent and structured branch of environmental science amid the often frightening complexity of ecosystems and the formidable number of chemicals with which it is contaminated.

Donald Mackay is indebted to his former colleagues at the University of Toronto, especially Wan Ying Shiu and Sally Paterson, whose collaboration has been crucial in developing the fugacity approach. We are grateful to our more recent colleagues at Trent University, and our industrial and government partners who have made the Canadian Environmental Modelling Centre a successful focus for the development, validation, and dissemination of mass-balance models. We especially acknowledge the contributions of Dr Eva Webster, who pioneered the dissemination of models from the Centre's website, and the work of Rodion Gordzevich, who transformed many of the models used in this work into spreadsheet format.

Without the support and diligent typing of Ness Mackay, the first two editions of this book, upon which this third is built, would not have been possible. Thank you, Ness!

We dedicate this book to our families in the hope that their lives will be spent in a cleaner, more healthful environment.

J. Mark Parnis
Donald Mackay

Authors

J. Mark Parnis is a physical chemist with a background in quantum mechanics, spectroscopy, and kinetics. Born in Calgary, Alberta, Mark graduated from the University of Toronto with a PhD in physical inorganic chemistry. After working in the NRC Laser Chemistry Group in Ottawa, he joined Trent University Chemistry Department in 1989, where he has taught inorganic, physical, and theoretical chemistry. Early on in his research career, he worked on the photochemistry and spectroscopic characterization of metal atom reaction products with small organic molecules in low-temperature matrices. This work led to studies of organic cation spectroscopy and the study of ion decomposition products in matrices. Later in his career, he joined forces with Donald Mackay to form the Chemical Properties Research Group, focusing on the techniques for estimating physico-chemical properties of molecular species with an emphasis on environmental modeling applications. He is currently the director of the Canadian Environmental Modelling Centre at Trent University.

Donald Mackay is an internationally renowned engineer and scientist, the acknowledged pioneer of fugacity-based modeling applications in environmental fate and exposure methodology. Born in Glasgow, Scotland, Don is a graduate of the University of Glasgow and is now professor emeritus in the School of the Environment and director emeritus of the Canadian Environmental Modelling Centre, both at Trent University. He is also professor emeritus in the Department of Chemical Engineering and Applied Chemistry of the University of Toronto where he taught for some 30 years and established himself as a pioneer of multimedia modeling in environmental science. Moving to Trent University in 1995, he contributed to the growth and maturation of the environmental science program and established the Canadian Environmental Modelling Centre, which he led until his official retirement in 2002. Since that time, Don has continued to work in the field, producing over 750 articles, many books, and numerous reports during his career. His principal research has been on the environmental fate of toxic substances and has included studies of numerous partitioning and transport processes in the environment, the focus being on organic contaminants. His recent work includes the extension of the environmental models to include food uptake and pharmacokinetic processes and their application as components of chemical risk assessments by regulatory agencies worldwide. The recipient of the Order of Canada and many other awards, Don is a leading figure in the field of environmental fate modeling.

1 Basic Concepts

THE ESSENTIALS

CLOSED SYSTEMS AT EQUILIBRIUM

Moles of chemical in phase "i" = phase "i" concentration $\left(\mathrm{mol\ m^{-3}}\right)$ × phase "i" volume $\left(\mathrm{m^3}\right)$

$$m_i = C_i V_i$$

Total moles of chemical in system = sum of moles of chemical in n phases

$$m_{\mathrm{Sys}} = \sum_{i=1}^{n} m_i = \sum_{i=1}^{n} C_i V_i$$

STEADY-STATE "OPEN" SYSTEMS

Total chemical input rate = total chemical output rate $\left(\mathrm{mol\ h^{-1}}\right)$

$$r_{\mathrm{Tot}}^{\mathrm{In}} = \overset{\text{All inputs}}{\sum_{i=1}} r_i = \overset{\text{All outputs}}{\sum_{j=1}} r_j = r_{\mathrm{Tot}}^{\mathrm{Out}}$$

Residence time for flow-only: $\tau_F = \dfrac{V_i}{G_i^{\mathrm{Outf}}}$

TIME-VARIABLE SYSTEMS

Rate of change in total moles = $\left(\text{total input rate}\right) - \left(\text{total output rate}\right)$

$$\frac{d\left(\text{total moles}\right)}{dt} = r_{\mathrm{Tot}}^{\mathrm{In}} - r_{\mathrm{Tot}}^{\mathrm{Out}}$$

$$t_{1/2} = \frac{0.693}{k^{\mathrm{Loss}}} \qquad t_{0.37} = \frac{1}{k^{\mathrm{Loss}}} \qquad t_{0.9} = \frac{2.303}{k^{\mathrm{Loss}}}$$

We recommend that the reader consult Appendix B to confirm the system of units used in this and later chapters. Of particular importance are units of concentration that differ from conventional usage in chemistry.

1.1 INTRODUCTION

Since the first edition of this book in 1991, there have been revolutions in environmental science. Climate change is now recognized as a global priority, requiring modeling and international actions to ensure that our planet remains habitable. The number of chemicals in commerce has greatly increased. We are now able to measure environmental concentrations at picogram $(10^{-12}\mathrm{g})$ levels. There have been enormous advances in toxicology, revealing increasingly subtle effects of chemicals on organisms, including ourselves. The explosion in information technology has revolutionized access to data on environmental systems and chemical properties. More than ever we need models to provide a quantitative accounting of chemical fate in our environment and to guide regulation and remediation efforts.

There is a common public perception and concern that when chemical substances are present in air, water, or food, there is a risk of adverse effects to human health. Assessment of this risk is difficult (a) because the exposure is usually (fortunately) well below levels at which lethal toxic effects and often sub-lethal effects cannot be measured with statistical significance against the "noise" of natural population variation and (b) because of the simultaneous multiple toxic influences of other substances, some taken voluntarily and others involuntarily. There is a growing belief that it is prudent to ensure that the functioning of natural ecosystems is unimpaired by these chemicals, not only because ecosystems have inherent value but also because they can act as sensing sites or early indicators of possible impact on human well-being.

Accordingly, there has developed a branch of environmental science concerned with describing, first qualitatively and then quantitatively, the behavior of chemicals in the environment. This science is founded on earlier scientific studies of the condition of the natural environment—meteorology, oceanography, limnology, hydrology, and geomorphology and their physical, energetic, biological, and chemical sub-sciences. This newer branch of environmental science has been variously termed *environmental chemistry*, *environmental toxicology*, or *chemodynamics*.

It is now evident that our task is to design a society in which the benefits of chemicals are enjoyed, while the risk of adverse effects from them is virtually eliminated. To do this, we must exert effective and cost-effective controls over the use of such chemicals, and we must have available methods of calculating their environmental behavior in terms of concentration, persistence, reactivity, and partitioning tendencies between air, water, soils, sediments, and biota. Such calculations are essential when assessing or implementing remedial measures to treat already contaminated environments. They become vital as the only available method for predicting the likely behavior of chemicals that may be (a) newly introduced into commerce or (b) subject to production increases or introduction into new environments.

In response to this societal need, this book develops, describes, and illustrates a framework and procedures for calculating the behavior of chemicals in the environment. It employs both conventional procedures that are based on manipulations of concentrations and procedures that use the concepts of activity and fugacity to characterize the equilibrium that exists between environmental phases such as air, water, and soil. Most of the emphasis is placed on organic chemicals, which are fortunately more susceptible to generalization than metals and other inorganic chemicals when assessing environmental behavior.

The concept of fugacity, which was introduced by G.N. Lewis in 1901 as a convenient thermodynamic equilibrium criterion, has been widely used in chemical process calculations. Its convenience in environmental chemical equilibrium or partitioning calculations has become apparent only in the last three decades. It transpires that fugacity is also a convenient quantity for describing mathematically the rates at which chemicals diffuse, or are transported, between phases; for example, volatilization of pesticides from soil to air. The transfer rate can be expressed as being driven by, or proportional to, the fugacity difference that exists between the source and destination phases. It is also relatively easy to transform chemical reaction, advective flow, and non-diffusive transport rate equations into fugacity expressions and build up sets of fugacity equations describing the quite complex behavior of chemicals in multiphase, non-equilibrium environments. These equations adopt a relatively simple form, which facilitates their formulation, solution, and interpretation to determine the dominant environmental phenomena.

We develop these mathematical procedures from a foundation of thermodynamics, transport phenomena, and reaction kinetics. Examples are presented of chemical fate assessments in both real and evaluative multimedia environments at various levels of complexity and in more localized situations such as at the surface of a lake. These calculations of environmental fate can be tedious and repetitive; thus, computer programs are described for many of the calculations discussed later in the text. These programs are freely available on the internet from the website of the Canadian Environmental Modelling Centre (CEMC) of Trent University (www.trentu.ca/cemc/resources-and-models). Many of these have been specifically designed and prepared to complement this text. The modern trend in model development is to implement them in spreadsheets, such as Microsoft Excel®, which have improved input and output features, including the ability to draw graphs and charts. Those familiar with the "logic" of spreadsheets should have no difficulty following the calculations in a step-by-step manner, parallel to the approach taken in the text. Sufficient information is given on each mass-balance model so that readers can write their own programs using the system of their choice.

Preparing a third edition of this book has enabled us to update, expand, and reorganize much of the material presented in the first two editions. We have benefited greatly from the efforts of those who have sought to understand environmental phenomena and who have applied the fugacity approach when deducing the fate of chemicals in the environment.

1.2 THE MASS-BALANCE CONCEPT

1.2.1 The Environment as Compartments

Our primary task in multimedia environmental modeling is to determine the ultimate fate of a chemical in the environment. We shall see that this is achieved by understanding the "choreography" of a chemical, i.e., the movement of a chemical through the environment. It is useful to view the environment as consisting of a number of interconnected *phases* or *media compartments*. Examples are the atmosphere, terrestrial soil, a lake, the bottom sediment under the waters of a lake, suspended sediment in a lake, and biota in soil or water, and humans. A phase may be continuous (e.g., water) or discrete, the latter consisting of a number of particles that are not in contact, but all of which reside in one phase (e.g., atmospheric aerosol particles or biota in water). In some cases, the phases may be similar chemically but physically distinct, e.g., the troposphere and the stratosphere which are principally distinct due to differences in pressure, temperature, and density. It may be convenient to lump all biota together as one phase or consider them as two or more classes each with a separate phase. Some compartments are in contact, such that a chemical may migrate between them (e.g., air and water at the surface of a lake), while others are not in contact, such that direct transfer is impossible (e.g., air and lake bottom sediment). Some phases are accessible in a short time to migrating chemicals (e.g., surface waters), whereas others may only be accessible at a very slow rate (e.g., deep lake or ocean waters), or effectively not at all (e.g., deep soil or rock).

We can consider these phases or parts of them as "compartments" about which we can write mass, mole, and, if necessary, energy balance equations. Lavoisier's Law of Conservation of Matter in chemical processes is a key concept that underlies such equations. We shall see that steady-state conditions will yield algebraic equations relating concentrations of chemicals distributed between these phases, whereas unsteady-state conditions will yield time-dependent differential equations. These equations may contain terms for discharges, flow (diffusive and non-diffusive) of material between phases, and for reaction or formation of a particular chemical. In doing so, we shall need to discriminate between equilibrium and steady-state conditions, as well as introduce the concepts of residence time and persistence.

1.2.2 Closed Systems at Equilibrium

The concepts introduced here are relatively simple and will be obvious to many readers. However, it is important to clarify the use of terms such as equilibrium, steady state, and unsteady state. The simplest system is one in which we aspire to describe how a given mass of a chemical will partition between various interacting phases of fixed volume once an equilibrium is achieved between them. This system is also closed or "sealed" in that no entry or exit of chemical is possible. Moreover, there are no inputs or losses associated with any phase or compartment. To describe such a system, we need only to generate a mass-balance equation that expresses the fact that the total amount of chemical present in the entire system must equal the sum of the amount in each phase, i.e., a statement of conservation of matter.

Total moles of chemical in system = sum of amount of chemical in each phase

The total molar quantity of chemical in each phase compartment "i" is generally given as a product of phase-specific volumes and concentrations:

$$m_i = C_i V_i$$

A variety of units may be used for these quantities, but here we express volumes in cubic metres (m^3), quantities of chemical in moles (mol), and concentrations in mol m^{-3}. The total moles in the whole system is the sum of the moles in each of "n" compartments:

$$\text{Total moles} = m_{Sys} = \sum_{i=1}^{n} m_i = \sum_{i=1}^{n} C_i V_i$$

For equilibrium and steady-state distribution of the chemical, the amount in any phase is understood to be unchanging. A typical calculation for such a three-phase system is illustrated in the following Worked Example.

Worked Example 1.1: A three-phase air-water-sediment system at equilibrium

A three-phase system consists of air ($V_A = 100\,m^3$), water ($V_W = 60\,m^3$), and sediment ($V_{Sed} = 3\,m^3$). To this is added 2 mol of a hydrocarbon such as benzene. What are the concentrations of benzene in each compartment at equilibrium?

The phase volumes are not affected by this addition, because the volume of hydrocarbon is small, as is usually the case for trace environmental contaminants. Defining the corresponding concentrations (mol m^{-3}) in the respective phases of air, water, and sediment as C_A, C_W, and C_{Sed}, we can write the following mass-balance equation for n phases at equilibrium:

$$\text{Total moles} = m_{Sys} = \sum_{i=1}^{n} C_i V_i$$

$$2 \text{ moles} = C_A V_A + C_W V_W + C_{Sed} V_{Sed}$$

To proceed further, we must have information about the relationships between C_A, C_W, and C_{Sed}. This could take the form of phase equilibrium relationships such as

$$K_{AW} = \frac{C_A}{C_W} = 0.4 \quad \text{and} \quad K_{Sed-W} = \frac{C_{Sed}}{C_W} = 100$$

Such ratios are referred to here as *partition ratios*, and quantify the equilibrium distribution of a single chemically distinct species between two phases. (Note that partition ratios are commonly referred to as "partition coefficients"; however, the use of this term is now considered inappropriate by the International Union of Pure and Applied Chemistry (IUPAC)). Here the partition ratio between air and water is designated K_{AW} and that between sediment and water is designated K_{Sed-W}. Rearranging to isolate expressions for C_A and C_{Sed},

$$C_A = 0.4 C_W \quad \text{and} \quad C_{Sed} = 100 C_W$$

We can now eliminate C_A and C_{Sed} by substitution to give an expression in which the sole unknown is C_W:

$$\text{Total moles} = 2 \text{ mol} = (0.4 C_W) V_A + C_W V_W + (100 C_W) V_{Sed}$$

Substituting the phase volumes and isolating C_W, we obtain

$$2 \text{ mol} = 0.4 C_W \times 100 \text{ m}^3 + C_W \times 60 \text{ m}^3 + 100 C_W \times 3 \text{ m}^3$$

$$= \left(40 \text{ m}^3 + 60 \text{ m}^3 + 300 \text{ m}^3\right) C_W = \left(400 \text{ m}^3\right) C_W$$

$$\therefore C_W = \frac{2 \text{ mol}}{400 \text{ m}^3} = 5 \times 10^{-3} \text{ mol m}^{-3}$$

It follows that

$$C_A = 0.4 C_W = 2 \times 10^{-3} \text{ mol m}^{-3}$$

and

$$C_{Sed} = 100 C_W = 0.5 \text{ mol m}^{-3}$$

The molar amounts in each phase m_i (mol) are the corresponding products of phase volume and molar concentration:

$$m_W = C_W V_W = 5 \times 10^{-3} \text{ mol m}^{-3} \times 60 \text{ m}^3 = 0.30 \text{ mol} \left(15\%\right)$$

$$m_A = C_A V_A = 2 \times 10^{-3} \text{ mol m}^{-3} \times 100 \text{ m}^3 = 0.20 \text{ mol} \left(10\%\right)$$

$$m_{Sed} = C_{Sed} V_{Sed} = 0.5 \text{ mol m}^{-3} \times 3 \text{ m}^3 = 1.5 \text{ mol} \left(75\%\right)$$

which sums to a total of 2.00 mol, as specified by the initial conditions.

Example 1.1 demonstrates that a simple algebraic procedure can establish the concentrations and amounts in each phase for a closed system at equilibrium, using only mass-balance equations and equilibrium relationships. The essential concept is that the total amount of chemical present must equal the sum of the individual amounts in each compartment. We will later refer to this as a "Level I" calculation. Such a calculation is useful because it is not always obvious where *concentrations* are high, as distinct from *amounts*.

1.2.3 OPEN SYSTEMS AT STEADY STATE

When a chemical can flow into and out of the system in question, the system is described as being "open." The chemical may enter or leave the system either by bulk transport (advection) between the system and its surroundings, direct emission into one or more compartments, or by generation or degradation of the chemical within a compartment of the system. When the concentrations within all compartments in the system are constant or static with respect to time, the system is said to be at "steady state." Thus, we may have a steady-state open system, an example of which might be a lake (the system) that has a river flowing both into and out of it. We will represent the bulk flow of matter into and out of a compartment with the symbol "G." For example, the water inflow and outflow rates for a lake would be represented by G_W^{Inf} and G_W^{Outf}, respectively, generally with units of $\text{m}^3 \text{h}^{-1}$.

The basic mass-balance requirement is then that the total rate at which a chemical enters the system must equal the total rate at which it leaves. In the absence of any loss due to degradation or reaction, the total input and output rates must necessarily balance.

$$\text{Total chemical input rate} = \text{total chemical output rate}$$

or

$$r_{\text{Tot}}^{\text{In}} = \overbrace{\sum_{i=1}}^{\text{All inputs}} r_i = \overbrace{\sum_{j=1}}^{\text{All outputs}} r_j = r_{\text{Tot}}^{\text{Out}}$$

The corresponding class of mass-balance equation must therefore account for the possibility of the chemical entering and leaving the system, as well as appearing or disappearing within one or more compartments. Mass-transfer rates are conventionally expressed either in moles or in grams per unit time, and whereas the basic unit in the closed system balance was mol or g, it is now mol h^{-1} or g h^{-1}. Such a calculation will later be referred to as a "Level II" calculation, of which Worked Example 1.2 is typical.

Worked Example 1.2: A pond with an inflow source containing a chemical of known concentration and outflow of concentration to be determined.

A thoroughly mixed pond of constant volume ($V = 10^4 \text{m}^3$) has constant water inflow and outflow rates of $G_W^{\text{Inf}} = G_W^{\text{Outf}} = 5\,\text{m}^3\,\text{h}^{-1}$. The inflow water contains $C_W^{\text{Inf}} = 0.01\,\text{mol m}^{-3}$ of a chemical. The same chemical is also discharged directly into the pond at a rate of $r_W^{\text{Em}} = 0.1\,\text{mol h}^{-1}$. Assuming that there is no reaction, volatilization, or other losses of the chemical, determine the concentration of chemical in the lake water and in the outflow water.

We designate the desired but unknown outflow concentration of the chemical in water as C_W^{Outf} with units of mol m^{-3}. To express the inflow rate of the chemical itself, r_W^{Inf}, we can state

$$r_W^{\text{Inf}} = G_W^{\text{Inf}} C_W^{\text{Inf}} = 5\,\text{m}^3\,\text{h}^{-1} \times 0.01\,\text{mol m}^{-3} = 0.05\,\text{mol h}^{-1}$$

The rate of chemical outflow r_W^{Outf} is then determined by balancing the molar transfer rates for chemical input and output:

$$r_W^{\text{Inf}} + r_W^{\text{Em}} = r_W^{\text{Outf}}$$

$$0.05\,\text{mol h}^{-1} + 0.1\,\text{mol h}^{-1} = 0.15\,\text{mol h}^{-1}$$

Thus, both the total inflow and outflow rates of chemical are 0.15 mol h^{-1}.

Given that the chemical is contained in the outflow water, which is flowing at a rate of $G_W^{\text{Outf}} = 5\,\text{m}^3\,\text{h}^{-1}$, we conclude that the concentration of the chemical in the outflow water is

$$C_W^{\text{Outf}} = \frac{r_W^{\text{Outf}}}{G_W^{\text{Outf}}} = \frac{0.15\,\text{mol h}^{-1}}{5\,\text{m}^3\,\text{h}^{-1}} = 0.03\,\text{mol m}^{-3}$$

Note that, as required for a steady-state system, the total inflow and outflow rates of chemical are both 0.15 mol h^{-1}.

A somewhat more involved calculation arises when one considers chemical degradation in one or more compartments, as illustrated in Worked Example 1.3.

Worked Example 1.3: A pond with an inflow source and outflow, and first-order degradation or reaction loss in the water.

Consider the same pond as in Worked Example 1.2, but assume that the chemical in question also decomposes or reacts in a first-order manner with a rate constant $k_W^{\text{Deg}} = 10^{-3}\,\text{h}^{-1}$ and use $V_W = 10^4\,\text{m}^3$. Determine the steady-state concentration of the chemical in the outflow water.

The rate of loss of chemical in the water volume by decomposition in mol h⁻¹ is given by the product of the rate constant for decomposition and the total molar amount of chemical. The total amount of chemical is given by $C_W V_W$. Thus, we have

$$r_W^{Deg}\left(\text{mol h}^{-1}\right) = k_W^{Deg} C_W V_W = 10^{-3}\,\text{h}^{-1} C_W \times 10^4\,\text{m}^3 = 10\,\text{m}^3\text{h}^{-1} C_W$$

Now, using the same mass flow balance approach, we can again write

$$\text{Total chemical input rate} = \text{total chemical output rate}$$

$$r_W^{Inf} + r_W^{Em} = r_W^{Outf} + r_W^{Deg}$$

From Worked Example 1.2, $r_W^{Outf} = G_i^{Outf} C_W^{Outf} = (5C_W)$ mol h⁻¹. As well, $r_W^{Em} = 0.1$ mol h⁻¹, and $r_W^{Inf} = G_W^{Inf} C_W^{Inf} = 0.05$ mol h⁻¹. Therefore, we can write

$$(0.05+0.1)\,\text{mol h}^{-1} = \left(5\,\text{m}^3\,\text{h}^{-1} \times C_W + 10\,\text{m}^3\,\text{h}^{-1} \times C_W\right) = 15\,\text{m}^3\,\text{h}^{-1}\,C_W$$

or

$$0.15\,\text{mol h}^{-1} = 15\,\text{m}^3\,\text{h}^{-1}\,C_W$$

Solving for C_W, we find

$$C_W = \frac{0.15\,\text{mol h}^{-1}}{15\,\text{m}^3\,\text{h}^{-1}} = 0.01\,\text{mol m}^{-3}$$

The total input of 0.15 mol h⁻¹ is again equal to the total output of 0.15 mol h⁻¹, now consisting of 0.05 mol h⁻¹ outflow and 0.10 mol h⁻¹ reaction.

In both worked examples above, an inherent assumption is that the prevailing concentration in the pond C_W is constant and equal to the outflow concentration C_W^{Outf}. This is the "well-mixed" or "continuously stirred tank" assumption. It may not always apply, but it greatly simplifies calculations when it does.

We may further increase the complexity of the calculation by considering additional losses due to volatilization from water to air, as demonstrated in Worked Example 1.4.

Worked Example 1.4: Steady-state concentration of a chemical in a lake with contaminated inflow, outflow, direct emission, first-order degradation loss in water, and volatilization loss to air.

A lake of area $A_{AW} = 10^6$ m² and depth $Y_W = 10$ m (therefore, $V_W = A_{AW} Y_W = 10^7$ m³) receives an input of $r_W^{Em} = 20.0$ mol h⁻¹ of chemical in an effluent discharge. The chemical is also present in the inflow water with $G_W^{Inf} = 420$ m³ h⁻¹ at a background concentration of $C_W^{Inf} = 0.01$ mol m⁻³. The chemical reacts with a first-order rate constant k_W^{Deg} of 10^{-3} h⁻¹, and it volatilizes with a flux of $L_W^{Vol} = 3.6 \times 10^{-2}\,C_W$ mol m⁻² h⁻¹, where C_W is the water concentration in mol m⁻³. (Note that flux is a rate associated with a unit of area, as reflected by division by m² in the units. Here it reflects the rate of volatilization per unit of the air–water interface area.) The outflow is $G_W^{Outf} = 320$ m³ h⁻¹, there being some loss of water by evaporation. Assuming that the lake water is well-mixed, calculate the lake water concentration and all the inputs and outputs in mol h⁻¹.

The rate of input of chemical to the lake from the inflow water is given by

$$r_W^{Inf} = G_W^{Inf} C_W^{Inf} = 420\,\text{m}^3\,\text{h}^{-1} \times 0.01\,\text{mol m}^{-3} = 4.2\,\text{mol h}^{-1}$$

The total molar input rate is given by the sum of all input rates:

$$\text{Total input rate } r = \text{inflow rate} + \text{discharge rate}$$

$$r_{\text{Tot}}^{\text{In}} = r_W^{\text{Inf}} + r_W^{\text{Em}}$$

or

$$r_{\text{Tot}}^{\text{In}} = (4.2 + 20.0) \ \text{mol h}^{-1} = 24.2 \ \text{mol h}^{-1}$$

In terms of change in content of the entire lake volume per day due to decomposition, the reaction rate is given by the rate constant (h^{-1}) times the number of moles present in the lake, expressed as $C_W \left(\text{mol m}^{-3}\right) V_W \left(\text{m}^3\right)$:

$$r_W^{\text{Deg}} = k_W^{\text{Deg}} C_W V_W = 10^{-3} \ \text{h}^{-1} \times C_W \times 10^7 \ \text{m}^3 = \left(10^4 \ \text{m}^3 \ \text{h}^{-1}\right) C_W$$

The volatilization flux is given as $L_W^{\text{Vol}} = 3.6 \times 10^{-2} \ C_W \ \text{mol m}^{-2} \ \text{h}^{-1}$. Given the surface area of the lake $A_{AW} = 10^6 \text{m}^2$, we can calculate the volatilization loss rate for the whole lake as

$$r_W^{\text{Vol}} = L_W^{\text{Vol}} \times A_{AW}$$

$$r_W^{\text{Vol}} \left(\text{mol h}^{-1}\right) = \left(3.6 \times 10^{-2} \ \text{mol m}^{-2} \ \text{h}^{-1} \times C_W\right) \times 10^6 \ \text{m}^2 = \left(3.6 \times 10^4 \ \text{m}^3 \ \text{h}^{-1}\right) C_W$$

The outflow rate expressed in terms of the unknown concentration C_W is given as the volume flow rate times the outflow water concentration, which for a well-mixed situation equals the lake water steady-state concentration:

$$C_W^{\text{Outf}} = C_W$$

$$r_W^{\text{Outf}} = G_W^{\text{Outf}} C_W^{\text{Outf}} = G_W^{\text{Outf}} C_W$$

$$r_W^{\text{Outf}} \left(\text{mol h}^{-1}\right) = 320 \ \text{m}^3 \ \text{h}^{-1} \times C_W$$

The total molar output rate is given by the sum of all output rates:

$$\text{Total output rate} = \text{outflow rate} + \text{degradation rate} + \text{volatilization rate}$$

$$r_{\text{Tot}}^{\text{Out}} = r_W^{\text{Outf}} + r_W^{\text{Deg}} + r_W^{\text{Vol}}$$

$$r_{\text{Tot}}^{\text{Out}} \left(\text{mol h}^{-1}\right) = \left(320 \ \text{m}^3 \ \text{h}^{-1}\right) C_W + \left(1.0 \times 10^4 \ \text{m}^3 \ \text{h}^{-1}\right) C_W + \left(3.6 \times 10^4 \ \text{m}^3 \ \text{h}^{-1}\right) C_W$$

$$= \left(4.632 \times 10^4 \ \text{m}^3 \ \text{h}^{-1}\right) C_W$$

Now equating the input and output rates, we obtain

$$r_{\text{Tot}}^{\text{In}} = r_{\text{Tot}}^{\text{Out}}$$

$$24.2 \ \text{mol h}^{-1} = \left(4.632 \times 10^4 \ \text{m}^3 \ \text{h}^{-1} \ C_W\right)$$

Solving for the unknown lake water concentration C_W, we find

$$C_W = \frac{24.2 \ \text{mol h}^{-1}}{4.632 \times 10^4 \ \text{m}^3 \ \text{h}^{-1}} = 5.225 \times 10^{-4} \ \text{mol m}^{-3}$$

With C_W now known, we can go back and determine the following molar rates:

$$r_W^{Deg} = 10^4 \text{ m}^3 \text{ h}^{-1} \times 5.225 \times 10^{-4} \text{ mol m}^{-3} = 5.22 \text{ mol h}^{-1}$$

$$r_W^{Vol} = 3.6 \times 10^4 \text{ m}^3 \text{ h}^{-1} \times 5.225 \times 10^{-4} \text{ mol m}^{-3} = 18.8 \text{ mol h}^{-1}$$

$$r_W^{Outf} = 3.2 \times 10^2 \text{ m}^3 \text{ h}^{-1} \times 5.225 \times 10^{-4} \text{ mol m}^{-3} = 0.167 \text{ mol h}^{-1}$$

Note the sum of these three losses is equal to the total influx, as it must be. In all these examples, the chemical is flowing or reacting, but the observed conditions in the compartment are not changing with time; thus, the steady-state condition applies.

1.2.4 Dynamic (Time-Variable) Open Systems

Whereas the mass balances lead to simple algebraic equations for closed systems at equilibrium and open steady-state systems, time-variant or "unsteady-state" conditions give rise to differential equations in time. The simplest method of setting up an "unsteady-state" mass-balance equation is in terms of a balance of rates of mass transfer:

$$\text{Rate of change in total moles} = \left(\text{total input rate}\right) - \left(\text{total output rate}\right)$$

or

$$\frac{d\left(\text{total moles}\right)}{dt} = r_{Tot}^{In} - r_{Tot}^{Out}$$

The total input and output rates (r_{Tot}^{In} and r_{Tot}^{Out}) are expressed as a change in the amount of chemical per time, typically in molar or mass amounts yielding units of mol h^{-1} or g h^{-1}, respectively. The contents must be expressed with the corresponding mole or gram units. The resulting differential equation can then be solved with an appropriate initial or boundary conditions, to give an algebraic expression for concentration as a function of time. For more complex situations, it may be preferable to integrate numerically.

The simplest example is the first-order decay process, in which an initial amount of a chemical m (mol) is present at time $t = 0$, and it decays by a simple first-order loss process such as chemical decomposition with a rate constant k^{Deg}. Balancing total input (0) and output ($k^{Deg}m$) rates, we can write

$$\frac{dm}{dt} = 0 - k^{Deg}m$$

If we separate variables, we can immediately integrate to yield the standard equation for first-order decay:

$$\frac{dm}{m} = -k^{Deg}dt$$

or

$$\ln m \big|_{m_0}^{m_t} = -k^{Deg}t \big|_0^t$$

Remembering that the initial amount of the chemical at time $t = 0$ is m_0, we can simplify and write

$$\ln\left(\frac{m_t}{m_0}\right) = -k^{\text{Deg}}t$$

or

$$m_t = m_0 \exp\left(-k^{\text{Deg}}t\right)$$

This expression is the well-known "first-order-decay" equation, commonly met by scientists for the first time in the context of elemental radioactive decay. The result shows that the concentration will decrease exponentially with a rate constant k^{Deg}. The time that it takes to reach half the initial concentration is called the "half-time" and is given by

$$\ln\left(\frac{0.5m_0}{m_0}\right) = -k^{\text{Deg}}t_{1/2}$$

where

$$t_{1/2} = \frac{\ln 0.5}{-k^{\text{Deg}}} = \frac{0.693}{k^{\text{Deg}}}$$

The half-time or half-life for a process is one of the key "characteristic times" that help in the estimation of the relative importance of different competing growth and decay processes in a complex system.

The treatment of a water body with a single "dose" of a chemical followed by its decomposition is a common example of a first-order decay situation, as illustrated in Worked Example 1.5.

Worked Example 1.5: Deduce the piscicide concentration in an isolated lake at a given time following a single application, after which the piscicide decomposes by a first-order degradation process.

A lake of $V_W = 10^6$ m^3 with no inflow or outflow is treated with $m_W^{\text{Em}} = 10$ mol of a piscicide (a chemical that kills fish), with a first-order reaction degradation rate constant of $k_W^{\text{Deg}} = 10^{-2}$ h^{-1}. What will the concentration in the water be after (i) 1 and (ii) 10 days, assuming no further input, and (iii) when will half the chemical have been degraded (i.e., what is the half-life time for decomposition of the piscicide)?

We seek the total concentration C_W at some time t. However, some of the quantities we want to work with require this quantity. Therefore, as with the steady-state situations above, we must work with expressions that contain the concentration as the unknown C_W.

i. If we assume the lake is well-mixed, the lake contents in moles at any time are always given by the product of lake volume V_W times the "well-mixed" concentration C_W at that time, V_W (m^3) C_W (mol m^{-3}).

$$m_W\,(\text{mol}) = C_W\left(\text{mol m}^{-3}\right)V_W\left(\text{m}^3\right)$$

The output is only by decomposition reaction, and the rate of loss can be expressed as

$$r_W^{\text{Out}}\,(\text{mol h}^{-1}) = r_W^{\text{Deg}} = k_W^{\text{Deg}}C_W V_W$$

Since there is no input, the rate of change in the content amount with time is

$$\frac{d\left(C_W V_W\right)}{dt} = r_W^{\text{Inf}} - r_W^{\text{Outf}} = 0 - k_W^{\text{Deg}}C_W V_W$$

Since we are seeking the concentration at selected times, we can simplify this expression by bringing the constant V_W outside the differential, and then dividing both sides by this quantity to obtain

$$V_W \frac{dC_W}{dt} = -k_W^{\text{Deg}} C_W V_W$$

or

$$\frac{dC_W}{dt} = -k_W^{\text{Deg}} C_W$$

We can again separate variables and integrate by inspection to yield the integrated equation for first-order decay:

$$\ln C_W \big|_{C_W^o}^{C_W^t} = -k_W^{\text{Deg}} t \big|_0^t$$

or

$$C_W^t = C_W^0 \exp\left(-k_W^{\text{Deg}} t\right)$$

Now, given that C_W^0 is 10 mol/10^6 m^3 or 10^{-5} mol m^{-3}, and $k_W^{\text{Deg}} = 10^{-2}$ h^{-1}, we can write

$$C_W^t = 10^{-5} \text{ mol m}^{-3} \times \exp\left(-10^{-2} \text{ h}^{-1} \, t\right)$$

After 1 day (24 h),

$$C_W^{t=24} = 10^{-5} \text{ mol m}^{-3} \times \exp\left(-10^{-2} \text{ h}^{-1} \times 24 \text{ h}\right) = 0.79 \times 10^{-5} \text{ mol m}^{-3}$$

The fraction remaining is given as a percentage by

$$\frac{0.79 \times 10^{-5} \text{ mol m}^{-3}}{1 \times 10^{-5} \text{ mol m}^{-3}} \times 100\% = 79\%$$

i.e., 79% remains in the lake after one day.

ii. After 10 days (240 h),

$$C_W^{t=240} = 10^{-5} \text{ mol m}^{-3} \times \exp\left(-10^{-2} \text{ h}^{-1} \times 240 \text{ h}\right) = 0.091 \times 10^{-5} \text{ mol m}^{-3}$$

In the same manner as in (i), we can show that after 10 days, 9.1% remains.

iii. As demonstrated above, the half-time for decay will be

$$t_{1/2} = \frac{-\ln(0.5)}{k_W^{Deg}} = \frac{0.693}{1 \times 10^{-2} \text{h}^{-1}} = 69.3 \text{ h}.$$

The decay curve for the water concentration in Worked Example 1.5 as a function of time is shown in Figure 1.1. The half-life concentrations and times are indicated by the dashed lines.

It is also possible to have concentration changes due to inflow and outflow processes as well as due to decomposition. Such a situation leads to exponential growth or decay in concentration to an asymptotic "steady-state" limit, as shown in the next example.

Worked Example 1.6: A pristine lake receives a constant inflow of water containing a chemical. The chemical decomposes and is also removed by outflow from the lake.

A well-mixed lake of volume $V_W = 10^6$ m^3, initially containing no chemical A, starts to receive an inflow of $G_W^{\text{Inf}} = 3.6 \times 10^4$ m^3 h^{-1} containing chemical A at a concentration $C_W^{\text{Inf}} = 0.2$ mol m^{-3}.

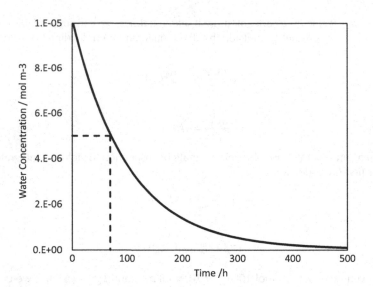

FIGURE 1.1 Concentration decay curve for Worked Example 1.5. The dashed lines indicate the position of the half-life time and concentrations.

The chemical reacts by first-order degradation with a rate constant of $k_W^{\text{Deg}} = 10^{-2}$ h^{-1}, and it also leaves with the outflow of $G_W^{\text{Outf}} = 3.6 \times 10^4$ m^3 h^{-1}. What will be the concentration of chemical in the lake one day after the start of the input of chemical?

The molar input rate is:

$$r_W^{\text{Inf}} = G_W^{\text{Inf}} C_W^{\text{Inf}}$$

$$= 3.6 \times 10^4 \text{ m}^3 \text{ h}^{-1} \times 0.2 \text{ mol m}^{-3}$$

$$= 7.2 \times 10^3 \text{ mol h}^{-1}$$

The molar loss rate due to reaction is

$$r_W^{\text{Deg}} \left(\text{mol h}^{-1} \right) = k_W^{\text{Deg}} C_W V_W$$

$$= 10^{-2} \text{ h}^{-1} \times C_W \left(\text{mol m}^{-3} \right) \times 10^6 \text{ m}^3$$

$$= 10^4 \text{ m}^3 \text{ h}^{-1} \, C_W$$

Remembering that the system is "well-mixed" so that the outflow water concentration is the same as the unknown lake water concentration that we seek here, the output by flow is

$$r_W^{\text{Outf}} \left(\text{mol h}^{-1} \right) = G_W^{\text{Outf}} C_W$$

$$= 3.6 \times 10^4 \text{ m}^3 \text{ h}^{-1} \times C_W$$

The total rate of loss or output for the system is the sum of all loss rates:

$$r_{\text{Tot}}^{\text{Out}} \left(\text{mol h}^{-1} \right) = r_W^{\text{Outf}} + r_W^{\text{Deg}}$$

$$= 3.6 \times 10^4 \text{ m}^3 \text{ h}^{-1} \, C_W + 1 \times 10^4 \text{ m}^3 \text{ h}^{-1} \, C_W$$

$$= 4.6 \times 10^4 \text{ m}^3 \text{ h}^{-1} \, C_W$$

We can now write the equation describing the rate of change in total moles in the system with time:

$$\frac{d\left(\text{total moles}\right)}{dt} = r_{\text{Tot}}^{\text{In}} - r_{\text{Tot}}^{\text{Out}}$$

or

$$\frac{d\left(V_W C_W\right)}{dt} = G_W^{\text{Inf}} C_W^{\text{Inf}} - \left(k_W^{\text{Deg}} V_W + G_W^{\text{Outf}}\right) C_W$$

After bringing the constant V_W out of the differential and dividing both sides by it, we have

$$\frac{dC_W}{dt} = \left[G_W^{\text{Inf}} C_W^{\text{Inf}} - \left(k_W^{\text{Deg}} V_W + G_W^{\text{Outf}}\right) C_W \right]\left(\frac{1}{V_W}\right)$$

As before, we separate variables to obtain

$$\frac{dC_W}{\left[G_W^{\text{Inf}} C_W^{\text{Inf}} - \left(k_W^{\text{Deg}} V_W + G_W^{\text{Outf}}\right) C_W \right]} = \left(\frac{1}{V_W}\right) dt$$

Given the following relationship from calculus:

$$\int \frac{dx}{\left(Ax + B\right)} = \frac{1}{A}\ln\left(Ax + B\right)$$

we can write

$$\left(\frac{1}{-\left(k_W^{\text{Deg}} V_W + G_W^{\text{Outf}}\right)}\right)\ln\left[G_W^{\text{Inf}} C_W^{\text{Inf}} - \left(k_W^{\text{Deg}} V_W + G_W^{\text{Outf}}\right) C_W \right]\Bigg|_{C_{t=0}}^{C_t} = \left(\frac{1}{V_W}\right) t\Big|_o^t$$

or

$$\ln\left[G_W^{\text{Inf}} C_W^{\text{Inf}} - \left(k_W^{\text{Deg}} V_W + G_W^{\text{Outf}}\right) C_W \right]\Bigg|_{C_{t=0}}^{C_t} = \left(\frac{-\left(k_W^{\text{Deg}} V_W + G_W^{\text{Outf}}\right)}{V_W}\right) t\Bigg|_o^t$$

Evaluating the integration limits explicitly, we obtain

$$\ln\left(\frac{\left[G_W^{\text{Inf}} C_W^{\text{Inf}} - \left(k_W^{\text{Deg}} V_W + G_W^{\text{Outf}}\right) C_W^t \right]}{\left[G_W^{\text{Inf}} C_W^{\text{Inf}} \right]}\right) = -\left(k_W^{\text{Deg}} + \left[G_W^{\text{Outf}} / V_W \right]\right) t$$

Taking antilogarithms and rearranging, the result is

$$\left[G_W^{\text{Inf}} C_W^{\text{Inf}} - \left(k_W^{\text{Deg}} V_W + G_W^{\text{Outf}}\right) C_W^t \right] = \left[G_W^{\text{Inf}} C_W^{\text{Inf}} \right]\exp\left(-\left(k_W^{\text{Deg}} + \left[G_W^{\text{Outf}} / V_W \right]\right) t\right)$$

or

$$\left(k^{\text{Deg}} V_W + G_W^{\text{Outf}}\right) C_W^t = \left[G_W^{\text{Inf}} C_W^{\text{Inf}} \right] - \left[G_W^{\text{Inf}} C_W^{\text{Inf}} \right]\exp\left(-\left(k_W^{\text{Deg}} + \left[G_W^{\text{Outf}} / V_W \right]\right) t\right)$$

or

$$C_W^t = \frac{\left[G_W^{Inf}C_W^{Inf}\right]}{\left(k_W^{Deg}V_W + G_W^{Outf}\right)}\left(1 - \exp\left(-\left(k_W^{Deg} + \left[G_W^{Outf}/V_W\right]\right)t\right)\right)$$

This has the form of $A(1 - \exp(-kt))$ which is an exponential rise to an asymptotic limit A. Note that when t is zero, $\exp(0)$ is unity and C_W is zero, as dictated by the initial condition. When t is very large, the exponential group becomes zero, and C_W approaches the asymptotic limit concentration of 0.157 mol m⁻³, which is equal to the pre-exponential term.

$$C_W^{\infty} = \frac{\left[G_W^{Inf}C_W^{Inf}\right]}{\left(k_W^{Deg}V_W + G_W^{Outf}\right)} = \frac{7.2\times10^3 \text{ mol h}^{-1}}{\left(10^4 \text{ m}^3 \text{ h}^{-1} + 3.6\times10^4 \text{ m}^3 \text{ h}^{-1}\right)} = 0.157 \text{ mol m}^{-3}$$

At such times, the total input of 7.2×10^3 mol h⁻¹ is equal to the total of the output by flow of 3.6×10^4 m³h⁻¹ $\times 0.157$ mol m⁻³ or 5.65×10^3 mol h⁻¹ plus the output by reaction of 10^{-2} h⁻¹ $\times 10^6$ m³ $\times 0.157$ mol m⁻³ or 1.57×10^3 mol h⁻¹. This is the steady-state solution, which the lake eventually approaches after a long period of time.

When t is 1 day or 24 h, C_W will be 0.105 mol m⁻³ or 67% of the way to its final value:

$$C_W^{t=24} = 0.157 \text{ mol m}^{-3}\left(1 - \exp\left(-\left(4.6\times10^{-2} \text{ h}^{-1}\right)\times 24 \text{ h}\right)\right) = 0.157 \text{ mol m}^{-3}\times 0.668 = 0.105 \text{ mol m}^{-3}$$

C_W will be halfway to its final value when the asymptotic term has the value of 0.5 or when $t = 15$ h:

$$t_{1/2} = \frac{-\ln(0.5)}{4.6\times10^{-2} \text{ h}^{-1}} = 15 \text{ h}$$

The growth half-time here is largely controlled by the residence time of the water in the lake, which is $(10^6$ m³)/(10 m³s⁻¹) or 10^5 s or 27.8 h, as discussed in Section 1.3. The growth curve for the water concentration in Worked Example 1.6 as a function of time is shown in Figure 1.2. The half-time concentrations and times are indicated by dashed lines, and the asymptotic limit concentration is indicated with a dotted line.

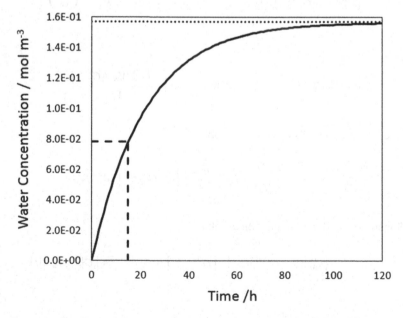

FIGURE 1.2 Concentration growth curve for Worked Example 1.6. The dashed lines indicate the position of the half-time time and concentrations, and the dotted line indicates the asymptotic limit concentration.

Worked Example 1.7: Calculate the time for a lake that is contaminated by a single emission event to eliminate 90% of the chemical by outflow and degradation reaction.

A well-mixed lake of $V_W = 10^5 \, m^3$ is initially contaminated with a chemical that results in a lake water concentration of $C_W^0 = 1 \, mol \, m^{-3}$. The chemical leaves by the outflow rate of $G_W^{Outf} = 1.8 \times 10^3 \, m^3 \, h^{-1}$, and it decomposes by reaction with a rate constant of $k_W^{Deg} = 10^{-2} \, h^{-1}$. What will be the chemical concentration after (i) 1 and (ii) 10 days, and (iii) when will 90% of the chemical have left the lake?

The inflow rate is 0, and after the initial contamination event, the total input rate is also 0. The output rate is the sum of the rate of loss by outflow and by reaction. Therefore, our differential equation is

$$\frac{d(\text{total moles})}{dt} = r_{Tot}^{In} - r_{Tot}^{Out}$$

or

$$\frac{d(V_W C_W)}{dt} = 0 - \left[k_W^{Deg} C_W V_W + G_W^{Outf} C_W \right]$$

or

$$\frac{dC_W}{dt} = -\left[k_W^{Deg} + \left(G_W^{Outf} / V_W \right) \right] C_W$$

Again separating variables, we have

$$\frac{dC_W}{C_W} = -\left[k_W^{Deg} + \left(G_W^{Outf} / V_W \right) \right] dt$$

Integrating from time 0, we arrive at a result similar to Worked Example 1.5, i.e., exponential decay. In this case, the decay is due to both decomposition and outflow losses:

$$\int_{C_W^0}^{C_W^t} \frac{dC_W}{C_W} = -\left[k_W^{Deg} + \left(G_W^{Outf} / V_W \right) \right] \int_0^t dt$$

or

$$\ln(C_W) \Big|_{C_W^0}^{C_W^t} = -\left[k_W^{Deg} + \left(G_W^{Outf} / V_W \right) \right] t \Big|_0^t$$

or

$$C_W^t = C_W^0 \exp\left(-\left[k_W^{Deg} + \left(G_W^{Outf} / V_W \right) \right] t \right)$$

i. Since C_W^0 is 1.0 mol m^{-3}, after 1 day C_W will be 0.51 mol m^{-3}:

$$C_W^{t=24} = 1.0 \, mol \, m^{-3} \times \exp\left(-\left[10^{-2} \, h^{-1} + \left(1.8 \times 10^3 \, m^3 \, h^{-1} / 10^5 \, m^3 \right) \right] \times 24 \, h \right) = 0.51 \, mol \, m^{-3}$$

ii. After 10 days, $C_W = 0.0012$ mol m^{-3}

$$C_W^{t=240} = 1.0 \, mol \, m^{-3} \times \exp\left(-\left[10^{-2} \, h^{-1} + \left(1.8 \times 10^3 \, m^3 \, h^{-1} / 10^5 \, m^3 \right) \right] \times 240 \, h \right) = 1.2 \times 10^{-3} \, mol \, m^{-3}$$

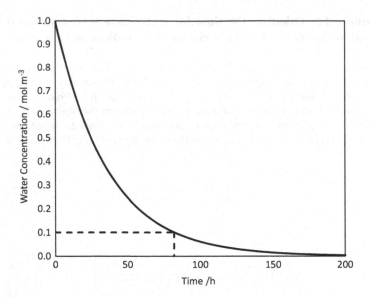

FIGURE 1.3 Concentration decay curve for Worked Example 1.7. The dashed lines indicate the position of the time and concentrations after 90% degradation.

iii. 10% of the initial concentration will remain when the exponential term equals 0.1. Solving for $t_{0.1}$ (10% remaining), we find

$$0.1 = \exp\left(-\left[10^{-2}\ \text{h}^{-1} + \left(1.8 \times 10^{3}\ \text{m}^{3}\ \text{h}^{-1}/10^{5}\ \text{m}^{3}\right)\right]t_{0.1}\right)$$

or

$$t_{0.1} = \frac{\ln(0.1)}{-\left[10^{-2}\ \text{h}^{-1} + \left(1.8 \times 10^{3}\ \text{m}^{3}\ \text{h}^{-1}/10^{5}\ \text{m}^{3}\right)\right]} = 82\ \text{h} = 3.4\ \text{days}$$

The water concentration decay curve in Worked Example 1.7 is shown as a function of time in Figure 1.3. The concentrations and times at 90% decay are indicated by the dashed lines.

These unsteady-state solutions usually contain exponential terms of the form $\exp(-kt)$. The term k is a characteristic rate constant with units of reciprocal time, a non-intuitive unit. As noted above, it is convenient to calculate its reciprocal $1/k$, which is a characteristic time. This is the time required for the process to reach 37% of the final value since

$$\exp(-kt) = \exp\left(-k\left(1/k\right)\right) = \exp(-1) = 0.37$$

or conversely to 63% completion. Those working with radioisotopes prefer to use half-lives rather than k, i.e., the time for half completion. This occurs when the term $\exp(-kt)$ is 0.5 or kt is ln 2 or 0.693; thus, the half-time t is $0.693/k$. Another useful time is the 90% completion value, which is $2.303/k$, where 2.303 is the natural logarithm (ln) of 10.

In some cases, the differential equation can become quite complex, and there may be several of them applying simultaneously. Setting up these equations requires practice and care. There is a common misconception that solving the equations is the difficult task. On the contrary, it is setting them up that is most difficult and requires the most skill. If the equation is difficult to solve, tables of integrals may be consulted, computer-based mathematical tools may be used, or an obliging mathematician can be sought. For many differential equations, an analytical solution is not feasible, and numerical methods must be used to generate a solution. We discuss techniques for doing this later.

1.3 STEADY STATE AND EQUILIBRIUM

In the previous section, we introduced the concept of "steady state" as implying unchanging with time, i.e., all time derivatives are zero. There is a frequent confusion between this concept and that of "equilibrium," which can also be regarded as a situation in which no change occurs with time. The difference is very important, and regrettably, the terms are often used synonymously. This is entirely wrong and is best illustrated by an example.

Consider the vessel in Figure 1.4a, which contains $100\,m^3$ of water and $100\,m^3$ of air. It also contains a small amount of benzene, say $1000\,g$. If this is allowed to stand at constant conditions for a long time, the benzene present will equilibrate between the water and the air and will reach unchanging but different concentrations, possibly $8\,g\,m^{-3}$ in the water and $2\,g\,m^{-3}$ in the air, i.e., a factor of 4 difference in the partition ratio in favor of the water. There is thus $800\,g$ in the water and $200\,g$ in the air. In this condition, the system is at equilibrium and at a steady state. If, somehow, the air and its benzene were quickly removed and replaced by clean air, leaving a total of $800\,g$ in the water, and the volumes

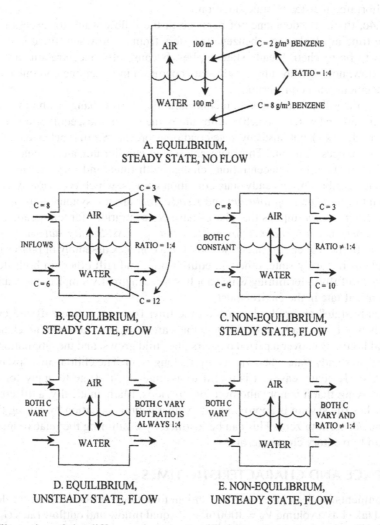

FIGURE 1.4 Illustration of the difference between equilibrium and steady-state conditions. *Equilibrium* implies that the benzene concentrations in the air and water achieve a partition ratio of 4. *Steady state* implies unchanging with time, even if flow occurs and regardless of whether or not equilibrium applies. (A) Equilibrium, steady state, no flow; (B) equilibrium, steady state, flow; (C) non-equilibrium, steady state, flow; (D) equilibrium, unsteady state, flow; (E) non-equilibrium, unsteady state, flow.

remained constant, the concentrations would adjust (some benzene transferring from water to air) to give a new equilibrium (and steady state) of $6.4\,\text{g m}^{-3}$ in the water (total $640\,\text{g}$) and $1.6\,\text{g m}^{-3}$ in the air (total $160\,\text{g}$), again with a factor of 4 difference. This factor is a partition ratio or distribution ratio or, as is discussed later, a form of Henry's law constant. During the adjustment period (for example, immediately after removal of the air when the benzene concentration in air is near zero and the water is still near $8\,\text{g m}^{-3}$), the concentrations are not at a ratio of 4, the conditions are non-equilibrium, and since the concentrations are changing with time, they are also of unsteady state in nature.

This correspondence between equilibrium and steady state does not, however, necessarily apply when flow conditions prevail. It is possible for air and water containing non-equilibrium quantities of benzene to flow into and out of the tank at constant rates as shown in Figure 1.4b. But equilibrium and a steady-state condition are maintained, since the concentrations in the tank and in the outflows are at a ratio of 1:4. It is possible for near equilibrium to apply in the vessel, even when the inflow concentrations are not in equilibrium, if benzene transfer between air and water is very rapid. Figure 1.4b thus illustrates a flow, equilibrium, and steady-state condition, whereas Figure 1.4a is a non-flow, equilibrium, and steady-state situation.

In Figure 1.4c, there is a deficiency of benzene in the inflow water (or excess in the air), and although in the time available some benzene transfers from air to water, there is insufficient time for equilibrium to be reached. Steady state applies, because all concentrations are constant with time. This is a flow, non-equilibrium, steady-state condition in which the continuous flow causes a constant displacement from equilibrium.

In Figure 1.4d, the inflow water and/or air concentration or rates change with time, but there is sufficient time for the air and water to reach equilibrium in the vessel; thus, equilibrium applies (the concentration ratio is always 4), but unsteady-state conditions prevail. Similar behavior could occur if the tank temperature changes with time. This represents a flow, equilibrium, and unsteady-state condition.

Finally, in Figure 1.4e, the concentrations change with time, and they are not in equilibrium; thus, a flow, non-equilibrium, unsteady-state condition applies, which is obviously quite complex.

The important point is that *equilibrium* and *steady state* are not synonymous; neither, either, or both can apply. *Equilibrium* implies that phases have concentrations (or temperatures or pressures) such that they experience *no tendency for net transfer of mass*. *Steady state* merely implies constancy with time. In the real environment, we observe a complex assembly of phases in which some are (approximately) in steady state, others in equilibrium, and still others in both steady state and equilibrium. By carefully determining which applies, we can greatly simplify the mathematics used to describe chemical fate in the environment.

Finally, "steady state" generally depends on the time frame of interest. Blood circulation in a sleeping child is nearly in steady state; the flow rates are fairly constant, and no change is discernible over several hours. But, over a period of years, the child grows, and the circulation rate changes; thus, it is not a true steady state when viewed in the long term. The child is in a "pseudo" or "short-term" steady state. In many cases, it is useful to assume steady state to apply for short periods, accepting or knowing that it is not valid over long periods. Mathematically, a differential equation that truly describes the system is approximated by an algebraic equation by setting the differential or the d(contents)/dt term to zero. This can be justified by examining the relative magnitude of the input, output, and inventory change terms.

1.4 RESIDENCE AND CHARACTERISTIC TIMES

In some environments, such as lakes, it is convenient to define a residence time (or detention time). If a well-mixed lake has a volume $V_W = 1000\,\text{m}^3$ and equal inflow and outflow rates $G_W^{\text{Outf}} = 2\,\text{m}^3\,\text{h}^{-1}$, then the flow residence time τ_F is $\left(V_W / G_W^{\text{Outf}}\right)$ (h), that is,

$$\tau_F = \frac{V_W}{G_W^{\text{Outf}}} = \frac{1000\ \text{m}^3}{2\ \text{m}^3\ \text{h}^{-1}} = 500\ \text{h}$$

That is, on average, the water spends 500 h (20.8 days) in the lake. This residence time may not bear much relationship to the actual time that a particular parcel of water spends in the lake, since some water may bypass most of the lake and reside for only a short time, and some may be trapped for years. The quantity is very useful, however, because it gives immediate insight into the time required to flush out the contents. Obviously, a large lake with a long residence time will be very slow to recover from contamination. Comparison of the residence time with a chemical reaction time (e.g., a half-life) indicates whether a chemical is removed from a lake predominantly by flow or by reaction.

If the lake were contaminated by a non-reacting (conservative) chemical at a concentration C_W^0 mol m^{-3} at zero time and there is no new emission, a balance of rates of mass transfer gives

$$\frac{d(V_W C_W)}{dt} = 0 - G_W^{\text{Outf}} C_W$$

or

$$\frac{dC_W}{C_W} = -\left(\frac{G_W^{\text{Outf}}}{V_W}\right) dt$$

which integrates to

$$C_W = C_W^0 e^{-\left(\frac{G_W^{\text{Outf}}}{V_W}\right)t}$$

Here we can see that the rate constant for loss without any reaction is

$$k_W^{\text{Outf}} = \frac{G_W^{\text{Outf}}}{V_W}$$

which is the reciprocal of the residence time for the lake. Therefore, the characteristic time for loss or decay in concentration is given by

$$\tau_F = \frac{1}{k_W^{\text{Outf}}} = \frac{V_W}{G_W^{\text{Outf}}}$$

The half-time for recovery occurs when t/τ_F or $k_W^{\text{Outf}} t$ is ln 2 or 0.693, i.e., when t is $0.693\tau_F$ or $0.693/k_W^{\text{Outf}}$.

If the chemical also undergoes a reaction with a rate constant k_W^{Deg} (h^{-1}), it can be shown that

$$C_W = C_W^0 \exp\left(-\left[k_W^{\text{Outf}} + k_W^{\text{Deg}}\right]t\right)$$

or

$$C_W = C_W^0 \exp(-k_T t)$$

where k_T is the "total" rate constant. As long as all individual processes are first order, k_T is simply the sum of the rate constants associated with each process, such that for this example:

$$k_T = k_W^{\text{Outf}} + k_W^{\text{Deg}}$$

Clearly, the larger rate constant dominates. Since the characteristic time is the inverse of k_T, it is also apparent that the characteristic times for individual compartments combine as reciprocals (as do electrical resistances in parallel), to give

$$\frac{1}{\tau_F} + \frac{1}{\tau_{Deg}} = \frac{1}{\tau_T}$$

We can now see that among compartment-specific characteristic times, the shorter time dominates. The term τ_{Deg} can be viewed as a reaction persistence. Characteristic times such as τ_{Deg} and τ_F are conceptually easy to grasp and are very convenient quantities to deduce when interpreting the relative importance of environmental processes. For example, if τ_F is 30 years and τ_{Deg} is 3 years, τ_T is 2.73 years, since

$$\frac{1}{30} + \frac{1}{3} = \frac{1}{2.73}$$

Thus, reaction dominates the chemical's fate and the overall characteristic time is only slightly less than that associated with the dominant degradation process. Ten out of every 11 molecules react, and only one leaves the lake by flow.

1.5   TRANSPORT MECHANISMS

In the air–water example in Section 1.3, it was argued that equilibrium occurs when the ratio of the benzene concentrations in water and air is 4. Thus, if the concentration in water is 4 mol m^{-3}, equilibrium conditions exist when the concentration in air is 1 mol m^{-3}. If the air concentration rises to 2 mol m^{-3}, we expect benzene to transfer by diffusion from air to water until the concentration in air falls, concentration in water rises, and a new equilibrium is reached. The equilibrium concentrations are easily calculated if the total amount of benzene is known. In a non-flow system, if the initial concentrations in air and water are C_A^0 and C_W^0 mol m^{-3}, respectively, and the volumes are V_A and V_W, then the total amount m_{Sys} is, by mass balance,

$$m_{Sys}(\text{mol}) = \left(C_A^0 V_A + C_W^0 V_W \right)$$

Here, C_A^0 is 2, and C_W^0 is 4 mol m^{-3}, and since the volumes are both 100 m^3, m is 600 moles:

$$m_{Sys}(\text{mol}) = 2 \text{ mol m}^{-3} \times 100 \text{ m}^3 + 4 \text{ mol m}^{-3} \times 100 \text{ m}^3 = 600 \text{ mol}$$

The system will eventually reach an equilibrated distribution, such that C_W is $4C_A$ or

$$600 \text{ moles} = C_A V_A + C_W V_W$$
$$= C_A V_A + \left(4C_A \right) V_W$$
$$= C_A \left(V_A + 4V_W \right)$$
$$= C_A \left(500 \text{ m}^3 \right)$$

Solving for C_A we find

$$C_A = \frac{600 \text{ mol}}{500 \text{ m}^3} = 1.2 \text{ mol m}^{-3}$$

$C_W = 4C_A$ or $4.8\,\text{mol m}^{-3}$. We see that the water concentration rises from 4.0 to $4.8\,\text{mol m}^{-3}$, while that of the air drops from 2.0 to $1.2\,\text{mol m}^{-3}$. If the concentration in water were instead increased to $10\,\text{mol m}^{-3}$, there would begin a corresponding new transfer from water to air until a new equilibrium state is reached.

Other transport mechanisms occur that are not driven by diffusion. For example, we could take $1\,\text{m}^3$ of water with its associated 1 mol of benzene and physically convey it into the air, forcing it to evaporate, thus causing the concentration of benzene in the air to increase. This non-diffusive, "bulk," or "piggyback" transfer occurs at a rate that depends on the rate of removal of the water phase and is not influenced by diffusion. Indeed, it may be in a direction opposite to that of diffusion.

In the environment, it transpires that there are many diffusive and non-diffusive processes operating simultaneously.

Examples of diffusive transfer processes include

1. Volatilization from soil to air
2. Volatilization from water to air
3. Absorption or adsorption by sediments from water
4. Diffusive uptake from water by fish during respiration

Some non-diffusive processes are

1. Fallout of chemical from air to water or soil in dustfall, rain, or snow
2. Deposition of chemical from water to sediments in association with suspended matter which deposits on the bed of sediment
3. Resuspension of sediment in the water column
4. Ingestion and egestion of food containing chemical by biota.

The mathematical expressions for these rates are quite different. For diffusion, the net rate of transfer or flux is written as the product of the departure from equilibrium and a kinetic quantity, and the net rate or flux becomes zero when the phases are in equilibrium. We examine these diffusive processes in Chapter 4. For non-diffusive processes, the rate is the product of the volume of the phase transferred per time unit (e.g., quantity of sediment or rain) and the concentration. We treat non-diffusive processes in Chapter 3.

We use the word *rate* as short form for transport rate. It has units such as mol h^{-1} or g h^{-1}. The word *flux* is used here to specify a rate of transfer per unit of area of interface, with units such as $\text{mol m}^{-2}\text{h}^{-1}$. Note that it is erroneous to use the term *flux rate* since flux, like velocity, already contains the "per time" term.

1.6   KEY CHEMICAL PROPERTIES

1.6.1   PARTITION RATIOS

There are several key chemical properties that are commonly used to characterize a chemical's partitioning tendencies. In the most simplified of views, the physical world may be considered to be comprised of air, water, and organic and inorganic matters. Of these four, the last is usually treated as being inert. As a result, we can learn a great deal about how a chemical partitions in the environment from its equilibrium partitioning behavior between air, water, and octanol (strictly 1-octanol), where the latter is commonly used to represent organic matter. There are three partition ratios (traditionally termed "partition coefficients"), K_{AW}, K_{OW}, and K_{OA}, but only two of these are independent, since K_{OA} must equal K_{OW}/K_{AW}:

$$K_{OA} = \frac{C_O}{C_A} = \frac{\left(C_O/C_W\right)}{\left(C_A/C_W\right)} = \frac{K_{OW}}{K_{AW}}$$

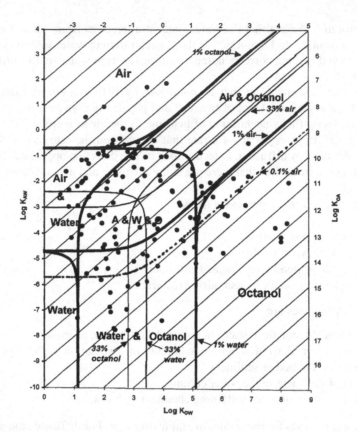

FIGURE 1.5 Plot of log K_{AW} versus log K_{OW} for the chemicals in Appendix A on which lines of constant K_{OA} are overlaid at 45° diagonally. The thicker lines represent constant percentages present at equilibrium in air, water, and octanol phases, assuming a volume ratio of 656,000:1300:1, respectively. (Modified from Gouin et al. 2000.) Note the large range in values of K_{AW} and K_{OW} for different chemicals.

These can be measured directly or estimated from vapor pressure, solubility in water, and solubility in octanol, but not all chemicals have measurable solubilities because of miscibility. Octanol is an excellent surrogate for natural organic matter in soils and sediments, lipids, or fats, and even plant waxes. It has approximately the same C:H:O ratio as lipids. Many correlations have been developed between, for example, soil–water and octanol–water partition ratios, as discussed in more detail below.

With any two of these three key partition ratios in hand, the partitioning behavior of any chemical is well defined. Any two may then be used as the basis for a "chemical space" plot, upon which a given chemical will have a unique or characteristic point that reflects its propensity for partitioning into air, water, or octanol. Figure 1.5 illustrates such a plot for the chemicals listed in Appendix A, based on their K_{OW} and K_{AW} values, which are the two more commonly available ratios of the three. Here, volatile compounds tend to lie to the upper left, water-soluble compounds to the lower left, and hydrophobic compounds to the lower right.

1.6.2 HYDROPHOBICITY

An important attribute of organic chemicals is the degree to which they are *hydrophobic*. This implies that the chemical is sparingly soluble in, or "hates," water and will preferentially partition into lipid, organic, fat, or other lower-polarity phases. A convenient descriptor of this hydrophobic tendency is K_{OW}, commonly expressed as log K_{OW} due to the large range of magnitude over which it typically varies. A high value of perhaps log $K_{OW} = 6$, as applies to dichlorodiphenyltrichloroethane

(DDT), implies that the chemical will achieve a concentration in an organic medium approximately a million times that of water with which it is in contact. In reality, most organic chemicals are approximately equally soluble in lipid or fat phases, but they vary greatly in their solubility in water. Thus, differences in hydrophobicity are largely due to differences of behavior in, or affinity for, the water phase, not differences in solubility in lipids. The word *lipophilic* is thus somewhat misleading, and the use of *hydrophobic* to describe chemicals with high log K_{OW} is preferable.

1.6.3 Vapor Pressure

A chemical's tendency to evaporate or partition into the atmosphere is primarily controlled by its vapor pressure, which is essentially the maximum pressure that a pure chemical can exert in the gas phase or atmosphere. It can be viewed as the *solubility* of the chemical in the gas phase. Indeed, if the vapor pressure in units of Pa is divided by the gas constant-temperature group RT, where R is the gas constant (8.314 Pa m^{3}mol^{-1} K^{-1}), and T is absolute temperature (K), then vapor pressure can be converted into a solubility with units of mol m^{-3}. Organic chemicals vary enormously in their vapor pressure and correspondingly in their boiling point. Some (e.g., the lower alkanes) that are present in gasoline are very volatile, whereas others (e.g., DDT) have exceedingly low vapor pressures. Partitioning from a pure chemical phase to the atmosphere is controlled by vapor pressure. Partitioning from aqueous solution to the atmosphere is controlled by K_{AW}, a joint function of vapor pressure and solubility in water. A substance may have a high K_{AW}, because its solubility in water is low. Partitioning from soils and other organic media to the atmosphere is controlled by K_{AO} (air/octanol), which is conventionally reported as its reciprocal, K_{OA}. Partitioning from water to organic media, including biotic tissues, is controlled by K_{OW}. Substances that have high vapor pressures and display a significant tendency to partition into the air phase over other phases are termed *volatile organic chemicals* or VOCs. As well, the term "semi-volatile organic chemicals" or SVOCs is also used to describe organic species that have more modest vapor pressures but remain subject to air transport.

1.6.4 Fugacity

Closely related to vapor pressure is the concept of fugacity, which is the same as partial pressure for a gas, but which also applies to a chemical in any environment. Fugacity plays a central role in this book and will be used extensively in the following chapters. First introduced by G.N. Lewis as a convenient thermodynamic equilibrium criterion, fugacity is a measure of the "escaping" or "flee-ing" tendency of a chemical from its immediate surroundings. It is identical to partial pressure in ideal gases and is linearly or nearly linearly related to concentration. Absolute values can be established because, at low partial pressures under ideal conditions, fugacity and partial pressure become equal. A chemical has an equilibrium distribution between two or more phases in contact when the fugacity of that chemical is equal in all equilibrating phases. If the fugacity of the chemical is lower in an adjacent compartment or phase to which the chemical has access, the chemical will move toward that adjoining phase, depleting the initial phase until the relative concentrations in the two interacting phases are equal to that given by the appropriate partition ratio, i.e., until they achieve equilibrium. At this point, the rates of transfer across the interface are equal, as is the fugacity in both phases. There is no longer any *net* transfer of chemical in either direction. Thus, when benzene migrates between water and air, it is "seeking to establish" an equal fugacity in both phases; its escaping tendency, or pressures, are equal in both phases. When a system is at equilibrium, and the fugacity is equal in all compartments, we shall refer to this overall or system fugacity as f_{Sys}.

Another useful quantity is the ratio of fugacity to some reference fugacity such as the vapor pressure of the pure liquid. This is a dimensionless quantity and is termed *activity*. Activity can also be used as an equilibrium criterion. Its use proves to be preferable for substances such as ions, metals, or polymers that do not appreciably evaporate and thus cannot establish measurable vapor phase concentrations and partial pressures.

Fugacity is directly related to concentration by a proportionality constant Z called a "fugacity capacity" or "Z-value", i.e.:

$$C = Zf$$

Here the Z-value is a reflection of the capacity of a certain volume of medium to "hold" fugacity. In this way, a volume of $1\,\text{m}^3$ with a concentration of $1\,\text{mol}\,\text{m}^{-3}$ would "contain" 1 Pa of that chemical's fugacity if Z were $1\,\text{Pa}\,\text{mol}^{-1}\text{m}^{-3}$. Alternatively, this same volume and concentration would hold only 1×10^{-3} Pa of fugacity, if Z were instead $1\times10^{-3}\,\text{Pa}\,\text{mol}^{-1}\text{m}^{-3}$. The fugacity of a chemical is determined by the properties of both the chemical and the medium in which it exists. For this reason, two different chemicals would be expected to have different Z-values for the same medium, and one chemical would be expected to have two different Z-values for two different media.

We assume that Z-values do not change with concentration. This is essentially a statement of Henry's law. Only at high concentrations does this become invalid. Z-values, like vapor pressure, are dependent on temperatures.

Since the fugacity of two connecting compartments at equilibrium is the same, we can redefine partition ratios in terms of fugacity capacities. Using the octanol–water partition ratio as an example, we may note that

$$K_{OW} = \frac{C_O}{C_W} = \frac{Z_O f_{\text{Sys}}}{Z_W f_{\text{Sys}}} = \frac{Z_O}{Z_W}$$

In this way, we can then see that a partition ratio is also a ratio of the two corresponding fugacity capacities, two numbers that reflect how much fugacity can be held per unit of molar concentration of the chemical within a particular medium or solvation environment.

Fugacity-based rate expressions for various processes are easily shown to be analogous to those of concentration-based rate theory. In the case of fugacity, the rate constant is replaced by a D-value. This relationship is easily demonstrated as follows for a first-order process with a rate r (mol h^{-1}), which is dependent on the concentration of the chemical C_i (mol m^{-3}) and the volume V_i (m^3):

$$r_i = k_1 V_i C_i$$

Substitution of concentration with Zf gives

$$r_i = \left(k_1 V_i Z_i\right) f_i = D_i f_i$$

Thus, we see immediately that a D-value is simply a rate constant with units of mol Pa^{-1}h^{-1} that relates the rate of change in the concentration to the prevailing fugacity. Note that the units of a D-value include the volume of the compartment in question, in order to give the rate in mol h^{-1}, i.e., the rate of change for the entire volume of the compartment. It is a simple matter to work in terms of rate of change in fugacity, as will be demonstrated below. Much of this book is dedicated to the estimation of various D-values for various physical and chemical transport processes in different simple and complex media.

1.6.5 IONIZATION

Many commercial chemicals are acids or bases, or both, and as such will ionize to greater or lesser extent in water as a function of pH. This behavior has a profound impact on the solubility of such chemicals in water, in that the neutral and ionized forms of an acid or base will generally have extremely different partition behaviors. For example, K_{OW} typically varies by $>10^4$

between neutral and ionized organic acids. As well, ionized chemicals have vanishingly low vapor pressures, such that the Henry's law constant for a neutral chemical will not be appropriate for use with the ionized form. Therefore, proper treatment of ionizing chemical partitioning requires use of a coupled-equilibrium approach that considered all relevant neutral and ionized forms. Organic chemicals may therefore be classified according to their dissociating tendencies in water solution.

The tendency to ionize is usually characterized by the acid dissociation constant K_A, often expressed as pKa, its negative base-ten logarithm. Basic compounds may be classified using the related base dissociation constant K_B. An example of an ionizing molecule is 2-nitrophenol, for which the pKa is 7.2. In acidic aqueous environments (pH < pKa), this molecule will be predominantly in its neutral form, and modeling its fate based on the properties of the neutral form will be successful. In contrast, the ionized (anionic) form will predominate in basic aqueous environments (pH > pKa), and treatment of the chemical in this form will be essential for meaningful modeling. When the pH is close to the pKa, both forms will be present and need to be considered explicitly. A more detailed discussion of ionization behavior and its impact on partitioning is given in Section 2.33.

1.6.6 REACTIVITY AND PERSISTENCE

In concert with partitioning characteristics, the other set of properties that determines environmental behavior is reactivity or persistence, usually expressed as a half-life. It is misleading to assign a single half-life to a chemical, because its value depends not only on the intrinsic properties of the chemical but also on the nature of its environment. Factors such sunlight intensity, hydroxyl radical concentration, the nature of the microbial community, and temperature vary considerably from place to place and time to time. Chemical half-lives may be semi-quantitatively grouped into classes, assuming that average environmental conditions apply, with different classes defined for air, water, soils, and sediments. The classification is that used in a series of "Illustrated Handbooks" by Mackay, Shiu, and Ma (2002) is shown in Table 1.1.

The half-lives are on a logarithmic scale with a factor of approximately 3 between adjacent classes. It is probably misleading to divide the classes into finer groupings; indeed, a single chemical may experience half-lives ranging over three classes, depending on environmental conditions such as seasonal variations in temperature, biodegradation processes, photodecomposition from sunlight, and other routes of conversion or loss.

TABLE 1.1
Classes of Chemical Half-Life or Persistence

Class	Mean Half-Life (h)	Range (h)
1	5	<10
2	17 (~1 day)	10–30
3	55 (~2 day)	30–100
4	170 (~1 week)	100–300
5	550 (~3 weeks)	300–1000
6	1700 (~2 months)	1000–3000
7	5500 (~8 months)	3000–10,000
8	17,000 (~2 years)	10,000–30,000
9	55,000 (~6 years)	>30,000

Adapted from the Handbooks of Mackay, D., Shiu, W.Y. and Ma, K.C. (2002).

PRACTICE PROBLEMS FOR CHAPTER 1

Problem 1.1: Equilibrium distribution of a pesticide between water, air, sediment, and biota.
0.04 mol of a pesticide of molar mass 200 g mol^{-1} is applied to a closed system at equilibrium consisting of 20 m^3 of water, 90 m^3 of air, 1 m^3 of sediment, and 2 L of biota (fish). If the concentration ratios are air/water 0.1, sediment/water 50, and biota/water 500, what are the concentrations and amounts in the fish in both gram and mole units?

Answer
The fish contains 0.1 g or 0.0005 mol at a concentration of 50 g m^{-3} or 0.25 mol m^{-3}.

Problem 1.2: Equilibrium distribution of a polychlorinated biphenyl (PCB) between water, suspended solids (SS), and biota.
A circular lake of diameter 2 km and depth 10 m contains SS with a volume fraction of 10^{-5}, i.e., 1 m^3 of SS per 10^5 m^3 water, and biota (such as fish) at a concentration of 1 mg L^{-1}. Assuming a density of biota of 1.0 g cm^{-3}, an SS/water partition ratio of 10^4, and a biota/water partition ratio of 10^5. Calculate the disposition and concentrations of 1.5 kg of a PCB in this system.

Answer
In this case, 8.3% is present in each of SS and biota and 83% in water with a concentration in water of 39.8 μg m^{-3}.

Problem 1.3: Steady-state CO_2 concentration in a building with an internal source.
A building, 20 m wide × 25 m long × 5 m high is ventilated at a rate of 200 m^{3}h^{-1}. The inflow air contains CO_2 at a concentration of 0.6 g m^{-3}. There is an internal source of CO_2 in the building of 500 g h^{-1}. What is the mass of CO_2 in the building and the exit CO_2 concentration?

Answer
7.75 kg and 3.1 g m^{-3}.

Problem 1.4: Steady-state pesticide concentration in field soil with periodic application, evaporative loss, and microbial degradation.
A pesticide is applied to a 10 ha field at an average rate of 1 kg ha^{-1} every 4 weeks. The soil is regarded as being 20 cm deep and well-mixed. The pesticide evaporates at a rate of 2% of the amount present per day, and it degrades microbially with a rate constant of 0.05 days^{-1}. What is the average standing mass of pesticide present at steady state? What will be the steady-state average concentration of pesticide (g m^{-3}), and in units of mg g^{-1} assuming a soil solids density of 2500 kg m^{-3}?

Answer
5.1 kg, 0.255 g m^{-3}, 0.102 mg g^{-1}.

Problem 1.5: Time-dependent change in CO_2 concentration following source input rate change.
If the concentration of CO_2 in Example 1.3 has reached a steady state of 3.1 g m^{-3} and then the internal source is reduced to 100 g h^{-1}, deduce the equation expressing the time course of CO_2 concentration decay and the new steady-state value.

Answer
New steady-state CO_2 concentration value is 1.1 g m^{-3} and $C_{CO_2}^t = 1.1 + 2.0 \exp(-0.08t)$.

Problem 1.6: Time for the lake to achieve a given concentration following pulse application of a piscicide that undergoes first-order loss by degradation.

A lake of volume $10^6\,m^3$ has an outflow of $500\,m^3h^{-1}$. It is to be treated with a piscicide, the concentration of which must be kept above $1\,mg\,m^{-3}$. It is decided to add $3\,kg$, thus achieving a concentration of $3\,mg\,m^{-3}$, and to allow the concentration to decay to $1\,mg\,m^{-3}$ before adding another $2\,kg$ to bring the concentration back to $3\,mg\,m^{-3}$. If the piscicide has a degradation half-life of $693\,h$ (29 days), what will be the interval before the second (and subsequent) applications are required?

Answer
30 days.

Problem 1.7: Relative efficiency of different application pulse intervals for a piscicide that undergoes first-order loss by degradation.

For the same lake as in Problem 1.6, Mr MacLeod, being economically and ecologically perceptive, claims that if he is allowed to make applications every 10 days instead of 30 days, he can maintain the concentration above $1\,mg\,m^{-3}$ but reduce the piscicide usage by 35%. Is he correct? What is the absolute minimum piscicide usage every 30 days to maintain $1\,mg\,m^{-3}$?

Answer
Yes. A total of $1.08\,kg$ would have to be added every 30 days.

SYMBOLS AND DEFINITIONS INTRODUCED IN CHAPTER 1

Symbol	Common Units	Description
A_{ij}	m^2	Area of interface between phases "i' and "j"
C_l	moles m^{-3}	Concentration of chemical in phase "i"
C_i^0	moles m^{-3}	Initial concentration of chemical in phase "i" at time $t = 0$
$C_i^{t=j}$	moles m^{-3}	Concentration of chemical in phase "i" at time $t = j$
C_i^{Inf}	moles m^{-3}	Inflow concentration of chemical associated with phase "i"
C_i^{Outf}	moles m^{-3}	Outflow concentration of chemical associated with phase "i"
f_i	Pa	Fugacity of chemical in phase "i"
f_{Sys}	Pa	Fugacity of chemical in entire system at equilibrium
Z_i	mol $Pa^{-1}m^{-3}$	Fugacity capacity for a chemical in phase "i"
G_i^{Inf}	m^3h^{-1}	Inflow rate associated with phase "i"
G_i^{Outf}	m^3h^{-1}	Outflow rate associated with phase "i"
K_{ij}	Various	Partition ratio between phases "i" and "j"
k_i	h^{-1}	First-order rate constant for a loss process in phase "i"
k_i^{Deg}	h^{-1}	First-order rate constant for chemical degradation process(es) in phase "i"
k_i^F	h^{-1}	First-order rate constant for flow residence in phase "i"
k_i^{Loss}	h^{-1}	First-order rate constant for loss or decay in phase "i"
L_i^{Vol}	mol $m^{-2}h^{-1}$	Flux (rate per area) of chemical by volatilization from phase "i"
m_i	moles	Moles of chemical present in phase "i"
m_0	moles	Moles of chemical present at time 0 or initially
m_{Sys}	moles	Moles of chemical present in the entire system in question
m_i^{Em}	moles	Amount of chemical emitted into phase "i"
n	—	Number of phases in the system

(*Continued*)

Symbol	Common Units	Description
r_i	mol h^{-1}	Molar flow rate associated with phase or process "i"
r_i^{Em}	mol h^{-1}	Molar emission flow rate into phase "i"
r_i^{Inf}	mol h^{-1}	Molar inflow rate associated with phase "i"
r_i^{Outf}	mol h^{-1}	Molar outflow rate associated with phase "i"
r_i^{Deg}	mol h^{-1}	Molar degradation rate of chemical in phase "i"
r_i^{Vol}	mol h^{-1}	Molar volatilization rate of chemical from phase "i"
r_{Tot}	mol h^{-1}	Total molar flow rate into and out of a steady-state system
r_{Tot}^{In}	mol h^{-1}	Total molar flow rate into a phase or system
r_{Tot}^{Out}	mol h^{-1}	Total molar flow rate out of a phase or system
τ_F	h	Flow residence time
τ_{Deg}	h	Degradation residence time
τ_{Loss}	h	Characteristic time for loss or decay
τ_T	h	Total residence time
$t_{1/2}$	h	Time to achieve half the theoretical equilibrium concentration
$t_{0.37}$	h	Time to achieve 0.37 of the theoretical equilibrium concentration
$t_{0.9}$	h	Time to achieve 0.9 of the theoretical equilibrium concentration
V_i	m^3	Volume of phase "i"
Y_i	m	Depth or thickness of phase "i"

2 Equilibrium Partitioning

THE ESSENTIALS

Partition ratios: $K_{12} = \dfrac{C_1}{C_2} \approx \dfrac{C_1^{\text{Sat}}}{C_2^{\text{Sat}}}$

Equilibrium fugacity condition: $f_{\text{Sys}} = f_1 = f_2$

Fugacity Ratio for Solids, accurate and approximate at 298 K, for a chemical with melting point T in Kelvin:

$$\ln F = -\frac{\Delta_f S}{RT}(T_M - T) + \frac{\Delta C_P}{RT}(T_M - T) - \frac{\Delta C_P}{R}\ln(T_M/T)$$

$$\ln F \simeq -\frac{\Delta_f S}{RT}(T_M - T) \simeq -\frac{\Delta_f S}{R}\left(\frac{T_M}{T} - 1\right)$$

$$\log F \simeq -0.01(T_M - 298) \quad \text{where} \quad \frac{\Delta_f S}{R} \simeq 6.79$$

Fugacity Capacities $(\text{mol}\,\text{Pa}^{-1}\,\text{m}^{-3})$ at temperature T:

Air: $Z_A \simeq \dfrac{1}{RT} = 4.00 \times 10^{-4}$ at 27.5°C

Solute in solvent "w": $Z_W = \dfrac{1}{v_W \gamma_i f_i^{\circ}} = \dfrac{1}{H} = \dfrac{C_L^{\text{Sat}}}{P_L^{\text{Sat}}}$

Ionizing acid in water: $Z = Z_W(I+1) = Z_W\left(10^{(\text{pH}-\text{pK}_a)}+1\right)$,

$$I = \frac{C_{A^-}}{C_{HA}} = 10^{(\text{pH}-\text{pK}_a)}$$

Solid sorbents or biota: $Z = \dfrac{K_P \rho_S}{H} = \dfrac{K_B \rho_S}{H}$

Pure solutes: $Z_P = \dfrac{1}{P_L^{\text{Sat}} v}$

Heterogeneous bulk mixtures: $Z_T = \displaystyle\sum_{i=1}^{n} v_i Z_i$

Multi-Compartment Equilibrium Partitioning:

Partition Ratio   Fugacity

Mass balance

$$m_{Sys} = C_1 \left[V_1 + \sum_{i=2}^{n} K_{i1} V_i \right]$$

$$m_{Sys} = f_{Sys} \sum_{i=1}^{n} V_i Z_i$$

$$f_{Sys} = \frac{m_{Sys}}{\sum_{i=1}^{n} V_i Z_i}$$

Concentration

$$C_1 = \frac{m_{Sys}}{\left[V_1 + \sum_{i=2}^{n} K_{i1} V_i \right]}$$

$$C_i = Z_i f_{Sys}$$

$$C_i = K_{i1} C_1$$

Amount of chemical

$$m_i = C_i V_i$$ $$m_i = V_i Z_i f_{Sys}$$

2.1 PARTITIONING THEORY

2.1.1 INTRODUCTION

Environmental scientists are often concerned with the distribution of a chemical between diverse phases. The essence of the partitioning calculation is to determine the equilibrium concentration of a chemical in a given phase if it is placed in contact with another phase with a known initial concentration of that chemical. The simplest situation of this type is when the chemical's distribution is static, that is, the system is at steady state and equilibrium. It is illustrative to consider a simple two-compartment system (Figure 2.1) in which a small volume of a non-aqueous phase is introduced into water that contains a dissolved chemical such as benzene. There is a tendency for some of the benzene to migrate into this new phase, with an equilibrium distribution of concentrations in the two phases being established in hours or days.

2.1.2 FUGACITY AND CONCENTRATION

It transpires that two approaches can be used to develop equations relating equilibrium concentrations to each other as shown in Figure 2.1. The simpler and most widely used is Nernst's Distribution law, which postulates that the concentration ratio C_1/C_2 is relatively constant and is equal to a partition or distribution ratio K_{12}. Thus, C_2 can be calculated as C_1/K_{12}. Ideally, K_{12} is expressed as a function of temperature and, if necessary, of concentration. To determine K_{12} by measurement, mixtures are equilibrated, both concentrations measured, and plotted as in Figure 2.1. Linear or nonlinear equations then can be fitted to the data.

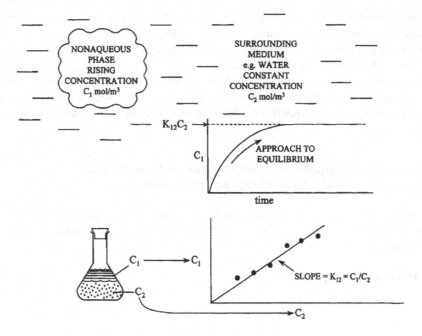

Experimental Determination of Partition Ratios

The Equilibrium Criterion Approach

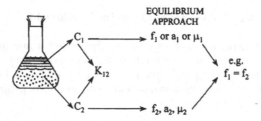

FIGURE 2.1   Some principles and concepts in phase equilibrium.

The second approach involves the introduction of the intermediate quantity, fugacity, which is a useful criterion of equilibrium. Fugacity is preferred for calculations involving most organic substances because of the simplicity of the corresponding equations that relate fugacity to concentration. It is linearly proportional to concentration via a constant of proportionality called a fugacity capacity or Z-value:

$$C = Zf$$

The advantage of the use of fugacity as an equilibrium criterion is that the properties of each phase are treated separately using a phase-specific equation. By contrast, partition ratios treat phases in pairs, which can obscure the nature of the underlying phenomena. We may detect a variability in K_{12} and not know from which phase the variability is arising. Further complications arise if we have a large number of phases to consider. For example, with ten phases, there are then 90 possible partition ratios, of which only nine are independent. By comparison, there are only ten Z-values needed to define this system at equilibrium.

Our task, then, is to start with a concentration of solute chemical in one phase, from this deduce the fugacity, argue that this equilibrium criterion will be equal in the other phase, and then calculate the corresponding concentration in the second phase. We therefore require recipes for deducing C from f and vice versa. This approach is depicted at the bottom of Figure 2.1.

The partition ratio approach sidesteps the use of fugacities. This is because of the inherent assumption that whatever the factors are that are used to convert C_1 to f_1 and C_2 to f_2, the ratio of these factors is constant over the range of concentration of interest. Thus, it is not actually necessary to calculate the fugacities. In the fugacity approach, no such assumption is made, and the individual calculations are undertaken. We can illustrate these different approaches with an example.

Worked Example 2.1: Equilibrium concentration of benzene in air in contact with water of a known benzene concentration.

Benzene is present in water at a concentration C_W of 1.0 mol m^{-3}. The partition ratio between air and water at the temperature of the water is $K_{AW} = 0.20$. What is the equilibrium concentration of benzene in the air C_A?

A. **Partition ratio approach**

$$K_{AW} = \frac{C_A}{C_W} = 0.20$$

Given that we know K_{AW} and the molar concentration of benzene in water C_W, we can isolate and solve for the molar concentration of benzene in air C_A:

$$C_A = K_{AW}C_W = 0.20 \times 1.0\,\text{mol m}^{-3} = 0.20\,\text{mol m}^{-3}$$

Partition ratios are generally unitless and are the same for any given partition ratio when expressed in any unit of concentration that is proportional to molarity. Since the molar mass of benzene is 78 g mol^{-1} the mass concentration of a 1.0 mol m^{-3} benzene solution in water is 78 g m^{-3}. The benzene mass concentration in air may be calculated in an analogous manner:

$$C_A = K_{AW}C_W = 0.20 \times 78\,\text{g m}^{-3} = 16\,\text{g m}^{-3}$$

B. **Fugacity approach**

For air and water under these conditions, it will be shown later that the equilibrium condition $f_{Sys} = f_A = f_W$ allows us to write

$$f_{Sys} = f_W = \frac{C_W}{Z_W} = \frac{C_A}{Z_A} = f_A$$

i.   Determine the fugacity f_A of the air at equilibrium, given the value of the fugacity capacities of water $Z_W = 2.0 \times 10^{-3}$ mol Pa^{-1}m^{-3} and air $Z_A = 4.0 \times 10^{-4}$ mol Pa^{-1}m^{-3}
     First, calculate the water fugacity from the water data:

$$f_W = f_{Sys} = \frac{C_W}{Z_W} = \frac{1.0\,\text{mol m}^{-3}}{2.0 \times 10^{-3}\,\text{mol Pa}^{-1}\text{m}^{-3}} = 500\,\text{Pa}$$

ii. Determine the concentration in the air, given Z_A, the fugacity capacity of air:

$$C_A = Z_A f_A = Z_A f_{Sys} = 4.0 \times 10^{-4}\,\text{mol Pa}^{-1}\,\text{m}^{-3} \times 500\,\text{Pa} = 0.2\,\text{mol m}^{-3}$$

Given that fugacity capacities Z are intrinsically molar quantities, we cannot sensibly do the calculation of mass concentration of benzene in air without defining mass-based fugacity capacities. However, as was demonstrated above, the ratio $Z_A/Z_W = K_{AW}$, allowing us to easily do the same calculation as in A above, but in the fugacity context.

Clearly, the principal challenge in working with the fugacity approach is the determination of the fugacity capacities Z_W and Z_A, or K_{AW}, which is their ratio. We therefore face the task of developing methods of estimating Z-values that relate concentration and fugacity, and partition ratios that are ratios of Z-values. The theoretical foundations are set out in Section 2.3, and result in a set of working equations applicable to the air–water–octanol system. The three solubilities (or pseudo-solubilities) in these media and the three partition ratios are then developed and discussed with regard to their application in problems involving environmental media such as soils and aerosols. We use the term "pseudo-solubility" in cases where the chemical, such as ethanol, is miscible with water, such that no solubility exists or can be measured.

2.2 PROPERTIES OF PURE SUBSTANCES

It is important to ascertain whether the substance of interest is solid, liquid, or vapor at the environmental temperature, for reasons discussed below. This is easily determined by comparing the ambient temperature with the melting and boiling point temperatures of the chemical in question, if known. The relationship between temperature, pressure, and phase of a given substance is compactly summarized in a pressure–temperature (P–T) diagram, of which Figure 2.2 is an example. Of particular interest for solids is the super-cooled (or sub-cooled) liquid vapor pressure line, shown as a dashed line. This is the vapor pressure that a solid (such as naphthalene, which melts at 80°C) would have *if* it were liquid at 25°C. The reason naphthalene is not liquid at 25°C is that it is able to achieve a lower free energy state by forming a solid crystal. Above 80°C, this lower energy solid state is less stable, and the substance remains liquid. Above the boiling point, the liquid state is abandoned in favor of a vapor state. It is not possible to measure the super-cooled liquid vapor pressure by direct experiment. It can be calculated as discussed shortly, and it can however be measured experimentally, though not directly, using gas chromatographic retention times. It is possible to measure the vapor pressure above the boiling point by operating at high pressures. Beyond the critical point, the vapor pressure cannot be measured, but it can be estimated.

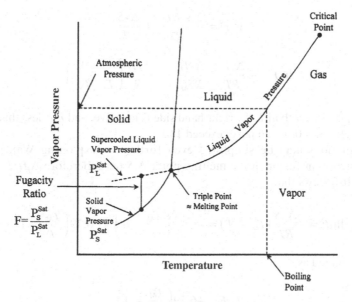

FIGURE 2.2 P–T diagram for a pure substance.

Correlations exist between the vapor pressure of solids and super-cooled liquids, which can be used to calculate the unmeasurable super-cooled liquid vapor pressure from that of the solid. The reason for this is that, when a solid such as naphthalene is present in a dilute, sub-saturated, dissolved, or sorbed state at 25°C, the molecules do not encounter each other with sufficient frequency to form a crystal. Thus, the low-energy crystal state is not accessible. The molecule thus behaves as if it were a liquid at 25°C. It "thinks" it is a liquid, because it has no access to information about the stability of the crystalline state, i.e., does not "know" its melting point. As a result, it behaves in a manner corresponding to the liquid vapor pressure. A similar phenomenon occurs above the critical point where a gas such as oxygen, when in solution in water, behaves as if it were a liquid at 25°C, not a gas. No liquid vapor pressure can be measured for either naphthalene or oxygen at 25°C; it can only be calculated. Later, we term this liquid vapor pressure the reference fugacity. We may need to know this fictitious vapor pressure for several reasons.

The ratio of the solid vapor pressure to the super-cooled liquid vapor pressure is termed the *fugacity ratio*, F. To estimate F, we need to know how much energy is involved in the solid–liquid transition, i.e., the enthalpy of melting or fusion. The rigorous equation for estimating F at temperature T (K) is (Prausnitz et al., 1986)

$$\ln F = -\frac{\Delta_f S}{RT}(T_M - T) + \frac{\Delta C_P}{RT}(T_M - T) - \frac{\Delta C_P}{R}\ln(T_M/T)$$

where $\Delta_f S$ (J mol^{-1} K^{-1}) is the entropy of fusion at the melting point T_M (K), ΔC_P (J mol^{-1} K^{-1}) is the difference in heat capacities between the solid and liquid substances, and R is the gas constant.

The heat capacity terms are usually small, and they tend to cancel, so the equation can often be simplified to

$$\ln F = -\frac{\Delta_f S}{RT}(T_M - T)$$

Since for the process of melting in which the Gibbs energies of both the solid and the liquid are the same and therefore $\Delta_f G = \Delta_f H - T_M \Delta_f S = 0$, we can expand this equation and equate the enthalpy of fusion $\Delta_f H$ (J mol^{-1}) with $T_M \Delta_f S$ to obtain

$$\ln F = -\frac{\Delta_f S T_M}{RT} + \frac{\Delta_f S}{R}$$

$$\ln F = -\frac{\Delta_f H}{RT} + \frac{\Delta_f H}{RT_M} = -\frac{\Delta_f H}{R}\left(\frac{1}{T} - \frac{1}{T_M}\right)$$

Note that since T_M is greater than T, the right-hand side is negative, and F is less than 1, except at the melting point, when it is 1.0. F can never exceed 1.0.

The above equations may be simplified even further by exploiting Walden's rule, which states that the entropy of fusion at the melting point $\Delta_f S$ (and therefore $\Delta_f H/T_M$) is often about 56.5 J mol^{-1} K^{-1}. It follows that

$$\ln F = -\frac{\Delta_f S}{RT}(T_M - T) = -\frac{\Delta_f S}{R}\left(\frac{T_M}{T} - 1\right) \approx -6.79\left(\frac{T_M}{T} - 1\right)$$

Thus, F is approximated as

$$\ln F \approx -6.79\left(\frac{T_M}{T} - 1\right)$$

If base ten logs are used and T is 298 K, this equation becomes

$$\log F = -\frac{6.79}{2.303}\left(\frac{T_M}{298\text{ K}} - 1\right) = -0.01(T_M(K) - 298) \text{ or } -0.01(T_M(^\circ C) - 25)$$

This is useful as a quick and easily remembered method of estimating F. If more accurate data are available for $\Delta_f H$ or $\Delta_f S$, they should be used, and if the substance is a high melting point solid, it may be advisable to include the heat capacity terms.

2.3 PROPERTIES OF SOLUTES

2.3.1 SOLUTION IN THE GAS PHASE

The easiest case is a solution in a gas phase (air) in which there are usually no interactions between molecules other than collisions. The basic fugacity equation as presented in thermodynamics texts (Prausnitz et al., 1986) is

$$f = y\phi P_T$$

where y is the mole fraction, ϕ is a fugacity coefficient, and P_T is the total (atmospheric) pressure. The partial pressure P_i is therefore yP_T and conversely $P_T = P_i/y$.

If the gas law applies,

$$P_T V = nRT \text{ and } P_i V = y(nRT)$$

Here, n is the total number of moles of all gases present, R is the gas constant, V is the volume (m^3), and T is the absolute temperature (K). Now the concentration of all gas molecules is given by $C = n/V$, and that of the solute $C_A = yC$ or $y(n/V)$. From the above equation, it follows that C_A also equals P_i/RT mol m^{-3} such that:

$$C_A = \frac{yP_T}{RT} = \left(\frac{1}{\phi RT}\right)f$$

and therefore

$$Z_A = \frac{1}{\phi RT}$$

Fortunately, the fugacity coefficient ϕ rarely deviates appreciably from unity under environmental conditions. The exceptions occur at low temperatures, high pressures, or when the solute molecules interact chemically with each other in the gas phase. Only this last class is important environmentally. This may be seen in cases involving chemicals such as certain gaseous molecules such as short-chain carboxylic acids and NO_2, which form dimers in the gas phase. The constant Z_A is thus usually $(1/RT)$ or about 4×10^{-4}mol Pa^{-1}m^{-3} near 298 K and is the same for all non-interacting substances.

The fugacity is thus numerically equal to the partial pressure of the solute P or yP_T under environmental conditions. This raises a question as to why we use the term *fugacity* in preference to *partial pressure*. The answers are that (1) under conditions when ϕ is not unity, fugacity and partial pressure are not equal, and (2) there is some conceptual difficulty about referring to a "partial pressure of DDT in a fish" when there is no vapor present for a pressure to be present in—even partially.

2.3.2 SOLUTION IN LIQUID PHASES

The fugacity equation for a solute is defined in the context of Raoult's law, which states that the partial pressure of a solute P_i is equal to the partial pressure of that solute in the pure liquid state P_i^0 times its mole fraction in the mixture x_i:

$$P_i = x_i P_i^0$$

The fugacity of the same solute (Prausnitz et al., 1986) is given by

$$f_i = x_i \left(\gamma_i f_i^0 \right)$$

where γ_i is an activity coefficient that accounts for observed deviations from Raoult's law and f_i^0 is the reference fugacity of the pure liquid.

Let us consider the solvent to be water, although the development here is completely general for any "well-behaved" solvent. The mole fraction of the solute x_i can be converted to concentration C (mol m^{-3}) using molar volumes v (m^{3}mol^{-1}), amounts m (mol), and volumes V (m^3) of solute (subscript i) and solution (solvent indicated here as water by a subscript w). Assuming that the solute concentration is small, i.e., $V_i \ll V_W$, we can write

$$C_i = \frac{m_i}{V_W + V_i} \simeq \frac{m}{V_W}$$

Expressing the solution volume in terms of the molar volume of the solvent water, we have

$$V_W = m_W v_W$$

At high dilution, we can approximate the mole fraction of the solute as

$$x_i = \frac{m_i}{m_W + m_i} \simeq \frac{m_i}{m_W}$$

or

$$m_i \simeq x_i m_W$$

Given these two expressions, we can deduce that

$$C_i = \frac{m_i}{V_W} \simeq \frac{x_i m_W}{m_W v_W} = \frac{x_i}{v_W}$$

or

$$x_i \simeq C_i v_W$$

Returning to our initial expression for fugacity of a solute, we can now state that

$$f_i = x_i \left(\gamma_i f_i^0 \right) = C_i v_W \left(\gamma_i f_i^0 \right)$$

and therefore

$$C_i = \frac{1}{v_W \left(\gamma_i f_i^0 \right)} f_i$$

Given that, in general, $C = Zf$, we have that the fugacity capacity of a solute in a solvent (water here) is given by

$$Z_W = \frac{1}{v_W \gamma_i f_i^0}$$

The reference fugacity f_i^0 is by definition the fugacity of the solute in the pure liquid state where x_i and γ_i are both 1.0 (the latter by definition), which corresponds to the vapor pressure of pure liquid solute at the temperature (and strictly the pressure) of the system.

The activity coefficient γ is defined here on a "Raoult's law" basis such that γ is 1.0 when x_i is 1.0. In most cases, γ-values exceed 1.0 and, for very hydrophobic chemicals, values may be in millions. An alternative convention, which we do not use here, is to define γ on a Henry's law basis such that γ is 1.0 when x_i is zero.

The activity coefficient is thus a very important quantity. It can be viewed as the ratio of the activity or fugacity of the solute in a given solvent, with respect to the activity or fugacity that this solute would have if it were in a solution consisting entirely of its own kind. Its dependence on the concentration of the solute often has the following general approximate form:

$$\log \gamma_i = \log \gamma_0 (1 - x_i)^2$$

where γ_0 is the activity coefficient at "infinite dilution," i.e., when the mole fraction x_i approaches zero.

Another useful way of viewing activity coefficients is that they can be regarded as an inverse expression of mole fraction solubility, i.e., an *insolubility*. A solute that is sparingly soluble in a solvent will have a high activity coefficient, an example being hexane in water. For a liquid solute such as hexane, at the solubility limit, when excess pure hexane is present, the fugacity equals the reference fugacity f_i^0 and

$$f_i = f_i^0 = x_i \gamma_i f_i^0$$

For this equality to hold, it is necessary that x_i and γ_i be inversely related, as

$$x_i = \frac{1}{\gamma_i} \text{ or } \gamma_i = \frac{1}{x_i}$$

The activity coefficient is thus the reciprocal of the solubility when expressed as a mole fraction. For solids at saturation, f_i is the fugacity of the pure solid f_S. Thus,

$$f_S = f_i^0 = x_i \gamma_i f_i^0$$

and

$$x_i = \frac{f_S}{\gamma_i f_i^0} = \frac{F}{\gamma_i}$$

where F is the fugacity ratio introduced earlier. Solid solutes of high melting point such as DDT thus tend to have low solubilities, because F is small and γ_i is large.

It is more common to express solubilities in concentration units such as g m^{-3}. Under dilute conditions, the solubility C_W^{Sat} mol m^{-3} is x_i/v_S, where v_S is the molar volume of the solution (m^3 mol^{-1}) and approaches the molar volume of the solvent. C_W^{Sat} is thus $1/\gamma_i v_S$ for liquids and $F/\gamma_i v_S$ for solids. In the gas phase, the solubility C_G^{Sat} is essentially the vapor pressure P^{Sat} in disguise, i.e.,

TABLE 2.1

Reference Table of Fugacity Capacity Definitions for Chemicals in Different Media

<table>
<tr><td colspan="3" align="center">Definitions of Fugacity Capacities (at Temperature T)</td></tr>
<tr><td>Compartment</td><td></td><td>Z (mol Pa^{-1}m^{-3})</td></tr>
<tr><td>Air</td><td>$1/(RT)$</td><td>$R = 8.314$ Pa m^{3}mol^{-1}</td></tr>
<tr><td></td><td></td><td>T = Temperature (K)</td></tr>
<tr><td>Water</td><td>$1/H$ or C_W^{Sat}/P^{Sat}</td><td>C_W^{Sat} = Aqueous solubility (mol m^{-3})</td></tr>
<tr><td></td><td></td><td>P^{Sat} = Vapor pressure (Pa)</td></tr>
<tr><td></td><td></td><td>H = Henry's law constant (Pa m^{3}mol^{-1})</td></tr>
<tr><td>Solid sorbent</td><td>$(K_P \rho_S)/H$</td><td>K_P = Partition ratio (L kg^{-1})</td></tr>
<tr><td></td><td></td><td>ρ_S = Density of solid (kg L^{-1})</td></tr>
<tr><td>Biota</td><td>$(K_B \rho_S)/H$</td><td>K_B = Bioconcentration factor (L kg^{-1})</td></tr>
<tr><td></td><td></td><td>ρ_S = Density of the biota (kg L^{-1})</td></tr>
<tr><td>Pure solute</td><td>$1/(P^{Sat}v)$</td><td>v = Solute molar volume (m^{3}mol^{-1})</td></tr>
</table>

$$C_G^{Sat} = \frac{n}{V} = \frac{P^{Sat}}{RT}$$

Returning to the definition of Z_W as $\left(1/\left(v_W \gamma_i f_i^0\right)\right)$, it is apparent that Z_W can also be expressed in terms of aqueous solubility, C_W^{Sat}, and vapor pressure. For liquid solutes, C_W^{Sat} is $1/(\gamma_i v_W)$, and for solid solutes, it is $F/(\gamma_i v_W)$. The reference fugacity f_R is the vapor pressure of the liquid, i.e., it is P^{Sat} for a liquid and $\left(P^{Sat}/F\right)$ for a solid. Substituting gives Z_W as C_W^{Sat}/P^{Sat} in both cases, the F canceling for solids. The ratio P^{Sat}/C_W^{Sat} is the Henry's law constant H in units of Pa m^{3}mol^{-1}; thus, Z_W is $1/H$:

$$\text{Liquid: } Z_W = \frac{1}{v_W \gamma_i f_i^0} = \frac{\left(1/v_W \gamma_i\right)}{f_i^0} = \frac{C_W^{Sat}}{P^{Sat}} = \frac{1}{\left(P^{Sat}/C_W^{Sat}\right)} = \frac{1}{H}$$

$$\text{Solid: } Z_W = \frac{1}{v_W \gamma_i f_i^0} = \frac{\left(1/v_W \gamma_i\right)}{f_i^0} = \frac{\left(C_W^{Sat}/F\right)}{\left(P^{Sat}/F\right)} = \frac{1}{\left(P^{Sat}/C_W^{Sat}\right)} = \frac{1}{H}$$

Polar solutes such as ethanol do not have measurable solubilities in water, because they are miscible. This generally occurs when γ is less than about 20. We can still use the concept of solubility and call it a "hypothetical or pseudo-solubility" if it is defined as $1/(\gamma_i v_S)$. For a liquid substance that behaves nearly ideally, i.e., γ_i is 1.0, the solubility approaches $1/v_S$, which is the density of the solvent in units of mol m^{-3}. For water, this is about 55,500 mol m^{-3}, i.e., 10^6 g m^{-3} divided by 18 g mol^{-1}. For a solid solute under ideal conditions, the solubility approaches F/v_S (mol m^{-3}).

These equations are general and apply to any non-ionizing chemical in solution in any liquid solvent, including water and octanol. The solution molar volumes and the activity coefficients vary from solvent to solvent. The Z-value for a chemical in octanol is, by analogy, $1/(v_O \gamma_i f_R)$, where v_O is the molar volume of octanol (Table 2.1).

2.3.3 Solutions of Ionizing Substances

Certain substances, when present in solution, adopt an equilibrium distribution between two or more chemical forms. Examples are acetic acid, ammonia, and pentachlorophenol (PCP), which ionize by virtue of association with water releasing H$^+$ (strictly H$_3$O$^+$(H$_2$O)n) or OH$^-$ ions.

Some substances dimerize or form hydrates. For ionizing substances, their distribution between phases is pH dependent, and so their solubilities and activities are also pH dependent. This behavior could be accommodated by defining Z as being applicable to the total concentration, but it then becomes pH dependent.

A more rigorous approach is to define Z for each chemical species, noting that, for ionic species, Z in air must be zero under normal conditions, because ions as such do not evaporate. It is useful to know the relative proportions of each species, because they will partition differently. This issue is critical for metals in which only a small fraction may be in free ionic form.

For acids, an acid dissociation constant K_a is defined as

$$K_a = \frac{C_{H^+} C_{A^-}}{C_{HA}}$$

where C_{H^+} is the hydrogen ion concentration, C_{A^-} is that of the dissociated anionic form, and C_{HA} is that of the undissociated parent acid. The ratio of ionic to non-ionic forms I is thus

$$I = \frac{C_{A^-}}{C_{HA}} = \frac{K_a}{C_{H^+}} = \frac{10^{\log K_a}}{10^{\log C_{H^+}}} = \frac{10^{-pKa}}{10^{-pH}} = 10^{(pH-pKa)}$$

where we have used the definitions of $pH = -\log(C_{H^+})$ and $pKa = -\log(K_a)$.

This latter expression is one form of the Henderson–Hasselbalch equation. For acids, when pKa exceeds pH by two units or more, $I < 0.01$, such that ionization can be ignored. When considering substances which have the potential to ionize it is essential to obtain pKa and determine the relative proportions of each form. The handbook by Lyman et al. (1982) and the text by Perrin et al. (1981) can be consulted for more details of estimation methods for pKa, and for applications. As well, on-line database resources such as EPI Suite may be consulted for experimental or Quantitative Structure–Property Relationship (QSPR)-estimated pKa values. See the on-line fugacity models and modeling resources tabulation from the Canadian Environmental Modelling Centre website (www.trentu.ca/cemc/resources-and-models) for a listing of on-line resources of this type.

Dissociation can be regarded as causing an increase in the Z-value of a substance in aqueous solution. The total Z-value is the sum of non-ionic and ionic contributions, which will have respective fractions $1/(I + 1)$ and $I/(I + 1)$:

$$\text{Neutral fraction} = \frac{1}{(I+1)}$$

$$\text{Ionized fraction} = \frac{I}{(I+1)}$$

The Z-value of the non-ionic form Z_W can be calculated by measuring solubility, activity coefficient, or another property under conditions when I is very small, i.e., pH $\ll$ pKa. The same Z_W value applies to the non-ionic form at all pH levels. The additional contribution of the ionic form is then calculated at the pH of interest as IZ_W, and the total effective Z-value is $Z_W(I+1)$, which can be used to calculate the total concentration. An inherent assumption here is that the presence of the ionic form does not affect Z_W.

Bringing these ideas together, we then have an equation for the effective fugacity capacity for an ionizing acid in water:

$$Z = Z_W(I+1) = Z_W \left(10^{(pH-pKa)} + 1\right)$$

To illustrate the relationship between pKa and ratio of ionic to non-ionic forms, consider phenol which has a pKa of 9.90, and pentachlorophenol (PCP) which has a pKa of 4.74 (Mackay et al., 1995). At a pH of 6.0, the corresponding values of the ratio of ionic to non-ionic forms I are 1.26×10^{-4} (phenol) and 18.2 (PCP):

$$I_{\text{Phenol}} = 10^{(\text{pH}-\text{pKa})} = 10^{(6.0-9.9)} = 10^{(-3.9)} = 1.26 \times 10^{-4}$$

$$I_{PCP} = 10^{(pH-pKa)} = 10^{(6.0-4.74)} = 10^{(+1.26)} = 18.2$$

For phenol, ionization can be ignored for pH values up to about 8. For PCP, the dominant species in solution is the ionic form, and the total Z-value in aqueous solution is 19.2 times that of the non-ionic form:

$$Z = Z_W(I+1) = Z_W(18.2+1) = 19.2 \times Z_W$$

As a result, partitioning of PCP from solution in water to other media, such as air or octanol, is pH dependent, and the issue of whether the ionic form also partitions must be addressed. K_{OW} is thus pH dependent if (as is usual) total concentrations are used to calculate it.

2.4   PARTITION RATIOS

2.4.1   FUGACITY AND SOLUBILITY RELATIONSHIPS

If we have two immiscible phases or media (e.g., air and water or octanol and water), we can conduct experiments by shaking volumes of both phases with a small amount of solute such as benzene to achieve equilibrium, then measure the concentrations of solute in each phase and plot one concentration with respect to the other, as was shown in Figure 2.1. It is preferable to use identical concentration units in each phase of amount per unit volume but, when one phase is solid, it may be more convenient to express concentration in units such as amounts per unit mass (e.g., $\mu g\ g^{-1}$) to avoid having to estimate phase densities.

The plot of the concentration data is often linear at low concentrations; therefore, we can write

$$C_2 \propto C_1 \text{ or } C_2 = K_{21}C_1$$

The constant of proportionality, which is also the slope of the line, is K_{21}, the partition ratio, which is constant at low concentrations for a given temperature. Rearrangement gives the conventional definition of the concentration-based partition ratio:

$$K_{21} = \frac{C_2}{C_1}$$

Now, since C_2 is $Z_2 f_2$ and C_1 is $Z_1 f_1$, and at equilibrium $f_1 = f_2 = f_{\text{Sys}}$, it follows that K_{21} is Z_2/Z_1:

$$K_{21} = \frac{C_2}{C_1} = \frac{Z_2 f_{\text{Sys}}}{Z_1 f_{\text{Sys}}} = \frac{Z_2}{Z_1}$$

It is apparent that a fugacity capacity "Z" value can be regarded as "half" a partition ratio. If we know Z for one phase (e.g., Z_1 as well as K_{21}), we can deduce the value of Z_2 as $K_{21}Z_1$. This proves to be a convenient method of estimating Z-values.

Partition ratios are widely available and used for systems of air–water, aerosol–air, octanol–water, lipid–water, fat–water, hexane–water, "organic carbon"–water, and various minerals with water. Applying the theory that was developed earlier and noting that, at equilibrium, the solute fugacities will be equal in both phases, we can define partition ratios for air–liquid and liquid–liquid systems in terms of solubilities.

Using air–water as an example, at a total atmospheric pressure P_T, the mole fractions in water and air are given by

$$f_i = x_i \gamma_i f_i^0 \text{ or } x_i = \frac{f_i}{\gamma_i f_i^0}$$

and

$$y_i P_T = P_i \text{ or } y_i = \frac{P_i}{P_T}$$

Bringing these together, and recalling that the gas-phase fugacity is equal to the partial pressure of the species in question, we have for the ratio of the mole fractions in the gas and liquid phases:

$$\frac{y_i}{x_i} = \frac{\left(\dfrac{P_i}{P_T}\right)}{\left(\dfrac{f_i}{\gamma_i f_i^0}\right)} = \frac{\gamma_i f_i^0}{P_T}$$

It is more common to work with concentrations represented as C_j^i (mol m^{-3}) rather than mole fractions. Here as elsewhere, "j" represents the solvation phase and "i" represents the chemical in question. The necessary conversion relationships are

$$y_i = C_A^i v_A = C_A^i \left(\frac{RT}{P_T}\right) \text{ and } x_i = C_W^i v_W$$

where v_A and v_W are the molar volumes of air and water, respectively, the latter taken as an approximation for the molar volume of the dilute aqueous solution. Substitution into the mole fraction ratio given above yields, after recalling that $f_i^0 = P^{\text{Sat}}$:

$$\frac{y_i}{x_i} = \frac{\gamma_i f_i^0}{P_T} = \frac{C_A^i \left(\dfrac{RT}{P_T}\right)}{C_W^i v_W}$$

$$\therefore K_{AW} = \frac{C_A^i}{C_W^i} = \frac{v_W \gamma_i P^{\text{Sat}}}{RT}$$

Invoking the expressions for solubility in a liquid or a gas developed above we can see that

$$\therefore K_{AW} = \frac{\left(P^{\text{Sat}}/RT\right)}{\left(1/v_W \gamma_i\right)} = \frac{C_A^{\text{Sat}}}{C_W^{\text{Sat}}}$$

That is, in terms of solubilities, K_{AW} is simply the ratio of the solubility of the solute in air to that in water, $C_A^{\text{Sat}}/C_W^{\text{Sat}}$.

For a liquid solute in a liquid–liquid system such as octanol–water,

$$f_i = x_W^i \gamma_W^i f_i^\circ = x_O^i \gamma_O^i f_i^\circ$$

where subscripts W and O refer to water and octanol phases. The reference state value f_i^0 is that of the pure liquid solute and is invariant with the solvent in question. This therefore conveniently cancels and leaves us with

$$\frac{\gamma_W^i}{\gamma_O^i} = \frac{x_O^i}{x_W^i}$$

and

$$K_{OW} = \frac{C_O^i}{C_W^i} = \frac{x_O^i/v_O}{x_W^i/v_W} = \frac{\gamma_W^i v_W}{\gamma_O^i v_O} = \frac{C_O^{\text{Sat}}}{C_W^{\text{Sat}}}$$

We again see that the partition ratio may be understood as a ratio of solute solubilities in the two partitioning media. The same final equation applies for a solid solute, since F cancels in the same manner as f_R in the analogous derivation. It is wise to check the dimensional consistency of all equations to ensure that no erroneous terms are introduced. Here C has units of mol m^{-3}, f and P are in Pa, v is in m^3 mol^{-1}, and γ and x are dimensionless.

Because v_W and v_O are relatively constant, the variation in K_{OW} between solutes is a reflection of variation in the ratio of activity coefficients γ_W^i/γ_O^i. Hydrophobic substances such as DDT have very large values of γ_W^i and low solubilities in water. The solubility in octanol is usually fairly constant for organic solutes, thus K_{OW} is approximately inversely proportional to C_W^{Sat}. Numerous correlations have been proposed between log K_{OW} and log C_W^{Sat}, which are based on this fundamental relationship.

Finally, for completeness, the octanol–air partition ratio can be shown to be

$$K_{OA} = \frac{C_O^{\text{Sat}}}{C_A^{\text{Sat}}} = \frac{RT}{\gamma_O^i v_O P_O^{\text{Sat}}}$$

Here P_O^{Sat} and γ_O^i apply to the octanol phase. It can be shown that Z_O is $1/\gamma_O^i v_O P_O^{\text{Sat}}$ and that K_{OA} is Z_O/Z_A.

Measurements of solubilities and partition ratios are subject to error, as is evident by examining the range of values reported in handbooks. An attractive approach is to measure the three partition ratios, K_{AW}, K_{OW}, and K_{OA}, and perform a consistency check that, for example, K_{OA} is K_{OW}/K_{AW}. Further checks are possible if solubilities can be measured to confirm that K_{AW} is $C_A^{\text{Sat}}/C_W^{\text{Sat}}$ or $P_W^{\text{Sat}}/\left(C_W^{\text{Sat}} RT\right)$. These checks are also useful for assessing the "reasonableness" of data. For example, if an aqueous solubility C_W^{Sat} is reported as 1 part per million or 1 g m^{-3} or (say) 10^{-2} mol m^{-3}, and K_{OW} is reported to be 10^7, then the solubility in octanol must be $C_W^{\text{Sat}} K_{OW}$ or 100,000 mol m^{-3}. Octanol has a solubility in itself, i.e., a density of about 820 kg m^{-3} or 6300 mol m^{-3}. It is inconceivable that the solubility of the solute in octanol exceeds the solubility of octanol in octanol by a factor of 100,000/6300 or 16; therefore, either C_W^{Sat} or K_{OW} or both are likely erroneous.

The relationships between the three solubilities and the partition ratios are shown in Figure 2.3. Two points are worthy of note: There are numerous correlations for quantities such as K_{AW}, K_{OW}, C_W^{Sat}, C_A^{Sat} as a function of molecular structure and properties. They are generally derived independently, so it is possible to estimate C_W^{Sat}, C_A^{Sat}, and K_{AW} and obtain inconsistent results, i.e., K_{AW} will not equal $C_A^{\text{Sat}}/C_W^{\text{Sat}}$. It is preferable, in principle, to correlate C_A^{Sat}, C_W^{Sat}, and C_O^{Sat} independently and use the values to estimate K_{AW}, K_{OW}, and K_{OA}. There is then no possible inconsistency. It must be easier to correlate C_i^{Sat} (which depends on interactions in only one phase) than K_{ij} (which depends on interactions in two phases).

An annoying snag arises here because the water in K_{AW} is always pure water, but the water in K_{OW} will contain some dissolved octanol. The octanol in K_{OW} also contains a significant mole fraction of dissolved water, whereas K_{OA} is generally measured with pure octanol. These issues are generally of little importance, but should be recognized.

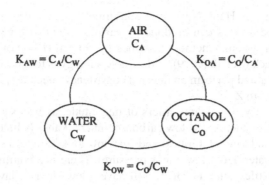

	Benzene	Hexachlorobenzene	DDT	Lindane
$K_{AW} =$	0.22	0.053	9.5×10^{-4}	6.0×10^{-5}
$K_{OW} =$	135	316000	1550000	5000

If $C_A = 1$ μg m⁻³, then the concentrations in water and octanol (μg m⁻³) are

$C_W =$	4.5	19	1050	16700
$C_O =$	614	5960000	1.6×10^9	8.3×10^7

FIGURE 2.3 Illustration of the relationships between the three solubilities, C_A, C_W, and C_O, and the three partition ratios, K_{AW}, K_{OW}, and K_{OA}, with values for four substances. Note the wide substance variation in concentrations corresponding to unit concentration in the air phase.

Finally, all activity coefficients, solubilities, and partition ratios are temperature dependent. The temperature coefficient is the enthalpy of phase transfer, e.g., pure solute to solution for solubility or from solution to solution for partition ratios. The enthalpies must be consistent around the cycle air–water–octanol such that their sum is zero. This provides another consistency check. It should be noted that the enthalpy change refers to the solubility or partition ratio variation when expressed in mole fractions, not mol m⁻³ concentrations. This is particularly important for partitioning to air, where a temperature increase causes a density decrease, thus C_i or C_i^{Sat} will fall while x_i remains constant. For details of the merits of applying the "three solubility" approach, the reader is referred to Cole and Mackay (2000). We discuss these partition ratios individually in more detail in the following sections.

2.4.2 AIR–WATER PARTITIONING

The nature of air–water partition ratios or Henry's law constants has been reviewed by Mackay and Shiu (1981), and estimation methods have been described by Mackay and Boethling (2000) and Baum (1997), and only a brief summary is given here. Several group contribution and bond contribution methods have been developed, and estimation methods are available through various on-line sites, such as those given in the on-line listing of fugacity models and other modeling resources on the CEMC website (www.trentu.ca/cemc/resources-and-models). As was discussed above, the simplest method of estimating Henry's law constants of organic solutes is as a ratio of vapor pressure to water solubility. It must be recognized that this contains the inherent assumption that water is not very soluble in the organic material, because the vapor pressure that is used is that of the pure substance (normally the pure liquid), whereas, in the case of solubility of a liquid such as benzene in water, the solubility is not actually that of pure benzene but is inevitably of benzene saturated with water. When the solubility of water in a liquid exceeds a few percent, this assumption may break down, and it is unwise to use this relationship. If a solute is miscible with water (e.g., ethanol), it is preferable to determine the Henry's law constant directly; that is, by measuring air and water concentrations at equilibrium. This can be done by various techniques, e.g., the EPICS (equilibrium partitioning in closed systems) method described by Gossett (1987) or a continuous stripping technique described by Mackay et al. (1979). A desirable strategy is to measure vapor pressure P^{Sat}, solubility C_i^{Sat}, and H or K_{AW} and perform an internal consistency check that H is indeed P^{Sat}/C_i^{Sat} or close to it. K_{AW} is also equal to Z_A/Z_W.

Henry's law constants vary over many orders of magnitude, as was shown in Figure 1.5, and tend to be high for substances such as the low carbon-number alkanes (which have high vapor pressures, low boiling points, and low solubilities) and very low for substances such as alcohols (which have a high solubility in water and a low vapor pressure). There is a common misconception that substances that are "involatile," such as DDT, will have a low Henry's law constant. This is not necessarily the case, because these substances also have very low solubilities in water, i.e., they are hydrophobic; thus, their low vapor pressure is offset by their very low water solubility, and they can have relatively large Henry's law constants. They are more logically termed "oleophilic" since they strongly prefer oily environments to either water or air. As a result, they may partition appreciably from water into the atmosphere through evaporation from rivers and lakes, despite being highly hydrophobic and of low water solubility.

The solubility and activity of a solute in water are affected by the presence of electrolytes and other *co-solvents;* thus, the Henry's law constant is also affected. The magnitude of the effect is discussed later in Section 2.4.5.

Worked Example 2.2: Air–water partitioning calculations with three chemicals with very different partitioning properties.

Deduce H and K_{AW} for benzene, DDT, and phenol (at low pH) given the following data at 25°C:

	Molar Mass (g mol⁻¹)	Solubility (g m⁻³)	Vapor Pressure (Pa)
Benzene	78.11	1780	12700
DDT	354.5	0.0055	0.00002
Phenol	94.11	88360	47

In each case, the molar solubility C_i^{Sat} (mol m⁻³) is the solubility in g m⁻³ divided by the molar mass, e.g., (1780 g m⁻³/78.11 g mol⁻¹) or 22.79 mol m⁻³ for benzene. H is then $P_{Bz}^{Sat}/C_{Bz}^{Sat}$ or (1.27 × 10⁴ Pa/22.79 mol m⁻³) = 557 Pa m³ mol⁻¹ for benzene. K_{AW} is H/RT or 557/(8.314 × 298) or 0.225.

	H	K_{AW}
Benzene	557	0.225
DDT	1.29	5.2 × 10⁻⁴
Phenol	0.050	2 × 10⁻⁵

Note that these substances have very different H and K_{AW} values because of their solubility and vapor pressure differences. The vapor pressure of DDT is about 600 million times less than that of benzene, but H is only 400 times less, because of DDT's very low water solubility. Phenol has a much higher vapor pressure than DDT, but it has a much lower H and K_{AW}. Benzene tends to evaporate appreciably from water into air, and DDT less so but still to a significant extent, while phenol does not evaporate significantly. Inherent in this calculation for phenol is the assumption that it does not ionize appreciably at the pH of determination.

2.4.3 OCTANOL–WATER PARTITIONING

The dimensionless octanol–water partition ratio (K_{OW}) is one of the most important and frequently used descriptors of chemical behavior in the environment. In the pharmaceutical and biological literature, K_{OW} is given the symbol P (for partition ratio), which we reserve here for pressure. The use of 1-octanol has been popularized by Hansch and Leo (1979), who have obtained correlations with many biochemical phenomena and have compiled extensive databases. Various methods are available for calculating K_{OW} from molecular structure, as reviewed by Lyman et al. (1982), Baum (1997), and Leo (2000). The US Environmental Protection Agency's EPI Suite tool provides the popular KOWWIN QSPR as part of its extensive estimation methods suite. Other extensive databases are also available as reviewed by Baum (1997). Octanol was selected because it has a similar carbon-to-oxygen ratio as lipids, is readily available in pure form, and is only sparingly soluble in water ($4.5 \, \text{mol m}^{-3}$). However, the solubility of water in octanol ($2300 \, \text{mol m}^{-3}$) is quite large (Baum, 1997). The molar volumes of these phases are $18 \times 10^{-6} \text{m}^3 \text{mol}^{-1}$ and $120 \times 10^{-6} \text{m}^3 \text{mol}^{-1}$, a ratio of 0.15. It follows that K_{OW} is $0.15 \, \gamma_W / \gamma_O$.

K_{OW} is a measure of hydrophobicity, i.e., the tendency of a chemical to "hate" or partition out of water. As was discussed earlier, it can be viewed as a ratio of solubilities in octanol and water but, in most cases of liquid chemicals, there is no real solubility, because octanol and the liquid are miscible. The "solubility" of organic chemicals in octanol tends to be fairly constant in the range of 2000–$4000 \, \text{mol m}^{-3}$; thus, variation in K_{OW} between chemicals is primarily due to variation in water solubility. It is therefore misleading to assert that K_{OW} describes lipophilicity or "love for lipids," because most organic chemicals "love" lipids equally, but they "hate" water quite differently. Viewed in terms of Z-values, K_{OW} is Z_O / Z_W. Z_O is (relatively) constant for organic chemicals; however, Z_W varies greatly and is very small (relatively) for hydrophobic substances. In this connection, it is noteworthy that Yalkowsky's General Solubility Equation (Jain and Yalkowski (2001), Sangvi et al., 2003) uses K_{OW} to predict water solubilities, by implicitly assuming equal solubilities of organic chemicals in octanol of approximately $3000 \, \text{mol m}^{-3}$.

Because K_{OW} varies over such a large range, from approximately 0.1 to 10^7, it is common to express it as log K_{OW}. K_{OW} is usually measured by equilibrating layers of water and octanol containing the solute of interest at sub-saturation conditions and analyzing both phases. If K_{OW} is high, the concentration in water is necessarily low, and even a small quantity of emulsified octanol in the aqueous phase can significantly increase the apparent concentration. A "slow stirring" method is usually adopted to avoid emulsion formation. An alternative is to use a generator column in which water is flowed over a packing coated with octanol containing the dissolved chemical.

2.4.4 OCTANOL–AIR PARTITIONING

This partition ratio is invaluable for predicting the extent to which a substance partitions from the atmosphere to organic media including soils, vegetation, and aerosol particles. It can be estimated as K_{OW}/K_{AW} or measured directly, usually by flowing air through a column containing a packing saturated with octanol with the solute in solution. Values of K_{OA} can be very large, i.e., up to 10^{12} for substances of very low volatility such as DDT, and values are especially high at low temperatures. Harner et al. (2000) have reported data for this coefficient and cite other data and measurement methods.

2.4.5 SOLUBILITY IN WATER

This property is of importance as a measure of the activity coefficient in aqueous solution, which in turn affects air–water and octanol–water partitioning. It can be regarded as a partition ratio between the pure phase and water, but the ratio of concentrations is not calculated. A comprehensive discussion is given in the text by Yalkowsky and Banerjee (1992), and estimation methods are described by Mackay (2000), Baum (1997), and Boethling and Mackay (2000). Extensive databases are available, for example, the handbooks by Mackay et al. (2002), which also give details of methods of experimental determination.

It is important to appreciate that solubility in water is affected by temperature and the presence of electrolytes and other solutes in solution. It is often convenient to increase the solubility of a sparingly soluble organic substance for aquatic toxicity testing by addition of a co-solvent to the water. Methanol and acetone are common co-solvents. To a first approximation, a "log-linear" relationship applies in that, if the solubility in water is C_W^{Sat} and that in pure co-solvent is C_C^{Sat}, then the solubility in a mixture C_M^{Sat} is given by

$$\log C_M^{Sat} = (1 - v_C^f) \log C_W^{Sat} + v_C^f \log C_C^{Sat}$$

where v_C^f is the volume fraction of the co-solvent in the solution.

Electrolytes generally decrease the solubility of organics in water, the principal environmentally relevant issue being the solubility in seawater. The Setschenow equation is usually applied for predictive purposes, namely

$$\log \frac{C_W^{Sat}}{C_E^{Sat}} = kC_S$$

where C_W^{Sat} is solubility in water, C_E^{Sat} is solubility in electrolyte solution, k is the Setschenow constant specific to the ionic species, and C_S is the electrolyte concentration (mol L^{-1}). Values of k generally lie in the range of 0.2–0.3 L mol^{-1}; thus, in seawater, which is approximately 33 g NaCl L^{-1} or C_S is about 0.5 mol L^{-1}, the solubility is about 70%–80% of that in water. Xie et al. (1997) have reviewed this literature, especially with regard to seawater.

2.4.6 Solubility in Octanol

There are relatively few data on this solubility, and for many substances, especially liquids, the low activity coefficients render the solute–octanol system miscible; thus, no solubility is measurable. Pinsuwan et al. (1995) have reviewed available solubility data and the relationships between K_{OW} and solubilities in octanol and water.

2.4.7 Solubility of a Substance "in Itself"

The fugacity of a pure solute is its vapor pressure P^{Sat}, and its "concentration" is the reciprocal of its molar volume v_S (m^3 mol^{-1}) (typically, 10^{-4} m^3 mol^{-1}). Thus,

$$C_P = \frac{1}{v_S} = Z_P f = Z_P P^{Sat}$$

and

$$Z_P = \frac{1}{P^{Sat} v_S}$$

Although it may appear environmentally irrelevant to introduce Z_P, the fugacity capacity of the pure solute, there are situations in which it is used. If there is a spill of a polychlorinated biphenyl (PCB) or an oil into water of sufficient quantity that the solubility is exceeded, at least locally, the environmental partitioning calculations may involve the use of volumes and Z-values for water, air, sediment, biota, and a separate pure solute phase. Indeed, early in the spill history, most of the solute will be present in this phase. The difference in behavior of this and other phases is that the pure phase fugacity (and, of course, concentration) remains constant, and as the chemical migrates out of the pure phase, the phase *volume* decreases until it becomes zero at total dissolution or evaporation. In the case of other phases, the *concentration* changes at approximately constant volume as a result of migration.

It can be useful to compare a set of calculated Z-values with Z_P to gain an impression of the degree of non-ideality in each phase. Rarely does a Z-value of a chemical in a medium exceed Z_P, but they may be equal when ideality applies and activity coefficients are close to 1.0.

2.4.8 PARTITIONING AT INTERFACES

Chemicals tend to adsorb from air or water to the surface of solids. An extensive literature exists on this subject as reviewed in texts in chemistry and environmental processes. A good review with environmental applications is given by Valsaraj (1995) in which the fundamental Gibbs equation is developed into the commonly used adsorption isotherms. These isotherms relate concentration at the surface to concentration in the bulk phase. Examples are the Langmuir, BET, and Freudlich isotherms. Generally, a linear isotherm applies at low concentrations as are usually encountered in the environment in relatively uncontaminated situations. Nonlinear behavior occurs at high concentrations in highly contaminated systems and in process equipment such as carbon adsorption units.

It is often not realized that partitioning also occurs at the air–water interface, where an excess concentration may exist. This is exploited in the solvent sublation process for removing solutes from water using fine bubbles. If an area of the surface is known, a surface concentration in units of mol m^{-2} can be calculated, but more commonly the concentration is given in moles per mass of sorbent, which is essentially the product of the surface concentration and a specific area expressed in m^2 surface per unit mass of sorbent. Solids such as activated carbon have very high specific areas and are thus effective sorbents. Partitioning to the air–water interface can become very important when the area of that interface is large compared with the associated volume of air or water. This occurs in fog droplets and snow where the ratio of area to water volume is very large, or in fine bubbles where the ratio of area to air volume is large. These ratios are (6/diameter) m^{2}m^{-3}. Note that a Z-value can be defined on an area basis (mol Pa^{-1}m^{-2}) or for the bulk phase by including the specific area.

Schwarzenbach et al. (1993) have reviewed mechanisms of sorption and have summarized reported data. This partitioning is important for ionizing substances but less important for non-polar compounds, which sorb more strongly to organic matter.

2.5 THEORETICAL APPROACHES TO ESTIMATING PARTITION RATIOS AND OTHER PROPERTIES

2.5.1 QUANTITATIVE STRUCTURE–PROPERTY RELATIONSHIPS

An invaluable feature of many series of organic chemicals is that their chemical properties vary systematically, and therefore predictably, with changes in molecular structure. Mathematical statements of such relationships are termed *Quantitative Structure-Property Relationships* or *QSPRs*. These relationships may be used to predict the properties of other chemical with structural features similar to those used to determine the QSPR. A typical simple QSPR relationship is illustrated for the chlorobenzenes in Figure 2.4. Figure 2.4a is a plot of log (sub-cooled liquid solubility) versus chlorine number from 0 (benzene) to 6 (hexachlorobenzene), which shows a steady drop in solubility (a change in the property) as a result of substituting a chlorine for a hydrogen (a change in the structure). The magnitude is a decrease in log solubility of about 0.65 units (factor of 4.5) per chlorine. Vapor pressure (Figure 2.4b) behaves similarly, with a drop of 0.72 units (factor of 5.2).

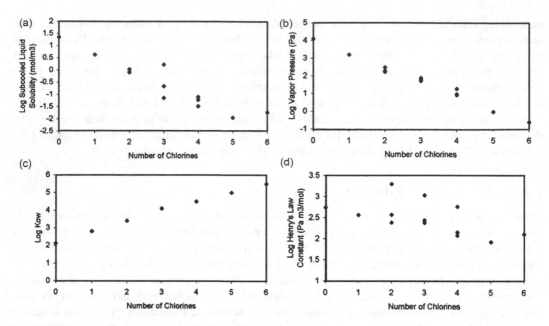

FIGURE 2.4 Illustration of quantitative structure property relationships for benzene and the chlorobenzenes showing the systematic changes in properties with chlorine substitution (a–d). (©2001 CRC Press LLC.)

K_{OW} (Figure 2.4c) shows an increase of 0.53 units (factor of 3.4). The Henry's law constant (Figure 2.4d) decreases by 0.16 units (factor of 1.4).

These plots are invaluable as a method of interpolating to obtain values for unmeasured compounds. They provide a consistency check for newly reported data. They form the basis of estimation methods in which these properties are calculated for a variety of atomic and group fragments.

QSPRs have their origin in the related relationships between molecular structural properties and biological "activity" such as toxicity. Such relationships are termed *Quantitative Structure–Activity Relationships* or *QSARs* and were introduced by Hansch and co-workers (1963), and Hansch (1964) wherein structural parameters associated with small organic molecules such as phenols were correlated with biological activity and toxicity. A good environmental example is the correlation of fish toxicity data expressed as LC_{50} (μ mol L^{-1}) versus K_{OW} as obtained by Konemann (1981).

$$\log\left(\frac{1}{LC_{50}}\right) = 0.87\log(K_{OW}) + 4.87$$

LC_{50} is the lethal concentration (usually expressed in μ mol L^{-1} or mmol L^{-1}) that causes death to 50% of the fish at a prescribed time, often 96 hours. This and other correlations have been reviewed and discussed by Veith et al. (1983), Kaiser (1984, 1987), Karcher and Devillers (1990), and Abernethy et al. (1986, 1988).

The fundamental relationship expressed by this correlation is best explained by an example: Consider two chemicals of log K_{OW} 3 and 5. The LC_{50} values (lethal concentration to 50% of the fish) will be 182 and 3.3 μmol L^{-1}, a factor of 55 different. If the target site for the toxic effect is similar to octanol in solvent properties, and equilibrium is reached, then the concentrations at the target site will be the product $LC_{50} \times K_{OW}$ or 182,000 and 330,000 μmol L^{-1}, only a difference of a factor of 1.8. The chemical of lower K_{OW} appears to be less toxic (it has a higher LC_{50}) when considered from the point of view of water concentration. When viewed from the target site concentration, it is slightly more toxic. The chemicals in the correlation display similar toxicities when evaluated from the target site concentrations. The correlation therefore expresses two processes, partitioning and toxicity, with most of the chemical-to-chemical variation being caused by partitioning difference. Many such non-biologically reactive chemicals are referred to as *narcotics* in which the effect appears to be induced by high lipid concentrations.

2.5.2 POLYPARAMETER LINEAR FREE ENERGY RELATIONSHIPS

Relationships between the structural features of a molecule and its activity or properties such as toxicity are implemented in QSARs and QSPRs may lack an underlying theoretical basis by which their results may be analyzed and interpreted. Rather, they are based on reasonably well-correlated relationships between a "countable" descriptor and one or more observable measurable properties. There is no direct causal relationship between the descriptors counted and the observed property. Rather, there is an implicit correlation between the observed property and the corresponding electrostatic properties of the molecular feature associated with the descriptor, the latter being the actual "causal" source of the change to which the initial property is correlated. For example, molecules with more methylene groups ($-CH_2-$) in an alkyl side chain will have a higher log K_{OW}, all other aspects being equal. The increase in log K_{OW} is strongly correlated with the increase in methylene unit count, but is actually *caused* by the increased hydrophobicity of the molecule that is due to the non-polar nature of the methylene unit in a simple alkyl chain. Thus, QSARs and QSPRs may work reasonably well, but one cannot "know" why without overlaying a certain amount of additional molecular bonding and property insight.

For this reason, much effort has been directed toward the elucidation of actual experimentally measurable electrostatic and physico-chemical properties of molecules that have a causal relationship with the observed solvation and partitioning properties. The most well known of these properties are the result of extensive work by Abraham and co-workers, and are often referred to as the "Abraham parameters." The Abraham parameters are designed to capture as completely as possible variations in the electrostatic interactions between solutes and solvents, as one moves from one solvent to another or one solute and another, or both. The Abraham solvation approach was first introduced in 1988 by Abraham and co-workers (Abraham et al., 1988), reviewed in 1993 (Abraham, 1993), and later described again in some detail in 2004 (Abraham et al., 2004).

The most commonly used Abraham parameters are measures of molecular (McGowan) volume (V), overall hydrogen bond acidity (A), overall hydrogen bond basicity (B), excess molar refraction (E), dipolarity/polarizability (S), and the gas–hexadecane partition ratio (L). These six properties are believed to span "solvation property space" as effectively and completely as possible. Note that the McGowan volume is not actually an experimental parameter, but a theoretical means by which the volume of a molecule is estimated from its structure.

Since the Abraham parameters are directly correlated with Gibbs (free) energy changes in molecular systems, correlations between these parameters and thermodynamic properties of molecules such as partition ratios and solubilities are termed "Linear Free Energy Relationships" or LFERs. Since in general a set of several parameters, commonly five, are used together to define these linear relationships, they are commonly termed "polyparameter Linear Free-Energy Relationships" (pp-LFERs) or "polyparameter Linear Solvation Energy Relationships (pp-LSERs)."

There are two typically used combinations of the Abraham parameters in the estimation of partition ratios:

$$\log K = c + eE + sS + aA + bB + vV$$

and

$$\log K = c + eE + sS + aA + bB + lL$$

Here the capital letters represent the Abraham parameters given above, which are specific to the molecular species of interest, i.e., the solute. The lowercase letters are solvent-specific parameters that reflect the influence of the partitioning solvent pair on the relative solvation energy and hence log K between two solvents. Determination of a robust and reliable set of solvent parameters requires data for a relatively diverse set of chemicals for which the molecular parameters are known accurately. Once developed, the pp-LFER may be used for any other molecular species that falls within the applicability domain of the chemical used to develop the relationship. Use of the relation outside of this domain is a risky venture and may result in significant errors. An example of such a catastrophe would be the use of a pp-LFER developed with and for neutral organic molecules to predict log K for a charged (ionized) species. Such an application would be doubly ill-advised, were the molecule to be a metal-containing species, for which both the charge and the presence the metal would place the chemical outside the applicability domain.

Goss has pioneered the application of pp-LFERs in the environmental context, publishing numerous studies in which he and co-workers demonstrate the application of this approach to such diverse media as organic carbon in soils and sediments (Nguyen et al., 2005), dry terrestrial aerosols (Arp et al., 2008), and membrane lipids. This work has culminated in the development of the UFZ-LSER database (Ulrich et al., 2017), wherein users may select combinations of numerous solutes and solvents including bovine serum, muscle protein, and lipid membranes, to allow the estimation of a significant number of solvation properties such as partition ratios, sorbed concentrations, and membrane permeabilities. DiToro and colleagues have pioneered the use of pp-LFERs for the estimation of the toxicity of narcotic chemicals, with the target lipid model (TLM) that has been applied to numerous chemicals and aquatic species (DiToro, 2001). This approach is particularly useful for hydrocarbons which are often present as complex mixtures. The Wania group at the University of Toronto has also applied pp-LFERs in a number of contexts including in several fugacity-based models, available from https://www.utsc.utoronto.ca/labs/wania/downloads/. Examples of such models are given in the on-line listing of fugacity models and other modeling resources on the CEMC website (www.trentu.ca/cemc/resources-and-models).

pp-LFERs are the most successful experimentally based means by which one may estimate solvation properties of molecules. Endo and Goss have critically reviewed the application of pp-LFER approaches in environmental chemistry in 2014 (Endo and Goss, 2014). Uncertainties of ±0.1–0.2 log units in log K for partitioning between two homogeneous phases are commonly obtained. Partitioning between heterogeneous phases such as biological mixtures have higher uncertainty, typically 0.3 log units or higher. When experimentally measured Abraham parameters are not available, theoretical estimates from QSAR approaches are available, but introduce significant uncertainty into the estimate. For example, use of estimates from the Absolv prediction module (ACD/Labs, Toronto, ON, Canada) generates uncertainties of about 1 log unit (Endo and Goss, 2014). It is therefore apparent that, in the absence of experimentally measured Abraham parameters for a given chemical species, the quality of the corresponding pp-LFER estimate is likely to be similar to or poorer than the result of a "first-principles approaches" based on quantum-mechanical molecular structure calculations outlined in the next section.

2.5.3 SOLVATION THEORY APPROACHES

Often researchers and especially regulatory policy makers are faced with the problem of judging the partitioning properties of new chemicals, many times from new classes of chemicals, for which experimental Abraham parameters are not yet available. As noted above, the uncertainty in the use of pp-LFERs with QSAR-estimated Abraham parameters can be high. When new families of chemicals are considered, this uncertainty may not even be estimable, since there is no evidence to demonstrate that these new chemicals fall within the applicability domain of QSARs used to generate the estimated Abraham parameters.

In such instances, "first-principles" approaches are an attractive alternative. Such approaches use the quantum-mechanically determined properties of a chemical as the basis for a calculation of solvation properties such as thermodynamic partitioning. The current approach that is most reliable and accurate is that of COSMO-RS theory (Klamt, 1995 and Klamt et al., 1998), a solvation theory based on density-functional theory (DFT) calculations of molecular structure and electronic distribution. As of 2016, the COSMO-RS approach was capable of an uncertainty of ±0.5 ln units in infinite-dilution partition ratios such as log K_{OW}, for example, which corresponds to about ±0.21 log units (Klamt, 2016). This uncertainty is believed to be partially due to experimental uncertainty, as well as the inherent uncertainty in DFT itself. The latter may be lowered by improvements in quantum-mechanical theoretical approaches in the future. When gas-phase properties are involved, the uncertainty is higher, generally of the order of ±0.4 log units for Henry's law constants, for example (Parnis et al., 2015). The significant increase in uncertainty in gas-phase related partitioning and other properties is due to the necessity to estimate the Gibbs energy of vaporization, which cannot be done as accurately as the condensed-phase determination described in brief below.

The COSMO-RS calculation involves several steps. First, a DFT calculation of molecular structure in the unsolvated (vacuum) state is performed. Then a "COSMO" calculation (Klamt and Schüürmann, 1993, Klamt, 2018) is performed, in which the electrostatic potential at slightly greater than the intermolecular collisional distance is calculated across the surface of the molecule. "Electrostatic screening" charges are then placed over the surface of the molecule and the minimum energy structure of the now-solvated molecule is recalculated. This new relaxed geometry has a different electrostatic potential, so the resetting of screening charges and re-optimization of the molecular geometry must be done iteratively until the fully relaxed geometry is obtained. At this point the "COSMO" calculation is done, and the COSMO-RS calculation may begin.

The actual COSMO-RS calculation starts by dividing the electrostatic potential surface of the fully solvated molecule into small segments. An empirical relationship that is globally parameterized with a large family of molecules is used to determine the energy of interaction between any pair of molecules of interest, such as solvent–solute and solvent–solvent. Using this relationship, the Gibbs energy of solvation for a molecule may be estimated, and the difference in Gibbs energy of solvation was used directly to estimate the partition ratio according to

$$\log K_{AB} = \exp(-\Delta G_{AB}/RT)$$

COSMO-RS theory can do much more than this calculation. For environmental applications, it is worth noting that it is capable of estimating partition ratios between a wide range of solvent mixtures, as well as vapor pressures, solubilities, boiling points, flash points, and many other properties. A particular strength of COSMO-RS is its ability to estimate properties over a wide temperature range with rigorous computation of the associated entropy changes. The applications of COSMO-RS theory in environmental applications such as partition ratio estimation have been described by Klamt (2005). This theory is also widely used in the process of drug design by the pharmaceutical industry. Regrettably, much of this work is confidential due to the highly competitive nature of research in that industry.

2.5.4 Key Environmental Thermodynamic Properties

The key thermodynamic properties of a pure substance for our purposes are its vapor pressure (i.e., its solubility in air), its solubility in water, its solubility in octanol, and the three partition ratios K_{AW}, K_{OW}, and K_{OA}. The magnitudes of these quantities are controlled by vapor pressure and activity coefficients.

We can also relate equilibrium concentrations in these phases using the three Z-values and fugacity. The partition ratios are simply the ratio of the respective Z-values; e.g., K_{AW} is Z_A/Z_W. The use of Z-values at this stage brings little benefit, but they become very useful when we calculate partitioning to other environmental media. We use Z_A, Z_W, and Z_O to calculate Z in phases such as soils, sediments, fish, and aerosol particles, and it proves useful to have Z_P (pure solute) as a reference point when examining the magnitude of Z-values. It is enlightening to calculate all the physical chemical properties of a solid and a liquid substance as shown in the following example.

Worked Example 2.3: Thermodynamic property calculations.

Deduce all relevant thermodynamic air–water partitioning properties for benzene (liquid at 25°C) and naphthalene (a solid of melting point 80°C) at 25°C.

BENZENE

Data: Vapor pressure (P^{Sat}) = 12,700 Pa, molar mass = 78 g mol^{-1}, solubility in water = 1780 g m^{-3}.

FUGACITY CAPACITY FOR BENZENE IN AIR

$$Z_A^{Bz} = \frac{1}{RT} = \frac{1}{8.314\,\text{Pa}\,\text{m}^3\,\text{mol}^{-1}\,\text{K}^{-1} \times 298\,\text{K}} = 4.04 \times 10^{-4}\,\text{mol}\,\text{Pa}^{-1}\,\text{m}^{-3}$$

MOLAR SOLUBILITY OF BENZENE IN WATER

$$C^{Sat} = \frac{1780\,\text{g}\,\text{m}^{-3}}{78\,\text{g}\,\text{mol}^{-1}} = 22.8\,\text{mol}\,\text{m}^{-3}$$

ACTIVITY COEFFICIENT OF BENZENE IN WATER

$$\gamma_W^{Bz} = \frac{1}{v_w C^{Sat}} = \frac{1}{18 \times 10^{-6}\,\text{m}^3\,\text{mol}^{-1} \times 23.1\,\text{mol}\,\text{m}^{-3}} = 2440$$

HENRY'S LAW CONSTANT FOR BENZENE

$$H = \frac{P^{Sat}}{C^{Sat}} = \frac{12,700\,\text{Pa}}{22.8\,\text{mol}\,\text{m}^{-3}} = 557\,\text{Pa}\,\text{mol}\,\text{m}^{-3}$$

FUGACITY CAPACITY FOR BENZENE IN WATER

$$Z_W^{Bz} = \frac{1}{H} = \frac{1}{557\,\text{Pa}\,\text{mol}\,\text{m}^{-3}} = 1.8 \times 10^{-3}\,\text{mol}\,\text{Pa}^{-1}\,\text{m}^{-3}$$

AIR–WATER PARTITION RATIO FOR BENZENE

$$K_{AW}^{Bz} = \frac{H}{RT} = \frac{557\,\text{Pa}\,\text{mol}\,\text{m}^{-3}}{8.314\,\text{Pa}\,\text{m}^3\,\text{mol}^{-1}\,\text{K}^{-1} \times 298\,\text{K}} = 0.22$$

NAPHTHALENE

Data: Solubility = 33 g m^{-3}, molar mass = 128 g mol^{-1}, vapor pressure = 10.9 Pa, melting point = 353 K.

FUGACITY CAPACITY FOR NAPHTHALENE IN AIR (AS WITH BENZENE)

$$Z_A^{Np} = \frac{1}{RT} = \frac{1}{8.314\,\text{Pa}\,\text{m}^3\,\text{mol}^{-1}\,\text{K}^{-1} \times 298\,\text{K}} = 4.04 \times 10^{-4}\,\text{mol}\,\text{Pa}^{-1}\,\text{m}^{-3}$$

FUGACITY RATIO OF NAPHTHALENE

$$F^{Np} = \exp\left(-6.79\left[\left(\frac{353\text{K}}{298\text{K}}\right)-1\right]\right) = 0.286$$

SUB-COOLED VAPOR PRESSURE OF NAPHTHALENE IN AIR ABOVE WATER AT 298 K

$$P^{Sat}_{SCL} = \frac{P^{Sat}}{F^{Np}} = \frac{10.9\,\text{Pa}}{0.286} = 38.1\,\text{Pa}$$

MOLAR SOLUBILITY OF SOLID NAPHTHALENE IN WATER AT 298 K

$$C^{Sat} = \frac{33\,\text{g\,m}^{-3}}{128\,\text{g\,mol}^{-1}} = 0.26\,\text{mol\,m}^{-3}$$

SUB-COOLED LIQUID MOLAR SOLUBILITY OF NAPHTHALENE IN WATER AT 298 K

$$C^{Sat}_{SCL} = C^{Sat}/F^{Np} = 0.26/0.286\,\text{mol\,m}^{-3} = 0.90\,\text{mol\,m}^{-3}$$

ACTIVITY COEFFICIENT OF NAPHTHALENE IN WATER

$$\gamma^{Np}_W = \frac{1}{v_w C^{Sat}_{SCL}} = \frac{1}{18\times10^{-6}\,\text{m}^3\,\text{mol}^{-1}\times0.90\,\text{mol\,m}^{-3}} = 6.2\times10^4$$

HENRY'S LAW CONSTANT FOR NAPHTHALENE

$$H = \frac{P^{Sat}_{SCL}}{C^{Sat}_{SCL}} = \frac{38.1\,\text{Pa}}{0.90\,\text{mol\,m}^{-3}} = 42\,\text{Pa\,m}^3\,\text{mol}^{-1} = P^{Sat}/C^{Sat} = 10.9\,\text{Pa}/0.26\,\text{mol\,m}^{-3}$$

FUGACITY CAPACITY OF NAPHTHALENE IN WATER

$$Z^{Np}_W = \frac{1}{H} = \frac{1}{42\,\text{Pa\,m}^3\,\text{mol}^{-1}} = 0.024\,\text{mol\,Pa}^{-1}\,\text{m}^{-3} = \frac{1}{v_w\gamma P^{Sat}}$$

AIR–WATER PARTITION RATIO FOR NAPHTHALENE

$$K^{Np}_{AW} = \frac{H}{RT} = \frac{42\,\text{Pa\,mol\,m}^{-3}}{8.314\,\text{Pa\,m}^3\,\text{mol}^{-1}\,\text{K}^{-1}\times298\,\text{K}} = 0.017 = \frac{4\times10^{-4}}{0.024} = \frac{Z_A}{Z_W}$$

Note that naphthalene has a higher activity coefficient in water corresponding to its lower solubility and greater hydrophobicity.

2.6 ENVIRONMENTAL PARTITION RATIOS AND Z-VALUES

2.6.1 INTRODUCTION

Our aim is now to use the physical chemical data to predict how a chemical will partition in the environment. Information on air–water–octanol partitioning is invaluable and can be used directly in the case of air and pure water, but the challenge remains of treating other media such as soils, sediments, vegetation, animals, and fish. The general strategy is to relate partition ratios involving these media to partitioning involving octanol. We thus, for example, seek relationships between K_{OW} and soil–water or fish–water partitioning. A comprehensive review of such relationships for environmental organic carbon is given by Schwartzenbach (2002) and specific application to soils has been reviewed by Sposito (1989).

2.6.2 ORGANIC CARBON–WATER PARTITION RATIOS

Studies by agricultural chemists have revealed that hydrophobic organic chemicals tended to sorb primarily to the organic matter present in soils. Similar observations have been made for bottom sediments. Karickhoff (1981) showed that organic carbon was almost entirely responsible for the sorbing capacity of sediments and that the partition ratio between sediment and water expressed in terms of an organic carbon partition ratio (K_{OC}) was closely related to the octanol–water partition ratio. Indeed, the simple relationship was suggested in this work to be

$$K_{OC} = 0.41\, K_{OW}$$

Note that subsequent work has shown that there is significant variability in the proportionality "constant" 0.41 (see below). This relationship is based on experiments in which a soil–water partition ratio was measured for a variety of soils of varying organic carbon content (y) and chemicals of varying K_{OW}. The soil concentration was measured in units of μg g^{-1} or mg kg^{-1} (usually of dry soil) and the water in units of μg cm^{-3} or mg L^{-1}. The ratio of soil and water concentration (designated K_{Soil-w}) thus has units of L kg^{-1} or reciprocal density.

$$K_{Soil-W} = \frac{C_{Soil}\,(\text{mg kg}^{-1})}{C_W\,(\text{mg L}^{-1})} = \text{L kg}^{-1}$$

If a truly dimensionless partition ratio is desired, it is necessary to multiply K_{Soil-W} by the soil density in kg L^{-1} (typically 2.5), or equivalently multiply C_{Soil} by density to give a concentration in units of mg L^{-1}. A plot of K_{Soil-W} versus organic carbon content, y (g g^{-1}), proves to be nearly linear and passes close to the origin, suggesting the relationship

$$K_{Soil-W} = y K_{OC}$$

where K_{OC} is an organic carbon–water partition ratio.

In practice, there is usually a slight intercept; thus, the relationship must be used with caution when y is less than 0.01, and especially when less than 0.001. Since y is dimensionless, K_{OC}, like K_{Soil-W}, also has units of L kg^{-1}. Measurements of K_{OC} for a variety of chemicals show that K_{OC} is related to K_{OW} as discussed above. K_{OW} is dimensionless; thus, the constant 0.41 has dimensions of reciprocal density, L kg^{-1}.

The relationship between K_{OW} and K_{OC} has been the subject of considerable investigation, and it appears to be variable. For example, DiToro (1985) has suggested that for suspended matter in water, K_{OC} approximately equals K_{OW}. Other workers, notably Gauthier et al. (1987), have shown that the sorbing quality of the organic carbon varies and appears to be related to its aromatic content as revealed by nuclear magnetic resonance (NMR) analysis. Gawlik et al. (1997) have recently reviewed some 170 correlations between K_{OC} and K_{OW}, solubility in water, liquid chromatographic retention time, and various molecular descriptors. They could not recommend a single correlation as being applicable to all substances. Seth et al. (1999) analyzed these data and suggested that K_{OC} is best approximated as $0.35K_{OW}$ (a coefficient slightly lower than Karickhoff's 0.41) but that the variability is up to a factor of 2.5 in either direction (i.e., a value of 0.35 with an uncertainty upper limit of 0.89 and a lower limit of 0.14).

$$K_{OC} = \left(0.35^{+0.54}_{-0.21}\right)K_{OW}$$

The uncertainty is more "balanced" when expressed in log form, where

$$\log K_{OC} = \log K_{OW} - (0.46 \pm 0.39)$$

It is thus expected that, depending on the nature of the organic carbon, K_{OC} can be as high as $0.89K_{OW}$ and as low as $0.14K_{OW}$. Values outside this range may occur because of unusual combinations of chemical and organic matter. Clearly the "blunt instrument" use of either 0.35 or 0.41 as a "hard number" in this context is ill-advised and attempts should be made to more accurately establish the effects of the organic fraction content by direct measurement. Doucette (2000) and Schwartzenbach (2002) have given comprehensive reviews of this issue, and Baum (1997) and Boethling and Mackay (2000) have reviewed estimation methods.

In summary, Z-values can be estimated for soils and sediments containing organic carbon of 0.35 $Z_O y(\rho/1000)$ or 0.41 $Z_O y(\rho/1000)$, where Z_O is for octanol, y is the organic carbon content, and ρ is the solid density, typically $2500 \, kg \, m^{-3}$. If an organic matter content is given, the organic carbon content can be estimated as 56% of the organic matter content (Sposito, 1989). These relationships provide a very convenient method of calculating the extent of sorption of chemicals between soils or sediments and water, provided that the organic carbon content of the soil and the chemical's octanol–water partition ratio are known. This is illustrated in Worked Example 2.4.

Worked Example 2.4: Soil–water partition ratio calculation.

Estimate the partition ratio for benzene (K_{OW} of 135) between a soil containing 0.02 g g⁻¹ of organic carbon and water, and then determine the concentration of benzene in soil in equilibrium with water containing 0.001 g m⁻³ benzene, using the Karickhoff (0.41 L kg⁻¹) correlation. Do the same for DDT (log K_{OW} of 6.19).

K_{OC}, THE ORGANIC CARBON–WATER PARTITION RATIO

$$K_{OC}^{Bz} = 0.41 \, L \, kg^{-1} \times K_{OW} = 0.41 \, L \, kg^{-1} \times 135 = 55 \, L \, kg^{-1}$$

$$K_{OC}^{DDT} = 0.41 \, L \, kg^{-1} \times K_{OW} = 0.41 \, L \, kg^{-1} \times 10^{6.19} = 6.35 \times 10^5 \, L \, kg^{-1}$$

K_{Soil-W}^{i}, THE SOIL–WATER PARTITION RATIO

$$K_{Soil-W}^{Bz} = 0.02 \times K_{OC} = 0.02 \times 55 \, L \, kg^{-1} = 1.1 \, L \, kg^{-1}$$

$$K_{Soil-W}^{DDT} = 0.02 \times K_{OC} = 0.02 \times 6.35 \times 10^5 \, L \, kg^{-1} = 1.27 \times 10^4 \, L \, kg^{-1}$$

Bear in mind that both K_{OC} and K_{Soil-W} have units of L kg⁻¹, i.e., reciprocal density, so that the calculated value of the soil concentration C_{Soil} will have units of µg g⁻¹.

C_{SOIL}, THE SOIL CONCENTRATION FROM THE SOIL–WATER PARTITION RATIO

$$C_{Soil}^{Bz} = K_{Soil-W}^{Bz} C_W^{Bz}$$

$$= 1.1 \, L \, kg^{-1} \times 0.001 \, g \, m^{-3} \times \left(\frac{10^{-3} \, kg \, g^{-1} \times 10^6 \, \mu g \, g^{-1}}{10^3 \, L \, m^{-3}}\right) = 0.0011 \, \mu g \, g^{-1}$$

$$C_{Soil}^{DDT} = K_{Soil-W}^{DDT} C_W^{DDT} = 1.27 \times 10^4 \, L \, kg^{-1} \times 0.001 \, g \, m^{-3} = 12.7 \, \mu g \, g^{-1}$$

Note the much higher DDT concentration because of its hydrophobic character. The concentrations in the organic carbon are $C_W K_{OC}$ or 0.055 µg g^{-1} for benzene and 635 µg g^{-1} for DDT. If octanol was exposed to this water, similar concentrations of $C_W K_{OW}$ or 0.135 µg cm^{-3} for benzene and 1549 µg cm^{-3} for DDT would be established in the octanol.

2.6.3 LIPID–WATER AND FISH–WATER PARTITION RATIOS

Studies of fish–water partitioning by workers such as Spacie and Hamelink (1982), Neely et al. (1974), Veith et al. (1979), and Mackay (1982) have shown that the primary sorbing or dissolving medium in fish for hydrophobic organic chemicals is lipid or fat. A similar approach can be taken as for soils, but there is a more reliable relationship between K_{OW} and K_{LW}, the lipid–water partition ratio. For most purposes, they can be assumed to be equal although, for the very hydrophobic substances, Gobas et al. (1987) suggest that this breaks down, possibly because of the structured nature of lipid phases. It is thus possible to calculate an approximate fish-to-water bioconcentration factor or partition ratio if the lipid content of the fish is known. Mackay (1982) reanalyzed a considerable set of bioconcentration data and suggested the simple linear relationship,

$$K_{FW} = 0.048 \, K_{OW}$$

Here 0.048 is the volume fraction of lipid in the fish (equivalent to octanol), i.e., the fish is viewed as containing about 4.8% lipid by volume. Lipid contents vary considerably, and it is certain that there is some sorption to non-lipid material, but it appears that, on average, the fish behaves as if it is about 4.8% octanol by volume. This leads to the approximate "three 5s" rule, i.e., for a chemical of log K_{OW} = 5, in a fish of 5% lipid by volume, the ratio of fish-to-water concentrations is 5000. This bioconcentration factor of 5000 is viewed as the lowest boundary of bioaccumulative substances.

In summary, Z_L for lipids can be equated (at least for approximate cases) with Z_O for octanol. For a phase such as a fish of lipid volume fraction v_L^f, $Z_F = v_L^f Z_O$.

2.6.4 MINERAL MATTER–WATER PARTITION RATIOS

Partition ratios of hydrophobic organics between mineral matter and water are generally fairly low and do not appear to be simply related to K_{OW}. Typical values of the order of unity to ten are observed as reviewed by Schwarzenbach et al. (1993). A notable exception occurs when the mineral surface is dry. Dry clays display very high sorptive capacities for organics, probably because of the activity of the polar inorganic sorbing sites. This raises a problem that some soils may display highly variable sorptive capacities as they change water content as a result of heating, cooling, and rainfall during the course of diurnal or seasonal variations. Some pesticides are supplied commercially in the form of the active ingredient sorbed to an inorganic clay such as bentonite.

In the environment, most clay surfaces appear to be wet and thus of low sorptive capacity. Most mineral surfaces that are accessible to the biosphere also appear to be coated with organic matter probably of bacterial origin. They thus may be shielded from the solute by a layer of highly sorbing organic material. It is thus a fair (and very convenient) assumption that the sorptive capacity of clays and other mineral surfaces can be ignored. Notable exceptions to this are sub-surface environments in which there may be extremely low organic carbon contents and when the solute ionizes. In such cases, the inherent sorptive capacity of the mineral matter may be controlling.

2.6.5 AEROSOL–AIR PARTITION RATIOS

One of the most difficult partition ratios is that between air and aerosol particles. These particles have very high specific areas, i.e., area per unit volume. They also appear to be very effective sorbents. The partition ratio is normally measured experimentally by passing a volume of air through a filter, then measuring the concentrations before and after filtration, and also the concentrations of the trapped particles. Relationships can then be established between the ratio of gaseous to aerosol material and the concentration of total suspended particulates, C_{TSP}.

There has been a profound change over the years in our appreciation of this partitioning phenomenon. The pioneering work was done by Junge and later by Pankow resulting in the Junge–Pankow equation, which takes the form

$$\phi = \frac{C\theta}{(P_L^{Sat} + C\theta)}$$

where ϕ is the fraction on the aerosol, θ is the area of the aerosol per unit volume of air, C is a constant, and P_L^{Sat} is the liquid vapor pressure. This is a Langmuir type of equation, which implies that sorption is to a surface, and the maximum extent of sorption is controlled by the available area.

Experimental data were better correlated by calculating K_P. It can be shown that

$$\phi = \frac{K_P C_{TSP}}{(1 + K_P C_{TSP})}$$

from which

$$K_P = \frac{\phi}{(C_{TSP}(1-\phi))}$$

The units of the total suspended particle concentration C_{TSP} are usually $\mu g\ m^{-3}$; thus, it is convenient to express K_P in units of $m^3\ \mu g^{-1}$. K_P is usually correlated against P_L^{Sat} for a series of structurally similar chemicals using the relationship

$$\log K_P = m \log\left(P_L^{Sat}\right) + b$$

where m and b are fitted constants, and m is usually close to -1 in value. Bidleman and Harner (2000) list 21 such correlations and present a more detailed account of this theory.

Mackay et al. (1986), in an attempt to simplify this correlation, forced m to be -1 and obtained a one-parameter equation, using the liquid vapor pressure,

$$K_{QA} = \frac{6 \times 10^6}{P_L^{Sat}}$$

where K_{QA} is a dimensionless partition ratio, i.e., a ratio of $(mol\ m^{-3})/(mol\ m^{-3})$, and P_L^{Sat} has units of Pa. Here, we use subscript Q to designate the aerosol phase. This enables Z_Q, the Z-value of the chemical in the aerosol particle, to be estimated as $K_{QA}Z_A$. It can be shown that K_{QA} is 10^{12} $K_P(\rho/1000)$, where ρ is the density of the aerosol (kg m^{-3}), i.e., typically 1500 kg m^{-3}. The fraction on the aerosol φ can then be calculated as

$$\varphi = \frac{K_{QA} v_Q^f}{(1 + K_{QA} v_Q^f)}$$

where v_Q^f is the volume fraction of aerosol and is $10^{-12}\ C_{TSP}/(\rho/1000)$ when C_{TSP} has units of $\mu g\ m^{-3}$. C_{TSP} is typically about 30 $\mu g\ m^{-3}$, plus or minus a factor of 5; thus, v_Q^f is about 20×10^{-12}, plus or minus the same factor. Equipartitioning between air and aerosol phases occurs when φ is 0.5 or $K_P \cdot C_{TSP}$ and $K_{QA} v_Q^f$ equals 1.0. This implies a chemical with a K_P of 0.03 m^3 μg^{-1} or K_{QA} of 0.05×10^{12} and a vapor pressure of about 10^{-4} Pa.

It is noteworthy that Z_Q has a value of $K_{QA}Z_A$ or about $\left(6 \times 10^6/P_L^{Sat}\right)\left(4 \times 10^{-4}\right)$ or $\left(2400/P_L^{Sat}\right)$. This is comparable to Z_P, the pure substance Z-value of $1/v\,P_L^{Sat}$, where v is the chemical's molar volume and is typically 100 cm^{3}mol^{-1} or 10^{-4}m^{3}mol^{-1}, giving a Z_P of about $10,000/P_L^{Sat}$. This implies that the solute is behaving near-ideally in the aerosol, i.e., the solubility in the aerosol is about 24% of the solubility of a substance in itself. This casts doubt on the surface sorption model, since it seems a remarkable coincidence that the area is such that it gives this near-ideal behavior. It further suggests that Z_Q may correlate well with Z_O for octanol.

This was explored by Finizio et al. (1997), Bidleman and Harner (2000), and Pankow (1998), leading ultimately to a suggestion that

$$K_P = 10^{-12} K_{OA} y \left(\frac{1000}{820} \right) = 10^{-11.91} K_{OA} y$$

where $820 \, kg \, m^{-3}$ is the density of octanol, and y is the volume fraction of organic matter in the aerosol, which is typically 0.2.

This relation reduces to

$$K_P = 10^{-12.61} K_{OA} \quad \text{or} \quad 0.25 \times 10^{-12} K_{OA} \, m^3 \, mg^{-1}$$

The use of K_{OA} is advantageous, because it eliminates the need to deduce the fugacity ratio, F, when calculating the sub-cooled liquid vapor pressure. It is also possible that, for a series of chemicals, the activity coefficients in octanol and aerosol organic matter are similar, or at least have a fairly constant ratio.

This approach is appealing, because it mimics the Karickhoff method of calculating soil–water partitioning, except that partitioning is now to air instead of water; thus, K_{OA} replaces K_{OW}. K_{QA} thus can be calculated by the analogous equation:

$$K_{QA} = y K_{OA} \left(\frac{\rho}{1000} \right)$$

where y is the organic matter mass fraction. This is equivalent to $Z_Q = y Z_O (\rho/1000)$, where Z_O is the chemical's Z-value in octanol and ρ is the aerosol density.

In summary, Z_Q can be deduced using the above equation, using the simple one-parameter expression for K_{QA} or one of the two-parameter equations for K_P. Bidleman and Harner (2000) discuss the merits of these approaches in more detail.

2.6.6 OTHER PARTITION RATIOS

In principle, partition ratios can be defined and correlated for any phase of environmental interest, usually with respect to the fluid media of air or water. For example, vegetation or foliage–air partition ratios K_{FA} can be measured and correlated against K_{OA}. Since K_{FA} is Z_F/Z_A and K_{OA} is Z_O/Z_A, the correlation is essentially of Z_F versus Z_O. Hiatt (1999) has suggested that for foliage, K_{FA} is approximately $0.01 \, K_{OA}$, implying that Z_F is about $0.01 \, Z_O$, or foliage has a content of octanol-equivalent material of 1%.

It is thus possible to estimate Z-values for chemicals in any phase of environmental interest, provided that the appropriate partition ratio has been measured or can be estimated. Figure 2.5 summarizes the relationships between fugacity, concentrations, partition ratios, and Z-values.

2.7 MULTIMEDIA PARTITIONING CALCULATIONS

2.7.1 THE PARTITION RATIO APPROACH

In the following sections, we apply the theory developed in Sections 2.1–2.6 to the issue of environmental partitioning and fate. Environmental fate calculations involving only the equilibrium distribution of a chemical between two or more environmental compartments are relatively

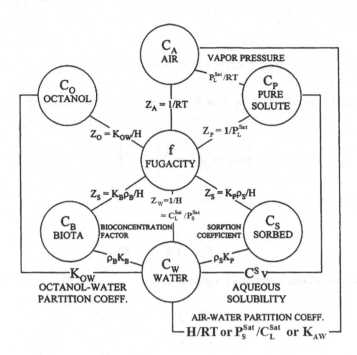

FIGURE 2.5 Relationships between Z-values and partition ratios and summary of Z-value definitions.

straightforward. As noted earlier, there are two approaches to the formulation of such problems, the concentration-based "partition ratio" approach and the "fugacity" approach. Of these, the calculation of one phase concentration from another by the use of a simple partition ratio is the most direct and convenient method. Care must be taken that the concentration units and the partition ratio dimensions are consistent, especially when dealing with solid phases. It is possible to deduce certain partition ratios from others; e.g., if an air/water and a soil/water partition ratio are available, then the air/soil or soil/air partition ratio can be deduced as follows:

$$K_{A-\text{Soil}} = \frac{K_{AW}}{K_{\text{Soil}-W}}$$

If we are treating ten phases, then it is possible to define nine independent interphase partition ratios, the tenth being dependent on the other nine. In principle, with ten phases, it is possible to define 90 partition ratios, half of which are reciprocals of the others. When dealing with very complex multi-compartment environmental media, extreme care must be taken to avoid underspecifying partition ratios and to ensure that the ratios are not inverted.

To illustrate the partition ratio approach with several interacting compartments consider a series of distinct and contacting phases of volume V_1–V_4, for which we know the partition ratios K_{12}–K_{14}. A known amount of chemical m_{Sys} (mol) is introduced into this hypothetical environment. By conservation of mass, the sum of the amount of the chemical in each compartment (C_iV_i) must equal the total amount introduced, so that we can write

$$m_{\text{Sys}} = C_1V_1 + C_2V_2 + C_3V_3 + C_4V_4$$

To solve this, we reduce the number of variables to one, using the relations between compartments captured by the partition ratios. Expressing the concentrations in each of phases 2–4 in terms of C_1 we can write

$$m_{\text{Sys}} = C_1V_1 + (K_{21}C_1)V_2 + (K_{31}C_1)V_3 + (K_{41}C_1)V_4$$

or

$$m_{Sys} = C_1 \left[V_1 + K_{21}V_2 + K_{31}V_3 + K_{41}V_4 \right]$$

Note that the general mass balance for a multiphase system of n phases expressed arbitrarily in terms of C_1 is

$$m_{Sys} = C_1 \left[V_1 + \sum_{i=2}^{n} K_{i1}V_i \right]$$

Solving for C_1, we find

$$C_1 = \frac{m_{Sys}}{\left[V_1 + K_{21}V_2 + K_{31}V_3 + K_{41}V_4 \right]}$$

The remaining unknown concentrations follow directly:

$$C_2 = K_{21}C_1$$

$$C_3 = K_{31}C_1$$

$$C_4 = K_{41}C_1$$

It is thus possible to calculate the concentrations in each phase, the amounts in each phase, and the percentages, and obtain a picture of the behavior of this chemical at equilibrium in an evaluative environment and a tentative picture of the chemical's likely fate in a real environment. This approach is illustrated in Worked Example 2.5.

Worked Example 2.5: Concentration-based partitioning of benzene between air, water, sediment, and fish, all at mutual equilibrium.

Suppose benzene partitions in a hypothetical environment consisting of air, water, sediment, and fish (subscripted A, W, Sed, and F), all of which are in equilibrium. Use the volumes of each phase and the dimensionless partition ratios given below to calculate the concentrations, amounts, and percentages in each phase, assuming that a total of 10 moles of benzene is introduced into this system.

$V_A = 1000$	$V_W = 20$	$V_{Sed} = 10$	$V_F = 0.050 \, \text{m}^3$
$K_{AW} = 0.20$	$K_{Sed-W} = 15$	$K_{FW} = 20$	

The analogous equation to the mass balance developed above is

$$C_W = \frac{m_{Sys}}{\left[V_W + K_{AW}V_A + K_{Sed-W}V_{Sed} + K_{FW}V_F \right]}$$

or

$$C_W = \frac{10 \, \text{mol}}{(20 + 0.20 \times 1000 + 15 \times 10 + 20 \times 0.050) \, \text{m}^3} = \frac{10 \, \text{mol}}{371 \, \text{m}^3} = 0.027 \, \text{mol m}^{-3}$$

Correspondingly, we have that the concentrations in each compartment are

$$C_A = K_{AW}C_W = 0.20 \times 0.027 \,\text{mol m}^{-3} = 0.0054 \,\text{mol m}^{-3}$$

$$C_{\text{Sed}} = K_{\text{Sed}-W}C_W = 15 \times 0.027 \,\text{mol m}^{-3} = 0.41 \,\text{mol m}^{-3}$$

$$C_F = K_{FW}C_W = 20 \times 0.027 \,\text{mol m}^{-3} = 0.54 \,\text{mol m}^{-3}$$

The amounts in each phase are given by the products C_iV_i:

$$m_A = C_A V_A = 0.0054 \,\text{mol m}^{-3} \times 1000 \,\text{m}^3 = 5.39 \text{ or } 53.9\%$$

$$m_W = C_W V_W = 0.027 \,\text{mol m}^{-3} \times 20 \,\text{m}^3 = 0.539 \text{ or } 5.4\%$$

$$m_{\text{Sed}} = C_{\text{Sed}} V_{\text{Sed}} = 0.41 \,\text{mol m}^{-3} \times 10 \,\text{m}^3 = 4.04 \text{ or } 41\%$$

$$m_F = C_F V_F = 0.54 \,\text{mol m}^{-3} \times 0.050 \,\text{m}^3 = 0.003 \text{ or } 0.3\%$$

It is clear that in this system, benzene partitions primarily into air, mainly because of the large volume of air. The concentration in the air is lower than in any other phase; thus, we must discriminate between phases in which the *concentration* is large and where the *amount* of chemical is large (which depends on the product of C_i and V_i). There is a high concentration in the fish, but only a negligible fraction of benzene is associated with fish. Such calculations are invaluable, because they establish the dominant medium into which the chemical is likely to partition, and they even give approximate concentrations.

2.7.2 THE FUGACITY-BASED APPROACH

We now repeat these calculations using the fugacity concept and replacing C_i by $Z_i f$. We know that Z_i will depend on

1. the nature of the solute (i.e., the chemical)
2. the nature of the medium or compartment
3. temperature
4. pressure (usually negligible)
5. concentration (negligible at low concentrations)

We have developed procedures by which Z-values can be estimated for any given environmental situation. Equilibrium concentrations can then be deduced using the fact that fugacity is equal for all phases at equilibrium, f_{Sys}. We can repeat the previous partitioning example using the fugacity method and demonstrate the equivalence of the two approaches as before, but now applying a common fugacity to all phases.

$$m_{\text{Sys}} = C_1V_1 + C_2V_2 + C_3V_3 + C_4V_4$$

$$m_{\text{Sys}} = Z_1 f_{\text{Sys}} V_1 + Z_2 f_{\text{Sys}} V_2 + Z_3 f_{\text{Sys}} V_3 + Z_4 f_{\text{Sys}} V_4$$

$$m_{\text{Sys}} = f_{\text{Sys}} \left[Z_1V_1 + Z_2V_2 + Z_3V_3 + Z_4V_4 \right]$$

Solving for the fugacity, we obtain

$$f_{\text{Sys}} = \frac{m_{\text{Sys}}}{\left[Z_1V_1 + Z_2V_2 + Z_3V_3 + Z_4V_4 \right]}$$

Once the fugacity is known, the concentrations in each phase follow immediately as the product of Z for that phase times the fugacity, $Z_i f_{Sys}$. The molar amount in each phase $m_i = C_i V_i = V_i Z_i f_{Sys}$

In general,

$$f_{Sys} = \frac{m_{Sys}}{\sum_{i=1}^{n} V_i Z_i}$$

$$C_i = Z_i f_{Sys}$$

$$m_i = V_i Z_i f_{Sys}$$

Worked Example 2.6: Fugacity-based partitioning of benzene between air, water, sediment, and fish, all at mutual equilibrium.

Using the data in Worked Example 2.5, recalculate the distribution using the fugacity approach, assuming that Z_A is 4.0×10^{-4} mol Pa^{-1}m^{-3}.

First, calculate the chemical's fugacity capacity in water from the value of Z_A and the partition ratio $K_{AW} = Z_A/Z_W$, then use that value to calculate the others using K_{SW} or K_{FW}. Watch for the change in order of the ratios!

$$Z_W = \frac{Z_A}{(Z_A/Z_W)} = \frac{Z_A}{K_{AW}} = \frac{4.0 \times 10^{-4} \text{ mol Pa}^{-1}\text{m}^{-3}}{0.20}$$

$$= 2.0 \times 10^{-3} \text{ mol Pa}^{-1}\text{m}^{-3}$$

$$Z_{Sed} = (Z_{Sed}/Z_W) Z_W = K_{Sed-W} Z_W$$

$$= 15 \times 2.0 \times 10^{-3} \text{ mol Pa}^{-1}\text{m}^{-3}$$

$$= 3.0 \times 10^{-2} \text{ mol Pa}^{-1}\text{m}^{-3}$$

$$Z_F = (Z_F/Z_W) Z_W = K_{FW} Z_W$$

$$= 20 \times 2.0 \times 10^{-3} \text{ mol Pa}^{-1}\text{m}^{-3}$$

$$= 4.0 \times 10^{-2} \text{ mol Pa}^{-1}\text{m}^{-3}$$

Now calculating the system fugacity:

$$f_{Sys} = \frac{m_{Sys} \text{ mol}}{\sum_{i=1}^{n} \left(V_i \left(\text{m}^3\right) Z_i \left(\text{mol Pa}^{-1}\text{m}^{-3}\right) \right)}$$

$$= \frac{10}{\left(1000 \times 4.0 \times 10^{-4} + 20 \times 2.0 \times 10^{-3} + 10 \times 3.0 \times 10^{-2} + 0.05 \times 4.0 \times 10^{-2}\right)}$$

$$= \frac{10}{0.742} = 13.48 \text{Pa}$$

The concentrations follow as the product of the system fugacity and the individual fugacity capacities, $Z_i f_{Sys}$:

$$C_A = Z_A f_{Sys} = 4.0 \times 10^{-4}\,\text{mol Pa}^{-1}\,\text{m}^{-3} \times 13.48\,\text{Pa} = 5.4 \times 10^{-3}\,\text{mol m}^{-3}$$

$$C_W = Z_W f_{Sys} = 2.0 \times 10^{-3}\,\text{mol Pa}^{-1}\,\text{m}^{-3} \times 13.48\,\text{Pa} = 2.7 \times 10^{-2}\,\text{mol m}^{-3}$$

$$C_{Sed} = Z_{Sed} f_{Sys} = 3.0 \times 10^{-2}\,\text{mol Pa}^{-1}\,\text{m}^{-3} \times 13.48\,\text{Pa} = 4.1 \times 10^{-1}\,\text{mol m}^{-3}$$

$$C_F = Z_F f_{Sys} = 4.0 \times 10^{-2}\,\text{mol Pa}^{-1}\,\text{m}^{-3} \times 13.48\,\text{Pa} = 0.54\,\text{mol m}^{-3}$$

The amounts and percentages are calculated as before.

The fugacity-based procedure involves the following steps:

1. Tabulate the compartment volumes.
2. Calculate and tabulate fugacity capacity Z-values.
3. Calculate and sum the products of volume and fugacity capacity VZ, i.e., ΣVZ.
4. Divide the total molar amount of material in the system as a whole by this sum and obtain the system fugacity.
5. Use the system fugacity f_{Sys} to calculate the concentration of chemical in each compartment as a product of the fugacity capacity and the fugacity, Zf_{Sys}, and the amounts as VZf_{Sys}.

Such calculations are readily done using computer spreadsheets, and there is no increase in mathematical complexity with the number of phase compartments. This approach will be reintroduced as a "Level 1"-type calculation in Chapter 5, in conjunction with associated spreadsheet-based programs that are available from the CEMC website (www.trentu.ca/cemc/resources-and-models).

2.7.3 SORPTION BY DISPERSED PHASES

A frequently encountered environmental calculation is the estimation of the fraction of a chemical that is present in a fluid that is sorbed to some dispersed sorbing phases within that fluid. This is a special case of multimedia partitioning involving only two phases. Examples are the estimation of the fraction of material attached to aerosols in air or associated with suspended solids or with biotic matter in water at equilibrium. The reason for this calculation is that the measured concentration is often comprised of the total of both the dissolved *and* the sorbed chemical, and it is useful to know what fractions are in each phase. This is particularly useful when subsequently calculating uptake of chemical by fish from water in which the partitioning may be only from the dissolved solute and not the sorbed fraction.

The general equations describing sorption in such cases are as follows. The continuous phase is designated by the subscript A and the dispersed phase by subscript B. The dispersed phase volume is typically a factor of 10^{-5} or less than that of the continuous phase.

- The volumes (m³) are denoted V_A and V_B, and usually V_A is much greater than V_B.
- The equilibrium concentrations are denoted C_A and C_B mol m⁻³.
- The dimensionless partition ratio K_{BA} is C_B/C_A.
- The total amount of solute m_{Sys} moles is distributed between the two phases.

$$m_{Sys} = V_T C_T = V_A C_A + V_B C_B$$

where V_T and C_T are the total volume and concentration, respectively. Since V_B is so much smaller than V_A, it can be generally assumed that V_T is approximately equal to V_A.

Now, as before, we express the mass balance in terms of one concentration only, using the partition ratio relation:

$$C_B = K_{BA} C_A$$

Our new mass balance is

$$m_{Sys} = C_A \left(V_A + V_B K_{BA} \right) = C_T V_A$$

Solving for our chosen unknown C_A

$$C_A = \frac{C_T V_A}{\left(V_A + V_B K_{BA} \right)} = \frac{C_T}{\left(1 + \dfrac{K_{BA} V_B}{V_A} \right)}$$

The ratio of V_B/V_A is essentially equal to V_B/V_T, the volume fraction of phase B, v_B^f. We may therefore write

$$C_A = \frac{C_T}{\left(1 + K_{BA} v_B^f \right)}$$

The fraction dissolved (i.e., in the continuous phase) is given by

$$\varphi_A = \frac{C_A}{C_T} = \frac{1}{\left(1 + K_{BA} v_B^f \right)}$$

The sorbed fraction is given by

$$\varphi_B = \left[1 - \left(\frac{C_A}{C_T} \right) \right] = \left| 1 - \frac{1}{\left(1 + K_{BA} v_B^f \right)} \right| = \frac{K_{BA} v_B^f}{\left(1 + K_{BA} v_B^f \right)}$$

The key quantity is thus $K_{BA} v_B^f$ or the product of the dimensionless partition ratio and the volume fraction of the dispersed sorbing phase. When this product is 1.0, half the solute is in each state. When it is smaller than 1.0, most is dissolved, and when it exceeds 1.0, more is sorbed. When the phase B is solid, it is usual to express the concentration C_B in units of moles or grams per unit mass of B in which case K_{BA} has units of volume mass^{-1} or reciprocal density. For example, it is common to use mg L^{-1} for C_A, mg kg^{-1} for C_B, and L kg^{-1} for K_P; then, with m_{Sys} in mg and V_A in L, then it can be shown that

$$m_{Sys} = C_A \left(V_A + m_B K_P \right) = C_T V_A$$

where m_B is the mass of sorbing phase (kg). Solving for C_A, we obtain

$$C_A = \frac{C_T V_A}{\left(V_A + m_B K_P \right)} = \frac{C_T}{\left(1 + K_P \left[\dfrac{m_B}{V_A} \right] \right)} = \frac{C_T}{\left(1 + K_P X_B \right)}$$

where X_B is the concentration of sorbent in kg L^{-1}. The units of the partition ratio K_{BA} or K_P and concentration of sorbent v_B^f or X_B do not matter as long as their product is dimensionless and consistent, i.e., the amounts of sorbing phase, continuous phase, and chemical are the same in the definition of both the partition ratio and the sorbent concentration.

Care must be taken when interpreting sorbed concentrations to ascertain if they represent the amount of chemical per unit volume or mass of sorbent, or the per unit volume of the environmental phase such as water.

The analogous fugacity equations are simply

$$\varphi_A = \frac{V_A Z_A}{(V_A Z_A + V_B Z_B)}$$

$$\varphi_B = \frac{V_B Z_B}{(V_A Z_A + V_B Z_B)}$$

In some cases, it is preferable to calculate a Z-value for a bulk phase consisting of other phases in equilibrium. Examples are air plus aerosols; water plus suspended solids; and soils consisting of solids, air, and water. If the total volume is V_T, the effective bulk Z-value is Z^{Bulk}, and equilibrium applies, then the total amount of chemical distributed over n phases must be

$$m_{Sys} = V_T Z^{Bulk} f = \sum_{i=1}^{n} V_i Z_i f$$

Thus,

$$Z^{Bulk} = \sum_{i=1}^{n} \left(\frac{V_i}{V_T} \right) Z_i = \sum_{i=1}^{n} v_i^f Z_i$$

where v_i^f is the volume fraction of each phase. The key point is that the component Z-values add in proportion to their *volume fractions*, that is the effective total fugacity capacity is a volume fraction weighted sum of the individual Z-values. Note that the use of bulk Z-values helps to simplify calculations by reducing the number of compartments, but it does assume that equilibrium exists within the bulk compartment.

This type of calculation is important when interpreting monitoring or analytical data, for example, for a water sample that may contain suspended solids. The total or bulk concentration may be quite different from the truly dissolved concentration.

Worked Example 2.7: Partitioning of a chemical between water and fish.

An aquarium contains $10\,\mathrm{m}^3$ of water and 200 fish, each of volume $1\,\mathrm{cm}^3$. How will $10\,\mathrm{mg}$ of (1) benzene and of (2) DDT partition between water and fish, given that the fish has a lipid content of 5% by volume? Use $\log K_{OW} = 2.13$ for benzene and 6.19 for DDT.

Given the lipid fraction of the fish, K_{FW} will be $0.05 K_{OW}$ or 6.7 for benzene and 7.7×10^4 for DDT:

$$K_{FW}^{Bz} = v_L^f K_{OW} = 0.05 \times 10^{2.13} = 6.7$$

$$K_{FW}^{DDT} = v_L^f K_{OW} = 0.05 \times 10^{6.19} = 7.7 \times 10^4$$

C_T is $1\,\mathrm{mg\,m}^{-3}$ in both cases, as demonstrated here, where we assume that the aquarium volume is acceptably approximated by the volume of water it contains, ignoring the volume of the fish:

$$C_T = \frac{m_{Sys}}{V_A} \simeq \frac{m_{Sys}}{V_W} = \frac{10\,\mathrm{mg}}{10\,\mathrm{m}^3} = 1\,\mathrm{mg\,m}^{-3}$$

The fraction dissolved φ_A for benzene is

$$\varphi_A^{Bz} = \frac{1}{\left(1 + K_{FW}^{Bz} v_F^f\right)} = \frac{1}{\left(1 + \left(6.7 \times \left[\dfrac{200 \times 10^{-6}\,\text{m}^3}{10\,\text{m}^3}\right]\right)\right)} = 0.99987$$

where v_F^f is the volume fraction of fish. By comparison, the same calculation for DDT, for which $K_{FW} v_F^f$ is 1.55, yields a dissolved fraction of 0.39:

$$\varphi_A^{DDT} = \frac{1}{\left(1 + K_{FW}^{DDT} v_F^f\right)} = \frac{1}{\left(1 + \left(7.7 \times 10^4 \times \left[\dfrac{200 \times 10^{-6}\,\text{m}^3}{10\text{m}^3}\right]\right)\right)} = 0.39$$

The dissolved concentrations are thus $9.9987 \times 10^{-4}\,\text{g m}^{-3}$ or $0.99987\,\text{mg m}^{-3}$ (benzene) and $3.9 \times 10^{-4}\,\text{g m}^{-3}$ or $0.39\,\text{mg m}^{-3}$ (DDT):

$$C_A^{Bz} = \varphi_A^{Bz} C_T = 0.99987 \times 1\,\text{mg m}^{-3} = 9.9987 \times 10^{-1}\,\text{mg m}^{-3}$$

$$C_A^{DDT} = \varphi_A^{DDT} C_T = 0.39 \times 1\,\text{mg m}^{-3} = 0.39\,\text{mg m}^{-3}$$

The sorbed concentrations per m^3 of fish are 0.0067 and $30\,\text{g m}^{-3}$, respectively.

$$C_B^{Bz} = K_{FW}^{Bz} \times C_A = 6.7 \times 9.9987 \times 10^{-1}\,\text{mg m}^{-3} = 6.7\,\text{mg m}^{-3} = 6.7 \times 10^{-3}\,\text{g m}^{-3}$$

$$C_B^{DDT} = K_{FW}^{DDT} \times C_A = 7.7 \times 10^4 \times 0.39\,\text{mg m}^{-3} = 3.03 \times 10^4\,\text{mg m}^{-3} = 30.3\,\text{g m}^{-3}$$

Note the huge disparity between the sorbed concentrations, which reflects the much greater hydrophobicity and therefore much higher value of K_{OW} for DDT.

2.7.4 Maximum Fugacity

When fugacities are calculated, it is advisable to check that the value deduced is lower than the fugacity of the pure phase, i.e., the solid or liquid fugacity or, in the case of gases, of atmospheric pressure. If these fugacities are exceeded, supersaturation has occurred, a "maximum permissible fugacity" has been exceeded, and the system will automatically "precipitate" a pure solute phase until the fugacity drops to the saturation value. It is possible to calculate inadvertently and use (i.e., misuse) these "over-maximum" fugacities. For example, a chemical may be spilled into a lake. The fugacity can be calculated as the amount spilled divided by VZ for water. If the resulting fugacity exceeds the vapor pressure, the water has insufficient capacity to dissolve all the chemical, and a separate pure chemical phase *must* be present. A similar situation can apply when a pesticide is applied to soils. Note that it is likely that the maximum Z-value that a solute can ever achieve is that of the pure phase Z_P. It may be useful to calculate Z_P to ensure that no mistakes have been made by grossly overestimating other Z-values.

2.7.5 Solutes of Negligible Volatility

A problem arises when calculating values of the fugacity and fugacity capacity of solutes that have a negligible or zero vapor pressure. Thermodynamically, the problem is that of determining the reference fugacity. The practical problem may be that no values of vapor pressure or air–water partition ratios are published or even exist. Examples are ionic substances, inorganic materials such as calcium carbonate or silica and polymeric, or high-molecular-weight substances including carbohydrates and proteins. Intuitively, no vapor pressure determination is needed (or may be possible), because the substance does not partition into the atmosphere, i.e., its "solubility" in air is effectively zero. Ironically, its air fugacity capacity can still be calculated as $(1/RT)$, but all the other (and the only useful) Z-values cannot be calculated, since H cannot be determined and may indeed be zero. Apparently, the other Z-values are infinite or at least are indeterminably large.

This difficulty is more apparent than real and is a consequence of the selection of fugacity rather than activity as an equilibrium criterion. There are two remedies: The first method, which is convenient but somewhat dishonest, is to assume a fictitious and reasonable, but small, value for vapor pressure (such as 10^{-6} Pa) and proceed through the calculations using this value. The result will be that Z for air will be very small compared with the other phases, and negligible concentrations will result in the air. It is obviously essential to recognize that these air concentrations are fictitious and erroneous. The relative values of the other concentrations and Z-values will be correct, but the absolute fugacity will be meaningless.

The second method, which is less convenient but more honest, is to select a new equilibrium criterion termed the *aquivalent concentration* (Mackay and Diamond, 1989) and can be used for metals in ionic form when the solubility is meaningless. In the "aquivalence" approach, Z is arbitrarily set to 1.0 in water, and all other Z-values are deduced from this basis using partition ratios. This approach is used in the equilibrium criterion (EQC) model for involatile substances (Mackay et al., 1996), and is based on the fact that all fugacity capacities are calculated from Z_W except for Z_W itself.

The concept of aquivalence is easy to demonstrate. Given that there is no vapor pressure at all for certain chemical species, the value of Z_A is undefined. However, we can still anticipate that such a species will partition between other media such as water and octanol, with "normal" fugacity capacities. Consider that if $K_{OW} = 100$ for a particular chemical, then the two fugacity capacities for water and octanol are defined, in relative terms:

$$K_{OW} = 100 = \frac{Z_O}{Z_W} \quad \text{or} \quad Z_O = 100 Z_W$$

Similarly, if $K_{AW} = 2$, then we can also show

$$K_{AW} = 2 = \frac{Z_A}{Z_W} \quad \text{or} \quad Z_A = 2 Z_W$$

It may now be demonstrated that the partition ratio between octanol and air K_{OA} is defined in absolute terms with two "relative" terms that contain Z_W:

$$K_{OA} = \frac{Z_O}{Z_A} = \frac{100 Z_W}{2 Z_W} = 50$$

One can see readily that the absolute value of Z_W chosen is arbitrary, and the choice of 1 Pa mol^{-1}m^{-3} is merely one of convenience. All other partition ratios determined with respect to Z_W will show the same cancelation and independence from the choice of Z_W.

2.7.6 SOME ENVIRONMENTAL IMPLICATIONS

Viewing the behavior of a solute in the environment in terms of Z introduces new and valuable insights. A solute tends to migrate into (or stay in) the phase of largest Z. Thus, SO_2 and phenol tend to migrate into water, freons into air, and DDT into sediment or biota. The phenomenon of bioconcentration is merely a manifestation of Z in biota, which is much higher (by the bioconcentration factor) than Z in the water. Occasionally, a solute such as inorganic mercury changes its chemical form becoming organometallic (e.g., methyl mercury). Its Z-values change, and the mercury now sets out on a new environmental journey with a destination of the new phase in which Z is now large. In the case of mercury, the ionic form will sorb to sediments or dissolve in water but will not appreciably bioconcentrate. The organic form experiences a large Z in biota and will bioconcentrate. The metallic form tends to evaporate.

Some solutes, such as DDT or PCBs, have very low Z-values in water because of their highly hydrophobic nature; i.e., they exert a high fugacity even at low concentrations, reflecting a large "escaping tendency." They will therefore migrate readily into any neighboring phase such as sediment, biota, or the atmosphere. Atmospheric transport should thus be no surprise, and the contamination of biota in areas remote from sites of use is expected. With this hindsight, it is not surprising that these substances are found in the tissues of Arctic bears and Antarctic penguins!

From the environmental monitoring and analysis viewpoint, it is preferable to sample and analyze phases in which Z is large, because it is in these phases that concentrations are likely to be large and thus easier to determine accurately. When monitoring for PCBs in lakes, it is thus common to sample sediment or fish rather than water, since the expected concentrations in water are very low. Likewise, those concerned with PCB behavior in the atmosphere may measure the PCBs on aerosols or in rainfall containing aerosols, since concentrations are higher than in the air.

In general, when assessing the likely environmental behavior of a new chemical, it is useful to calculate the various Z-values and from them identify the larger ones, since it is likely that the high Z compartments are the most important. It is no coincidence that solutes such as halogenated hydrocarbons, about which there is great public concern, have high Z-values in humans! It should be borne in mind that, when calculating the environmental behavior of a solute, Z-values are needed only for the phases of concern. For example, if no atmospheric partitioning is considered, it is not necessary to know the air–water partition ratio or H. An arbitrary value of H can be used to define Z for water and other phases, because H cancels. Intuitively, it is obvious that H, or vapor pressure, play no role in influencing water-fish-sediment equilibria.

There is an additional practical analytical benefit from these high Z-value phases. If the concentration in air or water is very low, it may be inconvenient to sample and extract the large volume of air or water necessary to obtain a reasonable mass of the chemical. A practical approach is to introduce a small volume of an artificial phase that has a high and known Z-value and which "soaks up" the chemical over time to approach equilibrium. This phase is then removed and analyzed. This is the principle of "passive sampling" as a method of measuring low concentrations. Such devices have been referred to as "fugacity meters."

In summary, in this chapter we have introduced the concept of equilibrium existing between phases and have shown that this concept is essentially dictated by the laws of thermodynamics. Fortunately, we do not need to use or even understand the thermodynamic equations on which equilibrium relationships are based. However, it is useful to use these relationships for purposes such as correlation of partition ratios. It transpires that there are two approaches that can be used to conduct equilibrium calculations. First is to develop and use empirical correlations for partition ratios. Using these ratios, it is possible to calculate the partitioning of the chemical in a multimedia environment.

The second approach, which we prefer and illustrate here, is to use the equilibrium criterion of fugacity which can be related to concentration for each chemical and for each medium using a proportionality constant or Z-value. The Z-value can be calculated from fundamental equations or from partition ratios. We have established simple expressions for the various Z-values in these media using information on the nature of the media and the physical chemical properties of the substance of interest. This enables us to undertake simple multimedia partitioning calculations.

SYMBOLS AND DEFINITIONS INTRODUCED IN CHAPTER 2

Symbol	Common Units	Description
C_i^{Sat}	mol m^{-3}	Solubility of chemical in phase "i"
C_i^j	mol m^{-3}	Concentration of chemical "j" in phase "i"
C_i^{j0}	mol m^{-3}	Initial concentration of chemical "j" in phase "i"
ΔC_P	J mol^{-1} K^{-1}	Thermodynamic heat capacity difference between solid and liquid at constant pressure
F	—	Ratio of solid vapor pressure to super-cooled vapor pressure
f_i°	Pa	Reference fugacity of solute "i" in pure liquid state
f_S	Pa	Fugacity of a pure solid
γ_i	—	Activity coefficient for solute "i"
γ_0	—	Activity coefficient for solute "i" at infinite dilution
γ_i^j	—	Activity coefficient for solute "j" solvated in phase "i"
ΔG_{AB}	kJ mol^{-1}	Gibbs energy difference of a chemical in phase A and B
I	—	Ratio of ionized form to neutral form of an acid
K_a	—	Acid dissociation constant
LC_{50}	—	Concentration lethal to half a population at a prescribed time
L_i	—	Decimal lipid fraction of medium "i"
m_i	—	Number of moles of chemical in phase "i"
ϕ	—	Fugacity coefficient (gas phase)
pH	—	$-\log$ (H$_3$O$^+$ concentration in water)
pK$_a$	—	$-\log$ (acid dissociation constant, Ka)
P_T	Pa	Total pressure
P_i	Pa	Partial pressure of solute "i"
P_i^0	Pa	Vapor pressure of pure liquid or super-cooled liquid solute "i"
P^{Sat}	Pa	Saturation or vapor pressure
P_S^{Sat}	Pa	Saturation or vapor pressure of solid
P_L^{Sat}	Pa	Saturation or vapor pressure of liquid or super-cooled liquid
ρ_S	kg L^{-1}	Density of phase or medium S
R	Pa m^3 K^{-1} mol^{-1}	Gas constant
φ_i		Fraction of sorbed chemical in phase "i"
$\Delta_f S$		Thermodynamic entropy of fusion of a solid
θ	m^2 m^{-3}	Area of an aerosol per unit volume of air
T	K	Temperature
T_M	K	Melting temperature of a pure solid
v_i	m^3 mol^{-1}	Molar volume of solvent "i"
v_i^f	—	Volume fraction of phase "i"
x_i	—	Mole fraction of solute "i" in a liquid solvent
X_i	kg L^{-1}	Mass concentration of chemical in phase "i"
y	—	Mass fraction of organic matter in a solid
y	—	Mole fraction of a chemical in the gas phase
Z_i	Pa mol^{-1} m^{-3}	Fugacity capacity of chemical in phase "i"
Z_P	Pa mol^{-1} m^{-3}	Fugacity capacity of a pure solute

PRACTICE PROBLEMS FOR CHAPTER 2

Problem 2.1: Equilibrium partitioning between water and a sorbing material.
A lake of volume $10^6 m^3$ contains $15\,mg\,L^{-1}$ of sorbing material. The total concentration of a chemical of K_P equal to $10^5 L\,kg^{-1}$ is $1\,mg\,L^{-1}$. What are the dissolved and sorbed concentrations?

Answer
$0.4\,mg\,L^{-1}$ dissolved and $0.6\,mg\,L^{-1}$ sorbed.

Problem 2.2: Equilibrium partitioning between three phases in tabular form.
A three-phase system has Z-values $Z_1 = 5 \times 10^{-4}$, $Z_2 = 1.0$, and $Z_3 = 20$ (all mol m^{-3} Pa^{-1}), and volumes $V_1 = 1000$, $V_2 = 10$, and $V_3 = 0.1$ (all m^3). Calculate the distributions, concentrations, and fugacity when 1 mol of solute is distributed at equilibrium between these phases. It is suggested that the calculations be done in tabular form.

Answer

Phase	Z	V	VZ	$C = Z_f$	CV	%
1	5×10^{-4}	1000	0.5	4×10^{-5}	0.04	4
2	1.0	10	10	0.08	0.80	80
3	20	0.1	2	1.6	0.16	16
Total			12.5		1.00	100

$$f_{Sys} = m_{Sys} \bigg/ \sum_{i=1}^{3} V_i Z_i = 1.0/12.5 = 0.08$$

3 Advection and Reaction

THE ESSENTIALS

Advection: Advection rate = volumetric flow rate of medium × concentration in medium

$$r \ (\text{mol h}^{-1}) = G \ (\text{m}^3 \ \text{h}^{-1}) \ C \ (\text{mol m}^{-3})$$

Fugacity with an emission or input E (mol h^{-1}) and n advective and reactive loss processes at steady state and equilibrium:

$$f_{\text{Sys}} = \frac{E}{\displaystyle\sum_{i=1}^{n} D_i}.$$

D-values (mol Pa^{-1}h^{-1}):
Advection: $D_i^{\text{Adv}} = G_i \ (\text{m}^3 \ \text{h}^{-1}) \ Z_i \ (\text{mol Pa}^{-1} \ \text{m}^{-3})$
Reaction: $D_i^{\text{Deg}} = k \ (\text{h}^{-1}) V \ (\text{m}^3) \ Z \ (\text{mol Pa}^{-1} \ \text{m}^{-3})$

	Concentration	**Fugacity**
Advection rate r^{Adv} (mol h^{-1})	$r^{\text{Adv}} = \displaystyle\sum_{i=1}^{n} G_i C_i^{\text{Adv}}$	$r^{\text{Adv}} = f_{\text{Sys}} \displaystyle\sum_{i=1}^{n} D_i^{\text{Adv}}$
Reaction rate r^{Deg} (mol h^{-1})	$r_i^{\text{Deg}} = \displaystyle\sum_{i=1}^{n} k_i C_i^{\text{Deg}} V_i$	$r_i^{\text{Deg}} = f_{\text{Sys}} \displaystyle\sum_{i=1}^{n} D_i^{\text{Deg}}$

Fugacity with non-steady-state advection and reaction processes at equilibrium:

$$f_{\text{Sys}} = f_{\text{Sys}}^0 \exp(-kt) \quad \text{where} \quad k = \frac{\displaystyle\sum_{i=1}^{n} D_i^{\text{Loss}}}{\displaystyle\sum_{i=1}^{n} V_i Z_i}$$

Characteristic times: (residence time for steady state, response time non-steady state.)

$$\tau = \frac{\displaystyle\sum_{i=1}^{n} V_i Z_i}{\displaystyle\sum_{i=1}^{n} D_i}$$

3.1 INTRODUCTION

In the simplest equilibrium partitioning (Level I) calculations, it is assumed that the chemical is conserved; i.e., it is neither destroyed by reactions nor conveyed out of the evaluative environment by flows in phases such as air and water. The conclusions drawn about the impact of a given chemical discharge or emission can be quite misleading when this assumption is made erroneously, such as in situations where a chemical degrades or is transported into or out of the system (Schwartzenbach, 2002, Thibodeaux and Mackay, 2011).

For example, suppose a chemical such as glucose degrades by rapid biodegradation and consequently survives for only 10 h in the environment. Such a chemical poses less of a threat with respect to bioaccumulation than PCBs, for example, which may survive for greater than 10 years. Nevertheless, if such reactivity is ignored, a Level I calculation will treat them identically. Alternatively, some chemicals may leave the area of discharge rapidly as a result of evaporation into air, and subsequently be removed from the system by wind. The corresponding contamination problem is solved locally, but only by shifting it to another location. Considerable attention is being paid to substances that are susceptible to long-range transport by just such a process. As well, it is possible that, in a given region, local contamination is largely a result of inflow of chemical from upwind or upstream regions. Local efforts to reduce contamination by controlling local sources may therefore be frustrated, because most of the chemical is inadvertently imported. This problem was at the heart of the Canada–U.S., and Scandinavia–Germany–U.K. squabbles over acid precipitation. It is also a concern in relatively pristine areas such as the Arctic and Antarctic, where ecosystems and residents have little or no control over the contamination of their environments.

In this chapter, we address these issues and devise methods of calculating the effect of advective inflow and outflow and degrading reactions on local chemical fate and subsequent exposures. It must be emphasized that, once a chemical is degraded, this does not necessarily solve the problem. Toxicologists rarely miss an opportunity to point out reactions, such as mercury methylation or benzo(a)pyrene oxidation, in which the product of the reaction is more harmful than the parent compound. For our immediate purposes, we will be content to treat only the parent compound. Assessment of degradation products is best done separately by having the degradation product of one chemical treated as formation of another new chemical.

If a chemical is introduced into a steady-state system at a rate of E mol h^{-1}, as shown in Figure 3.1a, then the rate of removal must also be E mol h^{-1}. Otherwise, net accumulation or depletion will occur. If the amount in the system is m mol, then, on average, the amount of time each molecule spends in the steady-state system will be m/E h. This time, τ, is a *residence time* and is also variously called a *lifetime*, *detention time*, or *persistence time*. Clearly, if a chemical persists longer, there will be more of it in the system. The key equation is

$$\tau = \frac{m}{E} \quad \text{or} \quad m = \tau E$$

This concept is routinely applied to retention time in lakes. If a lake has a volume of 100,000 m^3, and it receives an inflow of 1000 m^3 day^{-1}, then the retention time is 100,000/1000 or 100 days. A mean retention time of 100 days does not imply that all water will spend 100 days in the lake. Some will bypass in only 10 days, and some will persist in backwaters for 1000 days but, on average, the residence time will be 100 days.

For the purpose of evaluating the general features of chemical fate in the environment, it is convenient to define "evaluative" or hypothetical environments in which the modeler selects volumes, area, and flow rates that are reasonable but not specific to any actual system. Figure 3.1 depicts such an environment with corresponding inputs and outputs.

The reason that this concept is so important is that chemicals exhibit variable lifetimes, ranging from hours to decades. As a result, the amount of chemical present in the environment, i.e., the inventory of chemical, varies greatly between chemicals. We tend to be most concerned about persistent

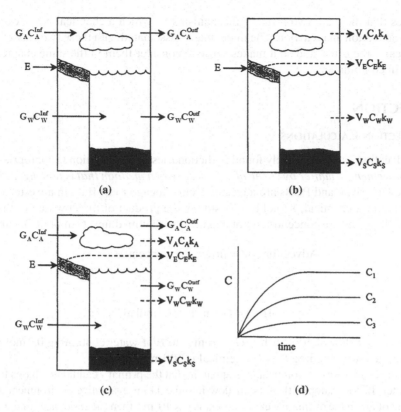

FIGURE 3.1 Diagram of a steady-state evaluative environment subject to (a) advective flow, (b) degrading reactions, (c) both, and (d) the time course to steady state.

and toxic chemicals, because relatively small emission rates (E) can result in large amounts (M) present in the environment since τ is large. This translates into high concentrations and possibly severe adverse effects. A further consideration is that chemicals that survive for prolonged periods in the environment have the opportunity to make long and often tortuous journeys. If applied to soil, they may evaporate, migrate onto atmospheric particles, deposit on vegetation, be eaten by cows, be transferred to milk, and then be consumed by humans. Chemicals may in this manner migrate up the food chain from water to plankton to fish to eagles, seals, and bears. Short-lived chemicals rarely survive long enough to undertake such adventures (or misadventures).

This lengthy justification leads to the conclusion that, if we are going to discharge a chemical into the environment, it is prudent to know

1. how long the chemical will survive, i.e., τ,
2. what causes its removal or "death," and
3. if it degrades to form other chemical species of concern.

This latter knowledge is useful, because it is likely that situations will occur in which a common removal mechanism does not apply. For example, a chemical may have the capacity to undergo rapid decomposition by photolysis, but this is of little relevance in long, dark Arctic winters or in deep, murky sediments.

In the process of quantifying this effect, we will introduce rate constants, advective flow rates, and, ultimately, using the fugacity concept, quantities called D-values, which prove to be immensely convenient. Indeed, armed with Z- and D-values, the environmental scientist has a powerful set of tools for calculation and interpretation.

It transpires that there are two primary mechanisms by which a chemical is removed from our environment: advection and reaction. The mathematics associated with these two classes of removal process are nearly the same, and so it makes sense to consider them in the same chapter, individually and then in combination.

3.2 ADVECTION

3.2.1 ADVECTION CALCULATIONS

Strangely, "advection" is a word rarely found in dictionaries, so a definition is appropriate. It means *the directed movement of matter by virtue of its presence in a medium that is flowing.* A floating leaf is advected down a river, and PCBs are advected from Chicago to Buffalo in a westerly wind. The rate of advection of a chemical, r (mol h^{-1}), is simply the product of the flow rate of the advecting medium, G (m^{3}h^{-1}), and the concentration of chemical in that medium, C (mol m^{-3}), namely,

$$\text{Advection rate} = \text{flow rate} \times \text{concentration}$$

or

$$r \,(\text{mol}\,\text{h}^{-1}) = G \,(\text{m}^3\,\text{h}^{-1})\,C\,(\text{mol}\,\text{m}^{-3})$$

Thus, if there is river flow of 1000 m^{3}h^{-1} (G_W) from A to B of water containing 0.3 mol m^{-3} (C_W) of chemical, then the corresponding flow of chemical is 300 mol h^{-1} (r_W).

Turning to the evaluative environment, it is apparent that the primary candidate phases for advection are air and water. If, for example, there is air flow into the 1 km^2 evaluative environment at 10^{9}m^{3}h^{-1}, and the volume of air in the evaluative environment is 6×10^9m^3, then the residence time will be 6 h, or 0.25 days. Likewise, a flow of 100 m^{3}h^{-1} of water into 70,000 m^3 of water results in a residence time of 700 h, or 29 days. It is easier to remember residence times than flow rates; therefore, we usually select or define a residence time and from it deduce the corresponding flow rate. Burial of bottom sediments can also be regarded as an advective loss, as can leaching of water from soils to groundwater. Advection of freons from the troposphere to the stratosphere is also of concern since it contributes to ozone depletion.

If we decree that our evaluative environment is at steady state, then inflow of air and water must equal outflows. Consequently, these inflow rates, designated G_i^{Inf} (m^{3}h^{-1}), must also be the corresponding outflow rates. If the concentrations of chemical in the phase of the evaluative environment is C_i (mol m^{-3}), then the outflow rate will be $G_i^{\text{Outf}}C_i$ (mol h^{-1}). This condition is often termed the *continuously stirred tank reactor*, or CSTR, assumption. The basic assumption is that, if a volume of phase (e.g., air) is well stirred, and some of that phase is removed, the air removed must have a concentration equal to that of the phase as a whole. If the chemical is introduced to the phase at a different concentration, it experiences an immediate change in concentration to that of the well mixed, or CSTR, value. The concentration experienced by the chemical then remains constant until the chemical is removed. The key point is that the outflow concentration equals the prevailing concentration. This concept greatly simplifies the algebra of steady-state systems. Essentially, we treat air, water, and other phases as being well-mixed CSTRs in which the outflow concentration equals the prevailing concentration. The fact that, in reality, the system will take a finite amount of time to achieve such "perfect" mixing, or that in the end it only closely approximates such mixing is ignored in the context of this assumption. Later we will examine situations in which this assumption is not valid, for which much more elaborate "Level IV" time-dependent computations are necessary.

If mixing in a compartment is sufficiently problematic such that overall chemical fate is affected, then the simplest expedient is to divide the compartment into two or more compartments that exchange both the chemical and the medium. This is routinely done for rivers and estuaries.

We can now consider an evaluative environment in which there is inflow and outflow of chemical in air and water. It is convenient at this stage to ignore the particles in the water, fish, and aerosols, and assume that the materials flowing into the evaluative environment are pure air and pure water. Since the steady-state condition applies, as shown in Figure 3.1a, the inflow and outflow rates are equal, and a mass balance can be assembled. The total inflow of the chemical is at a rate $G_A C_A^{\text{Inf}}$ in air, and $G_W C_W^{\text{Inf}}$ in water, the concentrations C_A^{Inf} and C_W^{Inf} being the "background" inflow values. There may also be emissions into the evaluative environment at a rate E. The total chemical inflow I is thus

$$I = \left[E + G_A C_A^{\text{Inf}} + G_W C_W^{\text{Inf}} \right] \text{ mol h}^{-1}$$

Now, the concentrations within the environment adjust instantly to the prevailing CSTR values of C_A and C_W in air and water. Thus, the outflow rates must be $G_A C_A$ and $G_W C_W$. These outflow concentrations could be constrained by equilibrium considerations; for example, they may be related through partition ratios or through Z-values to a common fugacity, as in Level I calculations.

This enables us to conceive of, and define, our first calculation in which we assume that equilibrium and steady-state conditions apply, *and* that inputs by emission and advection are balanced exactly by advective emissions. As well, we assume that equilibrium exists throughout the evaluative environment, such that all the phases are behaving like individual CSTRs. Later we will refer to this type of calculation as a "Level II" calculation.

Of course, starting with a clean environment and introducing these inflows, it would take the system some time to reach steady-state conditions, as shown in Figure 3.1d. At this stage, we are not concerned with how long it takes to reach a steady state, but only the conditions that ultimately apply at steady state. We can therefore develop the following mass-balance equations, using partition ratios and later fugacities.

$$I = E + G_A C_A^{\text{Inf}} + G_W C_W^{\text{Inf}} = G_A C_A + G_W C_W$$

This statement says that the total rate of chemical input equals the total rate of chemical output, a mass balance. As is our usual approach, we re-express this mass balance in terms of a single variable by way of partition ratios. Here we invoke the fact that $C_A = K_{AW} C_W$, to write

$$I = C_W \left[G_A K_{AW} + G_W \right]$$

and therefore

$$C_W = \frac{I}{\left[G_A K_{AW} + G_W \right]}$$

Other concentrations, amounts (m_i), and the total amount (m) can be deduced from C_W. The extension to multiple compartment systems is straightforward. For example, if soil is included, the concentration in soil will be in equilibrium with both C_A and C_W.

3.2.2 ADVECTION, FUGACITY, AND *D*-VALUES

The great advantage of working with fugacity becomes apparent when considering how to express the equilibrium environment mathematically. Since all compartments have the same fugacity at equilibrium, we assume a constant fugacity f_{Sys} to apply within the environment and to the outflowing media, such that

$$I = G_A C_A + G_W C_W = G_A Z_A f_{\text{Sys}} + G_W Z_W f_{\text{Sys}}$$

Solving for the prevailing fugacity, we find

$$f_{\text{Sys}} = \frac{I}{\left[G_A Z_A + G_W Z_W \right]}$$

Generally, for any number of phases in equilibrium, we have

$$f_{\text{Sys}} = \frac{I}{\displaystyle\sum_{i=1}^{n} G_i Z_i}$$

The grouping together of G- and Z-values, in the form of the product GZ (and other groups similar to it), appears so frequently in later calculations that it is convenient to designate them as D-values, i.e.:

$$D\left(\text{mol Pa}^{-1}\,\text{h}^{-1}\right) = G\left(\text{m}^3\,\text{h}^{-1}\right) Z\left(\text{mol Pa}^{-1}\,\text{m}^{-3}\right)$$

The advective process rate (mol h^{-1}) then equals Df. Such D-values are transport parameters, with units of mol Pa^{-1}h^{-1}. When multiplied by fugacity, they give molar rates of transport. They are thus similar in principle to rate constants, which, when multiplied by a mass of chemical, give a rate of reaction. Faster processes therefore have larger D-values.

In general, for any number of phases "n" in equilibrium contact, we have

$$f_{\text{Sys}} = \frac{I}{\displaystyle\sum_{i=1}^{n} G_i Z_i} = \frac{I}{\displaystyle\sum_{i=1}^{n} D_i^{\text{Adv}}}$$

Here, the superscript "Adv" refers to an advective transport process. Once the system fugacity is determined, one may then deduce all concentrations and amounts as before, using the Z-values for each phase and the total amount of chemical in the system as a whole. Such a calculation employing D-values is shown in Worked Example 3.1.

Worked Example 3.1: Advection in a well-mixed, three-phase system (air, water, sediment) with air and water inflow containing a chemical as well as with direct chemical emission.

An evaluative environment consists of 10^4 m^3 air, 100 m^3 water, and 1.0 m^3 sediment. There is an air inflow of 1000 m^3 h^{-1} and a water inflow of 1 m^3 h^{-1} at respective chemical concentrations of 0.01 mol m^{-3} and 1 mol m^{-3}. The Z-values are 4×10^{-4} (air), 0.1 (water), and 1.0 (sediment) mol Pa^{-1}m^{-3}. There is also an emission source with an associated rate of $E = 4$ mol h^{-1}. Calculate the fugacity, concentrations, amounts, outflow rates, and residence times.

Since we are assuming a well-mixed system, the overall system fugacity will be constant and the inflow and outflow rates must be equal. We can calculate the inflow (and therefore outflow) rate as the sum of all input rates:

$$I = E + G_A C_A^{\text{Inf}} + G_W C_W^{\text{Inf}}$$

or

$$I = 4\,\text{mol h}^{-1} + 1000\,\text{m}^3 \times 0.01\,\text{mol m}^{-3} + 1\,\text{m}^3 \times 1\,\text{mol m}^{-3}$$

$$= 15\,\text{mol h}^{-1}$$

The sum of the products of outflow rates and the corresponding Z-values is

$$\sum_{i=1}^{n} G_i Z_i = G_A Z_A + G_W Z_W$$

$$= 1000\,\mathrm{m^3 h^{-1}} \times 4 \times 10^{-4}\,\mathrm{mol\,Pa^{-1} m^{-3}} + 1\,\mathrm{m^3 h^{-1}} \times 1 \times 10^{-1}\,\mathrm{mol\,Pa^{-1} m^{-3}}$$

$$= 0.5\,\mathrm{mol\,Pa^{-1} h^{-1}}$$

Recasting the problem in terms of D-values, we have

$$D_A^{Adv} = G_A \left(\mathrm{m^3\ h^{-1}}\right) Z_A \left(\mathrm{mol\,Pa^{-1}\ m^{-3}}\right) = 0.4\,\mathrm{mol\,Pa^{-1}\ h^{-1}}$$

$$D_W^{Adv} = G_W \times Z_W = 0.1\,\mathrm{mol\,Pa^{-1}\ h^{-1}}$$

The sum of these D-values is $0.5\,\mathrm{mol\,Pa^{-1} h^{-1}}$, as was found using G- and Z-values above:

$$\sum_{i=1}^{n} D_i = D_A^{Adv} + D_W^{Adv}$$

$$= (0.4 + 0.1) = 0.5\,\mathrm{mol\,Pa^{-1}\ h^{-1}}$$

The prevailing fugacity is

$$f_{Sys} = \frac{I}{\displaystyle\sum_{i=1}^{n} G_i Z_i} = \frac{I}{\displaystyle\sum_{i=1}^{n} D_i} = \frac{15\ \mathrm{mol\,h^{-1}}}{0.5\,\mathrm{mol\,Pa^{-1}\ h^{-1}}} = 30\,\mathrm{Pa}$$

With the system fugacity in hand, and the corresponding Z-values, we can calculate the concentration in air as

$$C_A = Z_A f_{Sys} = 4 \times 10^{-4}\ \mathrm{mol\,Pa^{-1}\ m^{-3}} \times 30\,\mathrm{Pa}$$

$$= 1.2 \times 10^{-2}\,\mathrm{mol\,m^{-3}}$$

The analogous calculation for water and sediment results in $C_W = 3\,\mathrm{mol\,m^{-3}}$ and $C_{Sed} = 30\,\mathrm{mol\,m^{-3}}$. The molar amounts of chemical in each compartment are given by

$$m_i = V_i Z_i f_{Sys}$$

Therefore:

$$m_A = 1 \times 10^4\ \mathrm{m^3} \times 4 \times 10^{-4}\,\mathrm{mol\,Pa^{-1}\ m^{-3}} \times 30\,\mathrm{Pa} = 120\ \mathrm{moles}$$

$$m_W = 100\ \mathrm{m^3} \times 0.1\,\mathrm{mol\,Pa^{-1}\ m^{-3}} \times 30\,\mathrm{Pa} = 300\ \mathrm{moles}$$

$$m_{Sed} = 1\,\mathrm{m^3} \times 1.0\,\mathrm{mol\,Pa^{-1}\ m^{-3}} \times 30\,\mathrm{Pa} = 30\ \mathrm{moles}$$

The total amount of chemical in the system is therefore $m = 450$ moles. The outflow rates in air and water are given by product of the respective flow rates and the concentration of the flowing carrier medium, $G_i \times C_i$, such that

$$G_A C_A = 1000\,\mathrm{m^3\ h^{-1}} \times 1.2 \times 10^{-2}\,\mathrm{mol\,m^{-3}} = 12\,\mathrm{mol\,h^{-1}}$$

$$G_W C_W = 1\,\mathrm{m^3\ h^{-1}} \times 3.0\,\mathrm{mol\,m^{-3}} = 3\,\mathrm{mol\,h^{-1}}$$

Since there is no "flow" into or out of the sediment (such as would be present with deposition and sediment burial processes, respectively), the outflow rate for the sediment compartment is zero.

The total or overall residence time is given by

$$\tau = \frac{m}{I} = \frac{450\,\text{mol}}{15\,\text{mol}\,\text{h}^{-1}} = 30\,\text{h}$$

Residence times for the individual compartments are both m_i/G_iC_i or V_iZ_i/D_i:

$$\tau_A = \frac{m_A}{G_AC_A} = \frac{120\,\text{mol}}{12\,\text{mol}\,\text{h}^{-1}} = V_AZ_A/D_A = 10\,\text{h}$$

$$\tau_W = \frac{m_W}{G_WC_W} = \frac{300\,\text{mol}}{3\,\text{mol}\,\text{h}^{-1}} = V_WZ_W/D_W = 100\,\text{h}$$

Note that the overall residence time of the chemical is a weighted average of the residence times in the individual compartments, influenced by the extent to which the chemical partitions into the various phases. The sediment has no effect on the fugacity or the outflow rates, but it acts as a "reservoir" to influence the total amount present M and therefore the residence time or persistence.

It is apparent that the air D-value is larger and most significant. D-values can be added when they are multiplied by a common fugacity. Therefore, it becomes obvious which D-value, and hence which process, is most important. We can arrive at the same conclusion using partition ratios, but the algebra is less elegant.

Note also that, under the CSTR Level II assumption, the manner in which the chemical enters the environment is unimportant. All sources are combined or lumped into I, the overall input. This is because under this assumption, once in the environment, the chemical immediately achieves an equilibrium distribution, i.e., it "forgets" its origin.

3.2.3 ADVECTIVE PROCESSES

In an evaluative environment, there are several common advective flows that convey a chemical to and from the environment, namely,

1. inflow and outflow of air
2. inflow and outflow of water
3. inflow and outflow of aerosol particles present in air
4. inflow and outflow of particles and biota present in water
5. unidirectional transport of air from the troposphere to the stratosphere, i.e., vertical movement of air out of the environment
6. unidirectional sediment burial, i.e., sediment being conveyed out of the well-mixed layer to depths sufficient that it is essentially inaccessible
7. unidirectional flow of water from surface soils to groundwater (recharge)

It also transpires that there are several advective processes which can apply to chemical movement *within* the evaluative environment. Notable are rainfall, water runoff from soil, sedimentation, and food consumption, but we delay their treatment until later.

In situations 1–4, there is no difficulty in deducing the rate as $G_i \times C_i$ or $D \times f_{Sys}$, where G_i is the flow rate of the phase in question, C_i is the concentration of chemical in that phase, and the corresponding Z-values apply to the chemical in the phases in which it is dissolved or sorbed.

For example, aerosol may be transported to an evaluative world in association with the inflow of $10^{12}\,\text{m}^3\,\text{h}^{-1}$ of air. If the aerosol concentration is 10^{-11} volume fraction, then the flow rate of aerosol G_Q is $10\,\text{m}^3\,\text{h}^{-1}$.

$$G_Q = G_A v_Q^f = 10^{12}\,\text{m}^3\,\text{h}^{-1} \times 10^{-11} = 10\,\text{m}^3\,\text{h}^{-1}$$

The relevant concentration of chemical is that in the aerosol, not in the air, and is normally quite high, for example, $100 \, \text{mol m}^{-3}$. Therefore, the rate of chemical input in the aerosol is $1000 \, \text{mol h}^{-1}$. This can be calculated using the D and f route as follows, giving the same result.

If $Z_Q = 10^8$, then

$$f_{\text{Sys}} = \frac{C_Q}{Z_Q} = \frac{100 \, \text{mol m}^{-3}}{10^8 \, \text{mol Pa}^{-1}\text{m}^{-3}} = 10^{-6} \, \text{Pa}$$

$$D_Q^{\text{Adv}} = G_Q Z_Q = 10 \, \text{m}^3 \, \text{h}^{-1} \times 10^8 \, \text{mol Pa}^{-1} \, \text{m}^{-3} = 10^9 \, \text{mol Pa}^{-1} \, \text{h}^{-1}$$

Therefore, the molar flow rate due to the aerosol is

$$r_Q = D_Q^{\text{Adv}} f_{\text{Sys}} = 10^9 \, \text{mol Pa}^{-1} \, \text{h}^{-1} \times 10^{-6} \, \text{Pa} = 1000 \, \text{mol h}^{-1}$$

Treatment of transport to the stratosphere is somewhat more difficult. We can conceive of parcels of air that migrate from the troposphere to the stratosphere at an average, continuous rate, $G_A \, \text{m}^3\text{h}^{-1}$, being replaced by relatively clean stratospheric air that migrates downward at the same rate. We can thus calculate the D-value. This rate should correspond to a residence time of the troposphere of about 60 years, i.e., G_A is V_A/τ. Thus, if V_A is $6 \times 10^9 \text{m}^3$ and τ is $5.25 \times 10^5 \text{h}$, G_A is $11,400 \, \text{m}^3\text{h}^{-1}$. This rate is very slow and is usually insignificant, but there are situations in which it is important. For example, we may be concerned with the amount of chemical that actually reaches the stratosphere, for example, freons that catalyze the decomposition of ozone. This slow rate is then important from the viewpoint of the receiving stratospheric phase, although it is not an important loss from the delivering, or tropospheric, phase. Also, if a chemical is very stable and is only slowly removed from the atmosphere by reaction or deposition processes, then transfer *to* the troposphere may also be a significant mechanism of removal. Certain volatile halogenated hydrocarbons tend to be in this class. If we emit a chemical into the evaluative world at a steady rate by emissions and allow for no removal mechanisms whatsoever, its concentrations will continue to build up indefinitely. Such situations are likely to arise if we view the evaluative world as merely a scaled-down version of the entire global environment. There is certainly advective flow of chemical between, for example, the United States and Canada, but there is no advective flow of chemical out of the entire global atmospheric environment, except for the small amounts that transfer to the stratosphere. Whether advection is included depends upon the system being simulated. In general, the smaller the system, the shorter the advection residence time, and the more important advection becomes.

Sediment burial is the process by which the chemical is conveyed from the active mixed layer of accessible sediment into inaccessible buried layers. As was discussed earlier, this is a rather naive picture of a complex process, but at least it is a starting point for calculations. The reality is that the mixed surface sediment layer is rising, eventually filling the lake. Typical burial rates are 1 mm year^{-1}, the material being buried being typically 25% solids, 75% water. As it "moves" to greater depths, water is increasingly squeezed out, and the percentage of solids therefore increases.

Mathematically, the D-value for burial consists of two terms, the burial rate of solids and that of water. For example, if a lake has an area of 10^7m^2 and has a burial rate of 1 mm year^{-1}, the total rate of burial is $10,000 \text{m}^3\text{year}^{-1}$ or $1.14 \, \text{m}^3\text{h}^{-1}$, consisting of perhaps 25% solids, i.e., $0.29 \, \text{m}^3\text{h}^{-1}$ of solids (G_S) and $0.85 \, \text{m}^3\text{h}^{-1}$ of water (G_W). The rate of loss of chemical is then

$$G_S C_S + G_W C_W = G_S Z_S f_{\text{Sys}} + G_W Z_W f_{\text{Sys}} = f_{\text{Sys}} \left[D_S^{\text{Adv}} + D_W^{\text{Adv}} \right]$$

Usually, there is a large solid-to-pore water partition ratio; therefore, C_S greatly exceeds C_W or, alternatively, Z_S is very much greater than Z_W, and the term D_S^{Adv} dominates. A residence time of solids in the mixed layer can be calculated as the volume of solids in the mixed layer divided by G_S. For example, if the depth of the mixed layer is 3 cm, and the solids concentration is 25%, then the volume of solids is $75,000\,m^3$ and the residence time is $260,000\,h$, or 30 years. The residence time of water is probably longer, because the water content is likely to be higher in the active sediment than in the buried sediment. In reality, the water would exchange diffusively with the overlaying water during that time period. It is difficult to measure sediment burial rates and sediment residence times. The usual approach is to assess the sediment "chronology" by measurement of slowly decaying elemental isotopes such as lead 201 (^{210}Pb).

As discussed earlier in Section 2.7.3, there are occasions in which it is convenient to calculate a "bulk" Z-value for a medium containing a dispersed phase such as an aerosol. In this way, we may express two loss processes as one. In the case of aerosols, this would correspond to losses due to advection of the chemical dissolved in the aerosol particles, as well as losses due to advection of the chemical as a vapor in the gaseous air itself. The overall D-value is then GZ where G is the air flow rate and Z is the bulk fugacity capacity value for the air and the aerosol.

3.2.4 ADVECTION AS A PSEUDO REACTION

Examination of these equations shows that the group G/V plays the same role as a rate constant having identical units of h^{-1}. It may, indeed, be convenient to regard advective loss as a pseudo reaction with this rate constant and applicable to the phase volume of V. Note that the group V/G is the residence time of the phase in the system. Frequently, this is the most accessible and readily remembered quantity. For example, it may be known that the retention time of water in a lake is 10 days, or 240 h. The advective rate constant, k, is thus $1/240\,h^{-1}$, and the D-value is VZk, which is, of course, also GZ.

It is noteworthy that this residence time is not equivalent to a reaction half-time, which is related to the rate constant through the constant 0.693 or $\ln(2)$. Residence time is equivalent to $1/k$.

3.3 DEGRADATION REACTIONS

3.3.1 INTRODUCTION

The word *reaction* requires definition. We regard to reactions as processes that alter the chemical nature of the solute, i.e., change its *chemical abstract system (CAS)* number. For example, hydrolysis of ethyl acetate to ethanol and acetic acid is definitely a reaction, as is conversion of 1,2-dichlorobenzene to 1,3-dichlorobenzene, or even conversion of *cis*-but-2-ene to *trans*-but-2-ene. In contrast, processes that merely convey the unchanged chemical from one phase to another, or store it in inaccessible form, are not reactions. Uptake by biota, sorption to suspended material, or even uptake by enzymes are not reactions. A reaction may subsequently occur in these locations, but it is not until the chemical structure is actually changed that we consider reaction to have occurred. In the literature, the word *reaction* is occasionally, and wrongly, applied to these processes, especially to sorption.

When dealing with chemical reaction processes, we have two tasks at hand. The first is to assemble the necessary mathematical framework for treating reaction rates using rate constants, and the second is to devise methods of obtaining information on values of these rate constants.

3.3.2 REACTION RATE EXPRESSIONS

We prefer, when possible, to use a simple first-order kinetic expression for all reactions. For simplicity, we term all processes that result in the removal of the chemical of interest "degradations." The basic rate equation for the rate of change due to a degradation process for the total amount of a reactive species in a particular phase is

$$Rate = r^{Deg} = VCk = m \times k \ (mol\,h^{-1})$$

where V is the volume of the phase (m³), C is the concentration of the chemical (mol m⁻³), m is the amount of chemical in the whole system, and k is the first-order rate constant with units of reciprocal time. The group VCk thus has units of mol h⁻¹.

For any first-order process, the usual kinetic expression for rate of change in amount of chemical in the system is given by

$$\frac{dm}{dt} = -km \quad \text{or} \quad \frac{dC}{dt} = -kC$$

That is, the rate of change in amount of matter at a given time is proportional to the amount present at that time m. The use of C instead of m is permissible as long as V does not change with time.

Integrating from an initial condition concentration at zero time C_0 gives the following integrated equations that are familiar to most physical scientists:

$$\ln\left(\frac{C}{C_0}\right) = -kt \quad \text{or} \quad C = C_0 \exp(-kt)$$

Rate constants have units of frequency or reciprocal time and are therefore not easily grasped or remembered. It is more convenient to store and remember half-lives, i.e., the time, $t_{1/2}$, which is the time required for C to decrease to half of C_0, i.e., the time at which $C = 0.5C_0$. The rate constant is related to the half-life as

$$k = \frac{0.693}{t_{1/2}}$$

For example, a process with a half-life of 10 h has a corresponding rate constant, k, of 0.0693 h⁻¹.

3.3.3 Non-First-Order Kinetics

In reality, virtually all environmental reactions are second-order processes, i.e., the rate depends on the concentration of both the chemical and on a second term (B) that is characteristic of the reactive nature of the environment. This second term B could be the concentration of micro-organisms responsible for biodegradation of the chemical, or the intensity of light that causes photolytic decomposition. It is, however, convenient to combine this second term with the true second-order rate constant to give a simple, pseudo first-order rate expression, with a first-order rate constant, i.e.:

$$r = k_2BC = k_1'C \quad \text{where} \quad k_1' = k_2B$$

We can often circumvent complex reaction rate equations by expressing them in terms of a pseudo first-order rate reaction. The primary assumption is that the concentration of one reactant (B) is effectively constant and will not change appreciably as the reaction proceeds. Reactions between two chemicals can also be considered a pseudo first-order reaction when $C_A \ll C_B$, so the concentration of B does not change as the reaction proceeds.

A major advantage of forcing first-order kinetics on all reactions is that, if a chemical is susceptible to several reactions in the same phase, with rate constants k_A, k_B, k_C, etc., then the total rate constant for reaction is the sum of the individual rate constant $(k_A + k_B + k_C)$. Recall that half-lives for first-order processes add in a reciprocal manner, such that the combined effect of one mechanism with a half-life of 10 h, and another with a half-life of 20 h is a net process with a total half-life of 6.7 h, not 30 h, as one might naively suppose.

3.3.4 Partition Ratio Approach to Chemical Degradation

We are now able to perform certain calculations describing the behavior of chemicals in evaluative environments. The simplest is a "Level II-type" equilibrium steady-state reaction situation in which there is no advection, and there is a constant inflow of chemical in the form of an emission, as depicted in Figure 3.1b. When a steady state is reached, there must be an equivalent loss in the form of reactions. Starting from a clean environment, the concentrations would build up until they reach a level such that the rates of degradation or loss equal the total rate of input. We further assume that the phases are well-mixed and at equilibrium, i.e., transfer between them is very rapid.

As a result, the concentrations are related through partition ratios, or a common fugacity applies. The total emission rate E is given by the sum of all degradation rates over "n" equilibrating phases:

$$E = V_1 C_1 k_1 + V_2 C_2 k_2 + \cdots \sum_{i=1}^{n} V_i C_i k_i$$

Reducing this expression to one unknown C_W by employing partition ratios, we obtain

$$E = \sum_{i=1}^{n} V_i C_W K_{iW} k_i = C_W \sum_{i=1}^{n} V_i K_{iW} k_i$$

from which C_W can be deduced, followed by other concentrations, amounts, rates of reaction, and persistence.

> **Worked Example 3.2: A three-phase system (air, water, sediment) with an emissive source and reaction losses but not advection.**
>
> The same evaluative environment as in Worked Example 3.1 is subject to emission of 10 mol h^{-1} of a chemical, but no advection. The chemical reaction half-lives are 69.3 h (air), 6.93 h (water), and 693 h (sediment). Calculate the concentrations in each phase at steady-state assuming equilibrium between all phases. From before, $K_{AW} = 0.004$, $K_{Sed-W} = 10$, $V_A = 10^4$ m^3, $V_W = 100$ m^3, and $V_{Sed} = 1.0$ m^3.
>
> First, calculate the rate constant for reaction in air from the half-life:
>
> $$k_A = \frac{0.693}{\tau_{1/2}} = \frac{0.693}{69.3 \, \text{h}} = 0.01 \, \text{h}^{-1}$$
>
> The corresponding rate constants are 0.1 h^{-1} (water) and 0.001 h^{-1} (sediment). Now, the total loss due to reaction must equal the total emission rate at steady state, so we write
>
> $$E = V_A C_A k_A + V_W C_W k_W + V_{Sed} C_{Sed} k_{Sed}$$
>
> Again, employing partition ratios to reduce the expression to one unknown, we write
>
> $$E = C_W \left[V_A K_{AW} k_A + V_W k_W + V_{Sed} K_{Sed-W} k_{Sed} \right]$$
>
> Here
>
> $$E = 10 \, \text{mol h}^{-1} = C_W \left[10^4 \times 0.004 \times 0.01 + 100 \times 0.1 + 1 \times 10 \times 0.001 \right] \, \text{m}^3 \, \text{h}^{-1}$$
>
> or
>
> $$10 \, \text{mol h}^{-1} = C_W \left[0.4 + 10 + 0.01 \right] \, \text{m}^3 \, \text{h}^{-1}$$

Solving for C_W we obtain a steady-state water concentration of 0.9606 mol m^{-3}:

$$C_W = \frac{10\,\text{mol h}^{-1}}{[0.4+10+0.01]\ \text{m}^3\ \text{h}^{-1}} = 0.9606\,\text{mol m}^{-3}$$

The corresponding concentrations in air and sediment are obtained using partition ratios, that is,

$$C_A = K_{AW}C_W = 4\times10^{-3}\times0.9606\ \text{mol m}^{-3} = 0.00384\,\text{mol m}^{-3}$$

Similarly, we find $C_S = 9.606$ mol m^{-3}. The rates of reaction are given by VCk. For air, we have

$$r_A = V_A C_A k_A = 10^4\ \text{m}^3\times0.0038\,\text{mol m}^{-3}\times0.01\ \text{h}^{-1} = 0.38\,\text{mol h}^{-1}$$

The corresponding rates for water and sediment are 9.61 and 0.01 mol h^{-1}, respectively, for a total degradation rate of 10 mol h^{-1}, equal to the emissive input as required.

It is important to note that in the foregoing example, the reaction rate is controlled by the product V, C, and k and not any one of these three quantitates in isolation. A large value of only one of these quantities may convey the wrong impression that the reaction is important.

3.3.5 FUGACITY APPROACH TO CHEMICAL DEGRADATION

We can now follow the same process as used when treating advection and define D-values for reactions. If the rate is $VCk = VZf_{\text{Sys}}k$, we can also express it in "D-value" format by equating D^{Deg} with VZk:

$$D^{\text{Deg}}\left(\text{mol Pa}^{-1}\ \text{h}^{-1}\right) = V\left(\text{m}^3\right)Z\left(\text{mol Pa}^{-1}\ \text{m}^{-3}\right)k\left(\text{h}^{-1}\right),$$

such that the rate is also $D^{\text{Deg}}f_{\text{Sys}}$. Note that D^{Deg} has units of mol Pa^{-1}h^{-1}, identical to those of D^{Adv} or GZ, discussed earlier. If there are several reactions occurring to the same chemical in the same phase, then each reaction can be assigned a D-value, and these D-values can be added to give a total D-value. This is equivalent to adding the rate constants. The mass balance for a "Level II" calculation of an equilibrated system of n phases at steady state with loss due only to chemical reactions becomes

$$r^{\text{Deg}} = \sum_{i=1}^{n}V_iC_ik_i = \sum_{i=1}^{n}V_iZ_if_{\text{Sys}}k_i = f_{\text{Sys}}\sum_{i=1}^{n}V_iZ_ik_i = f_{\text{Sys}}\sum_{i=1}^{n}D_i^{\text{Deg}}$$

Thus, f_{Sys} can be deduced, followed by concentrations, amounts, the total amount m, and the rates of individual reactions as VCk or Df_{Sys}. We repeat Worked Example 3.2 in fugacity format as Worked Example 3.3.

Worked Example 3.3: Fugacity-based treatment of the three-phase system (air, water, sediment) from Worked Example 3.2 with an emissive source and reaction losses but no advection.

Phase	V_i (m³)	Z_i (mol Pa^{-1}m^{-3})	k_i (h^{-1})	D_i^{Deg} (mol Pa^{-1}h^{-1})
Air	10^4	4×10^{-4}	0.01	0.04
Water	100	0.1	0.1	1.0
Sediment	1.0	1.0	0.001	1×10^{-3}
			Total	1.041

The individual D-values for reaction in each phase are given by kVZ_i:

$$D_A^{\text{Deg}} = k_A^{\text{Deg}} V_A Z_A = 1.0 \times 10^{-2} \text{ h}^{-1} \times 1.0 \times 10^4 \text{ m}^3 \times 4 \times 10^{-4} \text{ mol Pa}^{-1} \text{ m}^{-3} = 4.0 \times 10^{-2} \text{ mol Pa}^{-1} \text{ h}^{-1}$$

$$D_W^{\text{Deg}} = k_W^{\text{Deg}} V_W Z_W = 1.0 \times 10^{-1} \text{ h}^{-1} \times 1.0 \times 10^2 \text{ m}^3 \times 1.0 \times 10^{-1} \text{ mol Pa}^{-1} \text{ m}^{-3} = 1.0 \text{ mol Pa}^{-1} \text{ h}^{-1}$$

$$D_{\text{Sed}}^{\text{Deg}} = k_{\text{Sed}}^{\text{Deg}} V_{\text{Sed}} Z_{\text{Sed}} = 1.0 \times 10^{-3} \text{ h}^{-1} \times 1.0 \text{ m}^3 \times 1.0 \text{ mol Pa}^{-1} \text{ m}^{-3} = 1.0 \times 10^{-3} \text{ mol Pa}^{-1} \text{ h}^{-1}$$

The sum of D-values is

$$\sum_{i=1}^{n} D_i^{\text{Deg}} = D_A^{\text{Deg}} + D_W^{\text{Deg}} + D_{\text{Sed}}^{\text{Deg}} = \left(4.0 \times 10^{-2} + 1.0 + 1.0 \times 10^{-3}\right) = 1.041 \text{ mol Pa}^{-1} \text{ h}^{-1}$$

The fugacity is

$$f_{\text{Sys}} = \frac{E}{\sum_{i=1}^{n} D_i^{\text{Deg}}} = \frac{10 \text{ mol h}^{-1}}{1.041 \text{ mol Pa}^{-1} \text{ h}^{-1}} = 9.606 \text{ Pa}$$

The corresponding concentrations in the equilibrated phases are given by the product Zf_{Sys} to yield

$$C_A = Z_A f_{\text{Sys}} = 4 \times 10^{-4} \text{ mol Pa}^{-1} \text{ m}^{-3} \times 9.606 \text{ Pa} = 3.84 \times 10^{-3} \text{ mol m}^{-3}$$

$$C_W = Z_W f_{\text{Sys}} = 0.1 \text{ mol Pa}^{-1} \text{ m}^{-3} \times 9.606 \text{ Pa} = 9.61 \times 10^{-1} \text{ mol m}^{-3}$$

$$C_{\text{Sed}} = Z_{\text{Sed}} f_{\text{Sys}} = 1.0 \text{ mol Pa}^{-1} \text{ m}^{-3} \times 9.606 \text{ Pa} = 9.61 \text{ mol m}^{-3}$$

The rates of loss due to chemical reaction in each compartment are given by Df_{Sys}, yielding

$$r_A^{\text{Deg}} = D_A^{\text{Deg}} f_{\text{Sys}} = 0.04 \text{ mol Pa}^{-1} \text{ h}^{-1} \times 9.606 \text{ Pa} = 3.84 \times 10^{-1} \text{ mol h}^{-1}$$

$$r_W^{\text{Deg}} = D_W^{\text{Deg}} f_{\text{Sys}} = 1.0 \text{ mol Pa}^{-1} \text{ h}^{-1} \times 9.606 \text{ Pa} = 9.606 \text{ mol h}^{-1}$$

$$r_{\text{Sed}}^{\text{Deg}} = D_{\text{Sed}}^{\text{Deg}} f_{\text{Sys}} = 1.0 \times 10^{-3} \text{ mol Pa}^{-1} \text{ m}^{-3} \times 9.606 \text{ Pa} = 9.606 \times 10^{-3} \text{ mol h}^{-1}$$

Worked Example 3.4: Three-phase system (air, water, sediment) at equilibrium and steady state with chemical input by emission and reactive loss in all three compartments.

An evaluative environment consists of $10,000 \text{ m}^3$ air, 100 m^3 water, and 10 m^3 sediment. There is input of $E = 25 \text{ mol h}^{-1}$ of chemical, which reacts with half-lives of 100 h in air, 75 h in water, and 50 h in sediment. Calculate the concentrations and amounts given the Z-values below:

Phase	Volume (m³)	Z (mol Pa⁻¹m⁻³)	k (h⁻¹)	VZk or D (mol Pa⁻¹h⁻¹)	C (mol m⁻³)	m (mol)	Rate (mol h⁻¹)
Air	1.0×10^4	4×10^{-4}	6.93×10^{-3}	0.0277	3.86×10^{-2}	386	2.68
Water	100	0.1	9.24×10^{-3}	0.0924	9.66	966	8.93
Sediment	10	1.0	1.39×10^2	0.139	96.6	966	13.4
			Total	0.2587	—	2318	25.0

The individual D-values for reaction in each phase are given by $k_i V_i Z_i$:

$$D_A^{\text{Deg}} = k_A^{\text{Deg}} V_A Z_A = 6.93 \times 10^{-3} \text{ h}^{-1} \times 1.0 \times 10^4 \text{ m}^3 \times 4 \times 10^{-4} \text{ mol Pa}^{-1} \text{ m}^{-3} = 2.77 \times 10^{-2} \text{ mol Pa}^{-1} \text{ h}^{-1}$$

$$D_W^{Deg} = k_W^{Deg} V_W Z_W = 9.24 \times 10^{-3} \text{ h}^{-1} \times 1.0 \times 10^2 \text{ m}^3 \times 1.0 \times 10^{-1} \text{ mol Pa}^{-1} \text{ m}^{-3} = 9.24 \times 10^{-2} \text{ mol Pa}^{-1} \text{ h}^{-1}$$

$$D_{Sed}^{Deg} = k_{Sed}^{Deg} V_{Sed} Z_{Sed} = 1.39 \times 10^{-2} \text{ h}^{-1} \times 10 \text{ m}^3 \times 1.0 \text{ mol Pa}^{-1} \text{ m}^{-3} = 1.39 \times 10^{-1} \text{ mol Pa}^{-1} \text{ h}^{-1}$$

The sum of D-values is

$$\sum_{i=1}^{n} D_i^{Deg} = D_A^{Deg} + D_W^{Deg} + D_{Sed}^{Deg} = \left(2.77 \times 10^{-2} + 9.24 \times 10^{-2} + 1.39 \times 10^{-1}\right) = 2.587 \times 10^{-1} \text{ mol Pa}^{-1} \text{ h}^{-1}$$

The rate constants in each case are 0.693/half-life. The fugacity is obtained by the emission rate E divided by the sum of the D-values:

$$f_{Sys} = \frac{E}{\sum_{i=1}^{n} D_i^{Deg}} = \frac{25 \text{ mol h}^{-1}}{0.2587 \text{ mol Pa}^{-1} \text{ h}^{-1}} = 96.6 \text{ Pa}$$

The corresponding concentrations in the equilibrated phases are given by the product Zf_{Sys}, to yield

$$C_A = Z_A f_{Sys} = 4 \times 10^{-4} \text{ mol Pa}^{-1} \text{ m}^{-3} \times 96.6 \text{ Pa} = 3.86 \times 10^{-2} \text{ mol m}^{-3}$$

$$C_W = Z_W f_{Sys} = 0.1 \text{ mol Pa}^{-1} \text{ m}^{-3} \times 96.6 \text{ Pa} = 9.66 \text{ mol m}^{-3}$$

$$C_{Sed} = Z_{Sed} f_{Sys} = 1.0 \text{ mol Pa}^{-1} \text{ m}^{-3} \times 96.6 \text{ Pa} = 96.6 \text{ mol m}^{-3}$$

The molar quantities in each compartment are given by VC

$$m_A = V_A C_A = 1 \times 10^4 \text{ m}^{-3} \times 3.86 \times 10^{-2} \text{ mol m}^{-3} = 3.86 \times 10^2 \text{ mol}$$

$$m_W = V_W C_W = 1 \times 10^2 \text{ m}^{-3} \times 9.66 \text{ mol m}^{-3} = 9.66 \times 10^2 \text{ mol}$$

$$m_{Sed} = V_{Sed} C_{Sed} = 10 \text{ m}^3 \times 96.6 \text{ mol m}^{-3} = 9.66 \times 10^2 \text{ mol}$$

$$m = \Sigma m_i = 386 + 966 + 966 = 2318 \text{ mol}$$

The rates of loss due to chemical reaction in each compartment are given by Df_{Sys}, yielding

$$r_A^{Deg} = D_A^{Deg} f_{Sys} = 2.77 \times 10^{-2} \text{ mol Pa}^{-1} \text{ h}^{-1} \times 96.6 \text{ Pa} = 2.68 \text{ mol h}^{-1}$$

$$r_W^{Deg} = D_W^{Deg} f_{Sys} = 9.24 \times 10^{-2} \text{ mol Pa}^{-1} \text{ h}^{-1} \times 96.6 \text{ Pa} = 8.93 \text{ mol h}^{-1}$$

$$r_{Sed}^{Deg} = D_{Sed}^{Deg} f_{Sys} = 1.39 \times 10^{-1} \text{ mol Pa}^{-1} \text{ m}^{-3} \times 96.6 \text{ Pa} = 13.4 \text{ mol h}^{-1}$$

$$\Sigma r_i^{Deg} = 2.68 + 8.93 + 13.4 = 25.0 \text{ mol h}^{-1}$$

It is clear that the D-value product kVZ controls the overall importance of each process. Despite its low volume and relatively slow reaction rate, the sediment provides a fairly fast-reacting medium because of its large Z-value. It is not until the calculation is completed that it becomes obvious where most reaction occurs. The overall residence time is given by the total quantity of chemical divided by the loss rate:

$$\tau = \frac{m = \Sigma m_i}{\Sigma r_i^{Deg}} = \frac{2318 \text{ mol}}{25.0 \text{ mol h}^{-1}} = 92.7 \text{ h}$$

Note that the persistence or m/E is a weighted mean of the persistence or reciprocal rate constants in each phase. It is also $\Sigma VZ / \Sigma D$.

3.4 ADVECTION AND DEGRADATION

3.4.1 COMBINING ADVECTION AND DEGRADATION PROCESSES

Advective and reaction processes can be included in the same calculation as shown in the example below, which is similar to those presented earlier for reaction. We now have inflow and outflow of air and water at rates given below and with background concentrations as shown in Figure 3.1c.

The mass-balance equation now becomes

$$I = E + G_A C_A^{Inf} + G_W C_W^{Inf} = G_A C_A^{Outf} + G_W C_W^{Outf} + \sum_{i=1}^{n} V_i C_i k_i$$

$$I = \text{emission} + \text{air and water inflow rates}$$

$$= \text{air and water outflow rates} + \text{total reaction loss rates}$$

This can be solved either by substituting $K_{iW}C_W$ for all other concentrations and solving for C_W, or calculating the advective D-values as GZ and adding them to the reaction D-values equal to VZk, then solving for the system fugacity.

> **Worked Example 3.5: Three-phase system (air, water, sediment) at equilibrium and steady-state concentration, with both advective and emissive inputs, and reactive losses.**
>
> The environment in Worked Example 3.4 has advective flows of $1000\,m^3\,h^{-1}$ in air and $1\,m^3\,h^{-1}$ in water as in Worked Example 3.1 and reaction D-values as in Worked Example 3.3, with a total input by advection and emission of $40\,mol\,h^{-1}$. Calculate the fugacity, concentrations, amounts, and chemical residence time.
>
> | Phase | Volume (m³) | Z (mol Pa⁻¹m⁻³) | D_A (mol Pa⁻¹h⁻¹) | D_R (mol Pa⁻¹h⁻¹) | C (mol m⁻³) | m_{Sys} (mol) | Rate (mol h⁻¹) $f(D_A + D_R)$ |
> |---|---|---|---|---|---|---|---|
> | Air | 10,000 | 4×10^{-4} | 0.4 | 0.0277 | 0.021 | 210 | 22.55 |
> | Water | 100 | 0.1 | 0.1 | 0.0924 | 5.27 | 527 | 10.14 |
> | Sediment | 10 | 1.0 | 0.0 | 0.1386 | 52.7 | 527 | 7.31 |
> | | | Total | 0.5 | 0.2587 | — | 1264 | 40 |
>
> The total of all D-values is 0.7587. Given that the emission rate is $E = 40\,mol\,h^{-1}$, the fugacity is given by
>
> $$f_{Sys} = \frac{E}{\sum_i D_i} = \frac{40\,mol\,h^{-1}}{0.7587\,mol\,Pa^{-1}\,h^{-1}} = 52.7\,Pa$$
>
> The total amount is 1264 moles, giving a mean residence time of 31.6 h:
>
> $$\tau = \frac{m = \sum_i m_i}{\sum_i r_i} = \frac{1264\,mol}{40\,mol\,h^{-1}} = 31.6\,h$$

The most important loss process is advection in air, which accounts for $21.08\,mol\,h^{-1}$. Next is sediment reaction at $7.31\,mol\,h^{-1}$, the water advection at $5.27\,mol\,h^{-1}$, etc. Each individual rate is $Df\,mol\,h^{-1}$.

3.4.2 RESIDENCE TIME FOR ADVECTIVE AND REACTIVE SYSTEMS

Confusion may arise when calculating the residence time or persistence of a chemical in a system in which advection and reaction occur simultaneously. The overall residence time in Worked Example 3.5 is 31.6 h and is a combination of the advective residence time and the reaction time. The presence of advection does not influence the rate constant of the reaction; therefore, it cannot affect the persistence of the chemical. But, by removing the chemical, it does affect the amount of chemical that is available for reaction, and thus it affects the rate of reaction. It would be useful if we could establish a method of breaking down the overall persistence or residence time into the time attributable to reaction and the time attributable to advection. This is best done by modifying the fugacity equations as shown below for total input I:

$$I = \left[\sum_{i=1}^{All\ Adv} D_i^{Adv} + \sum_{i=1}^{All\ Deg} D_i^{Deg} \right] f_{Sys}$$

But $I = m/\tau_0$, where m is the amount of chemical and τ_0 is the overall residence time. Furthermore, the total amount of chemical is also given by

$$m = \sum_{i=1}^{n} V_i Z_i f_{Sys} \ or \ f_{Sys} \sum_{i=1}^{n} V_i Z_i$$

Dividing both sides by m and canceling f_{Sys} gives

$$\frac{1}{\tau_0} = \sum_{i=1}^{n} \frac{D_i^{Adv}}{V_i Z_i} + \sum_{i=1}^{n} \frac{D_i^{Deg}}{V_i Z_i} = \frac{1}{\tau_{Adv}} + \frac{1}{\tau_{Deg}}$$

The key point is that the advective and reactive residence times τ_A and τ_R add as reciprocals to give the reciprocal overall time. These are the residence times that would apply to the chemical if only that process applied. Clearly, the shorter residence time dominates, corresponding, of course, to the faster rate constant. It can be shown that the ratio of the amounts removed by reaction and by advection are in the ratio of the overall rate constants or the reciprocal residence times.

Worked Example 3.6: Individual and overall residence times for a three-phase system (air, water, sediment) at equilibrium and steady-state concentration, with both advective and emissive inputs, and reactive losses.

Calculate the individual and overall residence times in Worked Example 3.5. Each residence time is VZ/D and the rate constant is D/VZ.

	$V_i Z_i$	$\sum_{i=1}^{n} \frac{V_i Z_i}{D_i^{Adv}}$	$\sum_{i=1}^{n} \frac{V_i Z_i}{D_i^{Deg}}$
Air	4	60	866
Water	10	240	260
Sediment	10	∞	173
Total	24		

Adding the reciprocals, i.e., the rate constants, gives

$$\frac{1}{\tau_0} = \sum_{i=1}^{n} \frac{D_i^{\text{Adv}}}{V_i Z_i} + \sum_{i=1}^{n} \frac{D_i^{\text{Deg}}}{V_i Z_i} = \frac{1}{60} + \frac{1}{240} + \frac{1}{866} + \frac{1}{260} + \frac{1}{\infty} + \frac{1}{173}$$

$$= 0.0167 + 0.0042 + 0.0012 + 0.0038 + 0 + 0.0058$$

$$= 0.0209 + 0.0108 = 0.0317$$

$$= 1/31.5$$

The advection residence time is 1/0.0209 or 47.8 h, and for reaction it is 1/0.0108 or 92.6 h. Each residence time (e.g., 60, 866) contributes to give the overall residence time of 31.5 h, *reciprocally*.

In mass-balance models of this type, it is desirable to calculate the advection, reaction, and overall residence times. An important observation is that these residence times are independent of the quantity of chemical introduced; in other words, they are *intensive* properties of the system. Concentrations, amounts, and fluxes are dependent on emissions and are *extensive* properties.

These concepts are useful, because they convey an impression of the relative importance of advective flow (which merely moves the problem from one region to another) versus reaction (which may help solve the problem). These are of particular interest to those who live downwind or downstream of a polluted area.

3.4.3 Dynamic (Non-Steady State) Calculations

A related calculation can be done in an unsteady-state model in which we introduce an amount of a reactive chemical, m, into the evaluative environment at zero time, then allow it to decay in concentration with time, but maintain equilibrium between all phases at the same time. This is analogous to a batch chemical reaction system. Although it is possible to include emissions or advective inflow, we prefer to treat first the case in which only reaction occurs to an initial mass m. We assume that all volumes and Z-values are constant with time. The basic statement of the rate of change in total amount of material in time is given by the sum of the loss rates in all compartments:

$$\frac{dm}{dt} = -\sum_{i=1}^{n} V_i C_i k_i = -f_{\text{Sys}} \sum_{i=1}^{n} V_i Z_i k_i = -f_{\text{Sys}} \sum_{i=1}^{n} D_i^{\text{Loss}}$$

However, given that the total amount of chemical can also be expressed in fugacity terms as

$$m = \sum_{i=1}^{n} V_i Z_i f_{\text{Sys}} = f_{\text{Sys}} \sum_{i=1}^{n} V_i Z_i$$

Taking the derivative of m with time using this latter expression, we obtain

$$\frac{dm}{dt} = \frac{d\left(f_{\text{Sys}}\right)}{dt} \times \sum_{i=1}^{n} V_i Z_i + f_{\text{Sys}} \times \frac{d\left(\sum_{i=1}^{n} V_i Z_i\right)}{dt}$$

All products of V and Z are invariant with respect to f_{Sys}, so their derivatives are zero. We can now equate the two expressions for $\frac{dm}{dt}$ and isolate $\frac{d\left(f_{\text{Sys}}\right)}{dt}$:

$$\frac{dm}{dt} = -f_{\text{Sys}} \sum_{i=1}^{n} D_i^{\text{Loss}} = \frac{d\left(f_{\text{Sys}}\right)}{dt} \times \sum_{i=1}^{n} V_i Z_i$$

or

$$\frac{d\left(f_{Sys}\right)}{dt} = \frac{-f_{Sys}\sum_{i=1}^{n}D_i^{Loss}}{\sum_{i=1}^{n}V_iZ_i}$$

Setting the group of constants equal to a rate constant k in the following manner:

$$k = \frac{\sum_{i=1}^{n}D_i^{Loss}}{\sum_{i=1}^{n}V_iZ_i}$$

we are left with a simple, first-order expression for rate of change in fugacity with time:

$$\frac{d\left(f_{Sys}\right)}{dt} = -kf_{Sys}$$

This expression integrates to an exponential decay equation, which takes the following form where the fugacity at time 0 is f_{Sys}^0:

$$f_{Sys} = f_{Sys}^0 \exp(-kt)$$

Note that k, the overall rate constant, is the reciprocal of the overall residence time.

Worked Example 3.7: Times to a given level of decay for a first-order decomposition process after all emissions cease.

Calculate the time necessary for the environment in Example 3.3 to recover 50%, 36.7%, 10%, and 1% of the steady-state level of contamination after all emissions cease.

Here, the sum of all VZ terms is 24 and the sum of D-values is 0.2587. Thus,

$$f_{Sys}^t = f_{Sys}^0 \exp(-0.2587t/24) = f_{Sys}^0 \exp(-0.01078t)$$

Since m is proportional to f_{Sys}^t and f_{Sys}^0 is 96.6 Pa, we wish to calculate t at which f_{Sys}^t is 48.3, 35.4, 9.66, and 0.966 Pa. Substituting and rearranging gives

$$\frac{f_{Sys}^t}{f_{Sys}^0} = \exp(-0.01078t)$$

or

$$t = \frac{\ln\left(\frac{f_{Sys}^t}{f_{Sys}^0}\right)}{-0.01078}$$

Here, t is 64, 93, 214, and 427 h for 50%, 36.7%, 10%, and 1% of the steady-state level of contamination, respectively. For example:

$$t_{36} = \frac{\ln(0.367)}{-0.01078} \simeq \frac{1}{0.01078} = 93\,\text{h}$$

The 93-h time is significant as both the steady-state residence time and the time of decay to 36.7% or exp(–1) of the initial concentration.

The dependence of fugacity on time is shown in Figure 3.2, in which the time to achieve the first three critical fugacity values are illustrated. The simple exponential decay results in a relatively rapid drop in the initial stages which tapers off as time progresses.

It is possible to include advection and emissions with only slight complications to the integration. The input terms may no longer be zero.

Worked Example 3.7 raises an important point, which we will address later in more detail. The steady-state situations in Level II calculations are somewhat artificial and contrived. Rarely is the environment at a steady state; things are usually getting worse or better. A valid criticism of Level II calculations is that steady-state analysis does not convey information about the rate at which systems will respond to changes. For example, a steady-state analysis of salt emission into Lake Superior may demonstrate what the ultimate concentration of salt will be, but it will take 200 years for this steady state to be achieved. In a much smaller lake, this steady state may be achieved in 10 days. Detractors of steady-state models point with glee to situations in which the modeler will be dead long before steady state is achieved.

Proponents of steady-state models respond that, although they have not specifically treated the unsteady-state situation, their equations do contain much of the key "response time" information, which can be extracted with the use of some intelligence. The response time in the unsteady-state Example 3.7 was 93 h, which was $\Sigma VZ/\Sigma D$. This is identical to the overall residence time, τ, in Example 3.4. The response time of an unsteady-state Level II system is equivalent to the residence time in a steady-state Level II system. By inspection of the magnitude of groups, VZ/D, or the reciprocal rate constants that occur in steady-state analysis, it is possible to determine the likely

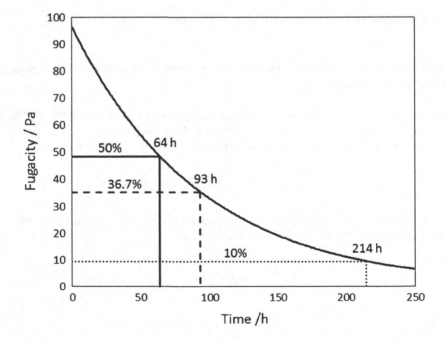

FIGURE 3.2 Dependence of fugacity on time in an environment that is recovering from an initial chemical release. The times and fugacities at 50%, 36.7%, and 10% remaining chemical are indicated by the solid, dashed, and dotted lines, respectively.

unsteady-state behavior. This is bad news to those who enjoy setting up and solving differential equations, because "back-of-the-envelope" calculations often show that it is not necessary to undertake a complicated unsteady-state analysis.

Indeed, when calculating D-values for loss from a medium, it is good practice to calculate the ratio VZ/D, where VZ refers to the source medium. This is the characteristic time of loss, or specifically the time required for that process to reduce the concentration to $\exp(-1)$ of its initial value if it were the only loss process. In some cases, we have an intuitive feeling for what that time should be. We can then check that the D-value is reasonable.

3.5 SUMMARY

In this chapter, we have learned to include advection and reaction rates in evaluative calculations involving equilibrated compartments under the steady-state assumption (Level II calculations). These calculations can be done using concentrations and partition ratios or fugacities and D-values. The concepts of residence time and persistence have been reintroduced, and are invaluable descriptors of environmental fate. Critics will be eager to point out a major weakness in these calculations. Environmental media are rarely in equilibrium; therefore, the use of a common fugacity or equilibrium partition ratios to relate concentrations between phases or media is often not valid. Treating non-equilibrium situations is the task of Chapter 4.

The susceptibility of a chemical in a specific medium to degrading reaction depends both on the inherent properties of the molecule and on the nature of the medium, especially temperature and the presence of candidate reacting molecules or enzymes. In this respect, environmental chemicals are fundamentally different from radioisotopes, which are totally unconcerned about external factors. Translation and extrapolation of reaction rates from environment to environment and laboratory to environment is therefore a challenging and fascinating task that will undoubtedly keep environmental chemists busy for many more decades.

SYMBOLS AND DEFINITIONS INTRODUCED IN CHAPTER 3

Symbol	Common Units	Description
C_i^{Inf}	mol m^{-3}	Inflow concentration of chemical in phase "i"
D_i^j	mol Pa^{-1} h^{-1}	"D-value" for process "j" in phase "i"
E	mol h^{-1}	Emission rate into system
f_{Sys}^t	Pa	Fugacity of chemical in entire system at time t
G_i	m^3 h^{-1}	Volume flow rate associated with phase "i"
I	mol h^{-1}	Total chemical influx or inflow rate
k_i^j	h^{-1}	Rate constant for process "j" in phase "i"
m	moles	Amount of chemical in a system
r_i^j	mol h^{-1}	Molar rate of reaction for process "j" in phase "i"
τ	h	Residence time
τ_i	h	Residence time of chemical in phase "i"

PRACTICE PROBLEMS FOR CHAPTER 3

Problem 3.1: Residence time for a system undergoing advective flow.

A pond of volume $1 \times 10^5 m^3$ is fed by a river of flow $150 m^3 h^{-1}$ and a stream of flow $10 m^3 h^{-1}$. A chemical is present at concentrations of $1 mg L^{-1}$ (river) and $10 mg L^{-1}$ (stream). Assuming the pond is well-mixed, (i) what is the concentration of the chemical exiting the pond at steady state? (ii) What percentage of chemical comes from each source? (iii) How much chemical is in the pond in kg? (iv) What is the residence time of water, and therefore the chemical, in the pond?

Answers

   (i) C: $1.56 mg L^{-1}$, (ii) Source apportionments: River: 60% Stream: 40%, (iii) Mass in pond: 156 kg, and (iv) Residence time: 625 h.

Problem 3.2: A pond with advective flow and degradation.

A pristine pond of volume $1 \times 10^4 m^3$ is exposed to an advective flow of $40 m^3 h^{-1}$ of inflow water with a concentration of $2 \times 10^{-6} mol m^{-3}$ of a chemical. The chemical decays by microbial degradation with a half time of 4 days. Assuming $Z = 10.0 mol Pa^{-1} m^{-3}$ for the chemical in water, (i) what is the steady-state fugacity and concentration of the chemical in the pond? (ii) What is the residence time of the water and the chemical in the pond? (iii) What fraction of the inflow chemical is degraded?

Answers

   (i) $f_{Sys} = 7.13 \times 10^{-8}$ Pa; $C_W = 7.13 \times 10^{-7} mol m^{-3}$, (ii) water res. time = 250 h chemical res. time = 89.1 h, and (iii) fraction degraded = 64.4%.

Problem 3.3: Time to a 90% decay for a first-order decomposition process.

A stagnant body of water with no inflow or outflow is contaminated with a chemical to an average concentration of $10^{-6} mol m^{-3}$. If the half-time for first-order degradation of this chemical in water is 120 days, how long will it take for 90% of the chemical to degrade in years? What fraction of the chemical will have degraded after 6, 12, and 18 months (use 30.5 days per month)?

Answer

   1.09 years for 90% degradation. Fractions remaining: 34.8%, 12.1%, 4.2%, respectively.

4 Transport within and between Compartments

Concentration	Fugacity

Non-diffusive advective molar flow rate $\left(\mathrm{mol\,h^{-1}}\right)$

$$r^{\mathrm{Adv}} = G^{\mathrm{Adv}}C \qquad\qquad r^{\mathrm{Adv}} = D^{\mathrm{Adv}}f$$

Diffusive molar flow rate within a phase $\left(\mathrm{mol\,h^{-1}}\right)$

$$r^{\mathrm{Diff}} = Ak^M\left(C_1 - C_2\right) \qquad\qquad r^{\mathrm{Diff}} = Ak^M Z\left(f_1 - f_2\right) = D^{\mathrm{Diff}}\left(f_1 - f_2\right)$$

Diffusive molar flow rate between phases $(1 \rightarrow 2)\left(\mathrm{mol\,h^{-1}}\right)$

$$r_{12}^{\mathrm{Diff}} = k_1^{\mathrm{Ov}}A\left(C_1 - \frac{C_2}{K_{21}}\right) = k_2^{\mathrm{Ov}}A\left(C_1 K_{21} - C_2\right) \qquad\qquad r_{12}^{\mathrm{Diff}} = D_{12}^{\mathrm{Ov}}\left(f_1 - f_2\right)$$

$$K_{21} = \frac{k_1^{\mathrm{Ov}}}{k_2^{\mathrm{Ov}}} = \frac{1/k_2^{\mathrm{Ov}}}{1/k_1^{\mathrm{Ov}}} \qquad\qquad K_{21} = \frac{k_1^{\mathrm{Ov}}}{k_2^{\mathrm{Ov}}} = \frac{Z_2}{Z_1}$$

$$\frac{1}{k_1^{\mathrm{Ov}}} = \left(\frac{1}{k_1^M} + \frac{1}{k_2^M K_{21}}\right) \qquad\qquad \frac{1}{D_{12}^{\mathrm{Ov}}} = \frac{1}{D_{12}^{\mathrm{Diff}}} + \frac{1}{D_{21}^{\mathrm{Diff}}} = \frac{1}{A_{12}}\left(\frac{1}{k_1 Z_1} + \frac{1}{k_2 Z_2}\right)$$

Combining D-values for series processes:

$$\frac{1}{D_{\mathrm{Tot}}} = \frac{1}{D_1} + \frac{1}{D_2} + \frac{1}{D_3} + \cdots$$

Combining D-values for parallel processes:

$$D_{\mathrm{Tot}} = D_1 + D_2 + D_3 + \cdots$$

4.1 STEADY-STATE DIFFUSIVE PROCESSES

4.1.1 INTRODUCTION

The steady-state "Level II" calculations described in Chapter 3 contain the major weakness that they assume environmental media to be at equilibrium. This is rarely the case in the real environment; therefore, the use of a common "system" fugacity (or concentrations related by equilibrium partition ratios) is usually, but not always, invalid. Reasons for this are best illustrated by an example.

Suppose we have air and water media as illustrated in Figure 4.1, with emissions of $100\,\mathrm{mol\,h^{-1}}$ of benzene into the water. There is only slow reaction in the water (say, $20\,\mathrm{mol\,h^{-1}}$), but there is rapid reaction (say, $80\,\mathrm{mol\,h^{-1}}$) in the air. This implies that benzene must be evaporating from water to air at a rate of $80\,\mathrm{mol\,h^{-1}}$. The question arises: is benzene capable of evaporating at $80\,\mathrm{mol\,h^{-1}}$, or will there be a resistance to transfer that prevents evaporation at this rate? If only $40\,\mathrm{mol\,h^{-1}}$ could evaporate, the evaporated benzene may react in the air phase at $40\,\mathrm{mol\,h^{-1}}$, but it will tend to build up in the water phase to a higher concentration and fugacity until the rate of reaction in the water increases to $60\,\mathrm{mol\,h^{-1}}$. The fugacity of benzene in the air will thus be lower

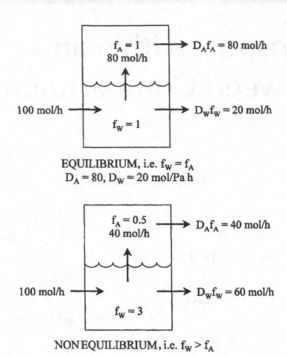

FIGURE 4.1   Illustration of non-equilibrium behavior in an air–water system. In the lower diagram, the rate of reaction in air is constrained by the rate of evaporation. Here $D_i = D_i^{\text{Deg}}$.

than the fugacity in water, and a non-equilibrium situation will have developed. The ability to calculate how fast chemicals can migrate from one phase to another is the challenging task of this chapter. The topic is one in which there still remains considerable uncertainty and scope for scientific investigation and innovation. We begin by considering all the transport processes that are likely to occur in an environmental context.

4.1.2   Non-Diffusive Processes

The first group of processes, which we have already met in Chapter 3, consists of *non-diffusive*, or *advective (piggyback)* processes. A chemical may move from one phase to another by "piggybacking" on the material that is making this journey, for reasons unrelated to the presence of the chemical. Examples include advective flows in air, water, or particulate phases; deposition of chemical in rainfall or sorbed to aerosols from the atmosphere to soil or water; and sedimentation of chemical in association with particles that fall from the water column to the bottom sediments.

Non-diffusive processes are usually unidirectional. As we saw in Chapter 3, the rate of chemical transfer is simply the product of the concentration of chemical in the moving medium, C mol m^{-3}, and the flow rate of that medium, G, m^{3}h^{-1}. We can thus treat *all* such non-diffusive processes as advection and calculate the D-value and rate as follows:

$$r^{\text{Adv}} = G^{\text{Adv}}C = G(Zf) = D^{\text{Adv}}f \left(\text{mol h}^{-1}\right)$$

Note that we may no longer use the symbol "f_{Sys}" for the fugacity in a given compartment, since this assumes an overarching system fugacity that only exists if the entire system is at equilibrium.

The usual problem is to measure or estimate G and the corresponding Z-values or partition ratio. We examine these rates in more detail later, when we focus on individual intermedia transfer processes.

4.1.3 Diffusive Processes

The second group of processes are *diffusive* in nature. Suppose that we have water that contains 1 mol m^{-3} of benzene and add some octanol to it as a second immiscible phase. Given that K_{OW} is 135 for benzene, it will diffuse from the water phase into the octanol phase until it reaches a concentration in octanol that is 135 times that in the water. We could rephrase this by stating that, initially, the fugacity of benzene in the water is (say) 500 Pa, and the fugacity in the octanol is 0. The benzene then migrates from water to octanol until both fugacities reach a common value of (say) 200 Pa. At this common fugacity, the ratio C_O/C_W is equal to Z_O/Z_W or K_{OW}. It is reasonable to conclude that diffusion will always occur from high fugacity (here f_W in water) to low fugacity (f_O in octanol). Therefore, it is tempting to write the transfer rate equation from water to octanol as a function of the difference in fugacity between the two phases, i.e., as

$$r_{WO}^{\text{Diff}}\left(\text{mol h}^{-1}\right) = D_{OW}^{\text{Ov}}\left(f_W - f_O\right)$$

Here D_{OW}^{Ov} is the D-value for "overall" diffusion between octanol and water, in either direction. This equation has the correct property that, when f_W and f_O are equal, there is no *net* diffusion. It also correctly describes the direction of diffusion. It transpires that this is indeed the correct form for this equation, as demonstrated below.

In reality, when the fugacities are equal, there is still active diffusion between octanol and water. Benzene molecules in the water phase do not "know" the fugacity in the octanol phase. At equilibrium, they diffuse from water to octanol at the same rate as they diffuse from octanol to water:

$$r_{WO}^{\text{Diff}}\left(\text{mol h}^{-1}\right) = D_{OW}^{\text{Ov}} f_W = D_{OW}^{\text{Ov}} f_O = r_{OW}^{\text{Diff}}\left(\text{mol h}^{-1}\right)$$

Note that at equilibrium, diffusion in both directions is described by the same "overall" D-value, D_{OW}^{Ov} as demonstrated below. The escaping tendencies from each phase to the other have become equal, and the net transport rate is zero. The term $(f_W - f_O)$ is termed a *departure from equilibrium* group, just as a temperature difference represents a departure from thermal equilibrium. It quantifies the diffusive *driving force.*

Our task is to devise means by which to calculate D for a number of processes involving diffusive interphase transfer. These include the following:

1. Evaporation of chemical from water to air and the reverse process of absorption. Note that we consider the chemical to be in solution in water and not present as a film or oil slick, or in sorbed form on an organic material suspended within the water column.
2. Sorption from water to suspended matter in the water column, and the reverse desorption.
3. Sorption from the atmosphere to aerosol particles, and the reverse desorption.
4. Sorption of chemical from water to bottom sediment, and the reverse desorption.
5. Diffusion within soils, and from soil to air.
6. Absorption of chemical by fish and other organisms by diffusion through the gills, following the same route traveled by oxygen.
7. Transfer of chemical across other membranes in organisms, for example, from air through lung surfaces to blood, or from gut contents to blood through the walls of the gastrointestinal tract, or from blood to organs in the body.

Armed with these D-values, we can set up mass-balance equations that are similar to the Level II calculations but allow for unequal fugacities between media.

To address these tasks, we return to first principles, quantify diffusion processes in a single phase, and then extend this capability to more complex situations involving two phases.

Chemical engineers have devoted considerable effort to quantifying diffusion rates, and especially to accomplishing diffusion processes inexpensively in chemical plants. We therefore exploit this body of profit-oriented information for the nobler purpose of environmental betterment.

4.1.4 DIFFUSIVE MIXING WITHIN A PHASE

In liquids and gases, molecules are in a continuous state of relative motion. If a group of molecules in a particular location is labeled at a point in time, as shown in the upper part of Figure 4.2, then at some time later, it will be observed that they have distributed themselves randomly throughout the available volume of fluid. Mixing has thereby occurred.

Since the number of molecules is large, it is exceedingly unlikely that they will ever return to their initial condition. This process is merely a manifestation of entropy-driven mixing in which one specific distribution of molecules gives way to one of many other statistically more likely mixed distributions. This phenomenon is easily demonstrated by combining salt and pepper in a jar, and then shaking it to obtain a homogeneous mixture. It is the rate of this mixing process that is at issue.

For simplicity, we will take the approach of working with the postulated equation above to describe this mixing, or diffusion, process. A more fundamental approach that seeks to understand the basic determinants of diffusion in terms of molecular velocities is pursued in most physical chemistry textbooks that demonstrate that such an equation form is correct. Most texts follow the mathematical approach and introduce a quantity termed *diffusivity* or *diffusion coefficient,* which

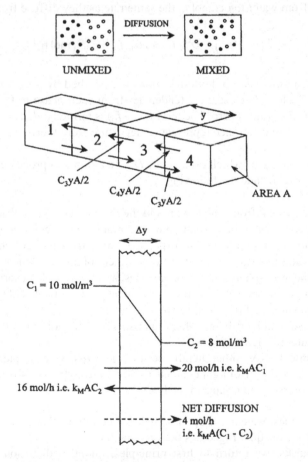

FIGURE 4.2   The fundamental nature of molecular diffusion.

has dimensions of $m^2 h^{-1}$, to characterize this process. It appears as the proportionality constant, B, in the equation expressing Fick's first law of diffusion, namely

$$r^{\text{Diff}} = -AB\left(\frac{dC}{dy}\right)$$

Here, r is the molar flow rate of chemical (mol h^{-1}), A is the area (m^2), B is the diffusivity ($m^2 h^{-1}$), and C is the concentration of the diffusing chemical (mol m^{-3}). y is the distance (m) in the direction of diffusion over which the diffusion occurs. (Note that we use the symbol B for diffusivity to avoid confusion with D-values, whereas most texts sensibly use the symbol D.) The group dC/dy is the concentration gradient, say between two adjacent points 1 and 2, and is characteristic of the degree to which the solution is unmixed or heterogeneous. The negative sign arises because the direction of diffusion is from high to low concentration, i.e., it is positive when dC/dy or $(C_2 - C_1)$ is negative. The equation is really a statement that the rate of diffusion is proportional to the concentration gradient and the proportionality constant is diffusivity. When Fick's law behavior appears not to be obeyed, it is generally assumed that such misbehavior is attributable to deviations or changes in the diffusivity, not to failure of the equation.

4.1.5 MASS-TRANSFER COEFFICIENTS

Diffusivity has some characteristics of a velocity but, dimensionally, it is the product of velocity and the distance to which that velocity applies. In many environmental situations, B is not known accurately, nor is y or Δy; therefore, the flow rate equation in finite difference form contains two unknowns, B and Δy. Substituting finite differences instead of derivatives, we can restate Fick's first law as

$$r^{\text{Diff}}\left(\text{mol h}^{-1}\right) = -AB\left(\frac{dC}{dy}\right) = -AB\left(\frac{(C_2 - C_1)}{\Delta y}\right)$$

$$= +AB\left(\frac{(C_1 - C_2)}{\Delta y}\right) = AB\left(\frac{\Delta C}{\Delta y}\right)$$

Note that by choosing to define $\Delta C = (C_1 - C_2)$, the concentration difference between the reference origin for diffusion and the proposed destination, we conveniently eliminate the negative sign.

To eliminate the dependence on distance traveled y and obtain an equation that depends only on a concentration difference, we can combine the two unknown quantities B and Δy in one term k^M, the ratio of diffusivity to distance traveled or $B/\Delta y$, with dimensions of velocity.

$$k^M\left(\text{m h}^{-1}\right) = \frac{\text{diffusivity}\left(m^2 h^{-1}\right)}{\text{distance traveled}\left(m\right)} = \frac{B}{\Delta y}$$

This allows us to write the more compact expression for net flow as

$$r^{\text{Diff}}\left(\text{mol h}^{-1}\right) = Ak^M\left(C_1 - C_2\right) = Ak^M \Delta C$$

The term "k^M" is a *mass-transfer coefficient* (MTC), with units of velocity (m h^{-1}). The MTC is widely used in environmental transport equations, and may be viewed as a net diffusion velocity. The rate r in one direction is then the product of the net diffusion velocity, area, and concentration. Note once again that, since we define $\Delta C = (C_1 - C_2)$, the sign of the net flow rate will be positive if the net flow is from point 1 to point 2.

Worked Example 4.1: Diffusion across a defined area A.

Assume that diffusion is occurring across an area of $1\,m^2$ from point 1 to 2, as in the lower section of Figure 4.2. Given that the concentrations at points 1 and 2 are $C_1 = 10\,mol\,m^{-3}$ and $C_2 = 8\,mol\,m^{-3}$, determine the net molar flow rate across the interface if the MTC for diffusion is $k^M = 2.0\,m\,h^{-1}$.

The net flow rate is given by

$$r^{Diff}\left(mol\,h^{-1}\right) = Ak^M\Delta C = Ak^M\left[C_1 - C_2\right]\left(mol\,h^{-1}\right)$$

or

$$r^{Diff} = 1\,m^2 \times 2.0\,m\,h^{-1}\times\left[10 - 8\right]mol\,m^{-3}$$

$$= 4.0\ mol\,h^{-1}$$

The group $k^M A$ is an effective volumetric flow rate and is equivalent to the term G^{Adv} $(m^3\,h^{-1})$, introduced for advective flow in Chapter 3.

4.1.6 DIFFUSION WITH D-VALUES IN FUGACITY FORMAT

To express diffusion using fugacity, we use the standard substitution of Zf for concentration C to express Fick's first law for diffusion in terms of a fugacity difference:

$$r^{Diff}\left(mol\,h^{-1}\right) = Ak^M\Delta C = Ak^M\Delta\left(Zf\right) = Ak^M Z\Delta f$$

Now, if we define $D^{Diff} = Ak^M Z$, and the flow rate expressions take the same form as was introduced above for advection and reaction:

$$r^{Diff}\left(mol\,h^{-1}\right) = D^{Diff}\Delta f = D^{Diff}\left(f_1 - f_2\right)$$

This equation is our basic working expression for diffusive flow in fugacity format, as was alluded to above. Note that the units of D^{Diff} are mol $Pa^{-1}h^{-1}$, identical to those of advection and reaction D-values.

Worked Example 4.2: Diffusion through a thin layer of water using concentration.

A chemical is diffusing through a layer of still water 1 mm thick, with an area of $200\,m^2$ and with concentrations on either side of 15 and $5\,mol\,m^{-3}$. If the diffusivity is $10^{-5}\,cm^2s^{-1}$, determine the net flow rate and the corresponding MTC?

The thickness is $y = 10^{-3}\,m$. Given the diffusivity value of $B = 10^{-5}\,cm^2s^{-1}$, the MTC is given by

$$k^M = \frac{B}{\Delta y} = \frac{10^{-5}\,cm^2\,s^{-1}\times1\times10^{-4}\,m^2\,cm^{-2}}{1\times10^{-3}\,m} = 1\times10^{-6}\ m\,s^{-1}$$

The molar flow rate r is then given by

$$r^{Diff} = Ak^M\left[C_1 - C_2\right]$$

$$= 200\,m^2\times10^{-6}\,m\,s^{-1}\times(15 - 5)\,mol\,m^{-3}$$

$$= 2\times10^{-3}\ mol\,s^{-1}$$

Worked Example 4.3: Rate of evaporation of a shallow volume of water to air using concentration.

Water is evaporating from a pan of area 1 m² containing 1 cm depth of water. The rate of evaporation is controlled by diffusion through a thin air film (assumed to be 2 mm thick) immediately above the water surface. The concentration of water in the air immediately at the surface is 25 g m⁻³ (this having been deduced from the water vapor pressure), and in the room the bulk air contains 10 g m⁻³. If the diffusivity B is 0.25 cm²s⁻¹, how long will the water take to evaporate completely?

The diffusivity is given as B = 0.25 cm²s⁻¹. This is equivalent to 0.09 m²h⁻¹. Using this and the assumed thickness of the barrier layer of Δy = 0.002 m, we can calculate the corresponding MTC:

$$k^M = \frac{B}{\Delta y} = \frac{0.09 \text{ m}^2 \text{ h}^{-1}}{2 \times 10^{-3} \text{ m}} = 45 \text{ m h}^{-1}$$

The net mass flow rate is therefore

$$r^M = A k^M [C_1 - C_2]$$

$$= 1 \text{ m}^2 \times 45 \text{ m h}^{-1} \times (25 - 10) \text{ g m}^{-3}$$

$$= 675 \text{ g h}^{-1}$$

The total mass of water initially in the pan, assuming a density of 1×10^6 g m⁻³ is given by the product of volume times density:

$$V \times \text{Density} = (A \times y) \times \text{Density}$$

$$= 1 \text{ m}^2 \times 1 \times 10^{-2} \text{ m} \times 1 \times 10^6 \text{ g m}^{-3}$$

$$= 1 \times 10^4 \text{ g}$$

Assuming the same flow rate holds for the entire period of evaporation, the time for complete evaporation will be

$$\frac{1 \times 10^4 \text{ g}}{675 \text{ g h}^{-1}} = 14.8 \text{ h}$$

To evaporate 10,000 g will take 14.8 h.

Note that the "amount" unit in r and C need not be moles. It can be another quantity such as grams, but it must be consistent in both. In this example, the 2 mm thick film is controlled by the air speed over the pan. Increasing the air speed could reduce this to 1 mm, thus doubling the evaporation rate. This Δy is rather suspect, so it is more honest to use an MTC, which, in the example above, is 0.09/0.002 or 45 m h⁻¹. This is the actual net velocity with which water molecules migrate from the water surface into the air phase. In cases such as this, the MTC depends on wind speed, so correlations have been developed to express this dependence, for example, for evaporation of volatilization from lakes.

4.1.7 DIFFUSION IN POROUS MEDIA

When a solute is diffusing in air or water, its movement is restricted only by collisions with other molecules. If solid particles or phases are also present, the solid surfaces will also block diffusion and slow the net velocity. Environmentally, this is important in sediments in which a solute may have been deposited at some time in the past, and from which it is now diffusing back to the overlying water. It is also important in soils from which pesticides may be volatilizing. It is therefore essential to address the question, "By how much does the presence of the solid phase retard diffusion?" We assume that the solid particles are in contact, but there remains a tortuous path for diffusion (otherwise, there is no access route, and the diffusivity would be zero).

The process of diffusion is shown schematically in Figure 4.3, in which it is apparent that the solute experiences two difficulties. First, it must take a more tortuous path, which can be defined by a unitless *tortuosity factor*, F_Y, the ratio of tortuous distance to direct distance. Second, it does not have available the full area for diffusion, i.e., it is forced to move through a smaller area, which can be defined using an *area factor*, F_A. This area factor F_A, is equal to the *void fraction* v^f_{Void}, i.e., the fraction of the total volume that is fluid, and thus is accessible to diffusion. It can be argued that the tortuosity factor, F_Y, is related to area factor or void fraction, raised possibly to the power of negative 0.5, that is

$$F_Y \simeq \left(F_A \text{ or } v^f_{\text{Void}} \right)^{-0.5}$$

Therefore, in total, we can postulate that the effective diffusivity in the porous medium, B_E, is related to the molecular diffusivity, B, by

$$B_E = B\left(\frac{F_A}{F_Y} \right) = B\left(\frac{\left(v^f_{\text{Void}} \right)^1}{\left(v^f_{\text{Void}} \right)^{-0.5}} \right) \simeq B\left(v^f_{\text{Void}} \right)^{1.5}$$

Such a relationship is found for packings of various types of solids, as discussed by Satterfield (1970). This equation may be seriously in error since (1) the effective diffusivity is sensitive to the shape and size distribution of the particles, (2) there may also be "surface diffusion" along the solid surfaces, and (3) the solute may become trapped in "cul-de-sacs" or become sorbed on active sites. At least the equation has the correct property that it reduces to intuitively correct limits that B_E equals B when v is unity, and B_E is zero when v is zero. There is no substitute for actual experimental measurements using the soil or sediment and solute in question.

For soils, it is common to employ the Millington–Quirk (MQ) expression for diffusivity as a function of air and water contents. An example is in the soil diffusion model of Jury et al. (1983). The MQ expression uses air and water volume fractions v^f_A and v^f_W and calculates effective air and water diffusivities as follows:

$$B_{AE} = \frac{B_A \left(v^f_A \right)^{10/3}}{\left(v^f_A + v^f_W \right)^2}$$

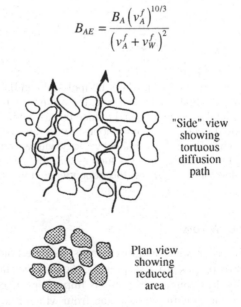

"Side" view
showing
tortuous
diffusion
path

Plan view
showing
reduced
area

FIGURE 4.3 Diffusion in a porous medium in which only part of the area is accessible, and the diffusing molecule must take a longer, tortuous path.

and

$$B_{WE} = \frac{B_w \left(v_W^f \right)^{10/3}}{\left(v_A^f + v_W^f \right)^2}$$

where B_A and B_W are the molecular diffusivities, and B_{AE} and B_{WE} are the effective diffusivities. Inspection of these equations shows that they reduce to a similar form to that presented earlier. If v_W^f is zero, B_{AE} is proportional to v_A^f to the power 1.33 instead of 1.5.

Occasionally, there is confusion when selecting the concentration driving force that is to be multiplied by B_E. This should be the concentration in the diffusing medium, not the total concentration including sorbed form. In sediments, the pore water concentration may be 0.01 mol m^{-3}, but the total sorbed plus pure water, i.e., bulk concentration, is 10 mol m^{-3}. B_E should then be multiplied by 0.01 not 10, which is appropriate since the latter is mostly made up of immobilized sorbed chemical. In some situations (regrettably), the total concentration (10) is used, in which case B_E must be redefined to be a much smaller "effective diffusivity," i.e., by a factor of 1000. The problem is that diffusivity is then apparently controlled by the extent of sorption.

In sediments, it is suspected that much of the chemical present in the pore or interstitial water, and therefore available for diffusion, is associated with colloidal organic material. These colloids can also diffuse; consequently, the diffusing chemical has the option of diffusing in solution or piggybacking on the colloid. From the Stokes–Einstein equation, the diffusivity B is approximately inversely proportional to the molecular radius. A typical chemical may have a molecular mass of 200. A colloid may have an equivalent molar mass of 6000 g mol^{-1}, i.e., a factor of 30 larger in mass and volume, but only a factor of $30^{0.33}$ or about 3 in radius. The colloid diffusivity will thus be about one-third that of the dissolved molecule. But if 90% of the chemical is sorbed, the colloidal diffusion rate will exceed that of the dissolved form. As a result, it is necessary to calculate and interpret the component diffusion processes, since it may not be obvious which route is faster.

4.2 DIFFUSION BETWEEN PHASES

4.2.1 Diffusivity and the MTC Approach

So far in this discussion, we have treated diffusion in only one phase, but in reality, we are most interested in situations where the chemical is migrating from one phase to another. It thus encounters two diffusion regimes, one on each side of the interface. Environmentally, this is discussed most frequently for air–water exchange, but the same principles apply to diffusion from sediment to water, soil to water and to air, and even to biota–water exchange.

An immediate problem arises at the interface, where the chemical must undergo a concentration "jump" from one equilibrium value to another. The chemical may even migrate across the interface from low to high concentration. Clearly, whereas concentration difference was a satisfactory "driving force" for diffusion *within* one phase, it is not satisfactory for describing diffusion *between* two phases. When diffusion is complete, the chemical's fugacities on both sides of the interface will be equal. Thus, we can use fugacity as a "driving force" or as a measure of "departure from equilibrium." Indeed, fugacity is the fundamental driving force in both cases, but it was not necessary to introduce it for one-phase systems, because only one Z applies, and the fugacity difference is proportional to the concentration difference.

Traditionally, interphase transfer processes have been characterized using the Whitman Two-Resistance MTC approach (Whitman, 1923), in which departure from equilibrium is characterized using a partition ratio, or in the case of air–water exchange, a Henry's law constant. We derive the flow rate equations for air–water exchange using the Whitman approach and following Liss and Slater (1974), who first applied it to transfer of gases between the atmosphere and the ocean, and Mackay and Leinonen (1975), who applied the same principles to other organic solutes. We will later derive the same equations in fugacity format. Unfortunately, the algebra is lengthy, but the conclusions are very important, so the pain is justified.

Figure 4.4 illustrates an air–water system in which a solute (chemical) is diffusing at steady state from solution in water at concentration C_W (mol m^{-3}) to the air at concentration C_A mol m^{-3}, or at a partial pressure P (Pa), equivalent to $C_A RT$. We assume that the solute is transferred relatively rapidly in the bulk of the water by eddies, thus the concentration gradient is slight and may be ignored. As it approaches the interface, however, the eddies are damped, diffusion slows, and a larger concentration gradient is required to sustain a steady diffusive flow. An MTC, k_W^M, applies over this water portion of the "interfacial" region. The solute reaches the air–water interface at a concentration C_W^I, then abruptly enters the air portion of the interfacial region, where its concentration changes to C_A^I, the air-phase value. At the same point, the relevant MTC abruptly becomes k_A^M.

The question arises as to whether there is a significant resistance to transfer at the interface. It appears that if it does exist, it is small and unmeasurable. In any event, we do not know how to estimate it, so it is convenient to ignore it and assume that equilibrium applies. We thus argue that there is no interfacial resistance, and C_W^I and C_A^I are in equilibrium, such that the partition ratio K_{AW} applies:

$$\frac{C_A^I}{C_W^I} = K_{AW} = \frac{Z_A}{Z_W} = \frac{H}{RT}$$

Once in the air interfacial region, the solute then diffuses in the air from C_A^I to C_A in the bulk air with an MTC k_A^M. We can write the flow equations for each phase, noting that all such flow rates r must be equal; otherwise, there would be net accumulation or loss at the interface.

$$r^{\text{Diff}} = k_W^M A\left(C_W - C_W^I\right) = k_A^M A\left(C_A^I - C_A\right) \text{mol h}^{-1}$$

Remember that the order of the concentrations is from the origin to the destination, i.e., we subtract from the concentration at the starting point of the diffusion the concentration at the endpoint.

Isolating each of the concentration differences in turn, we have

$$\left(C_W - C_W^I\right) = \frac{r^{\text{Diff}}}{k_W^M A}$$

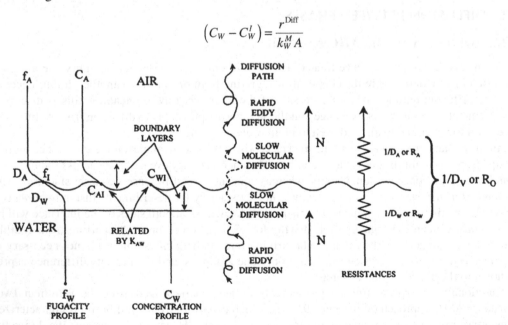

FIGURE 4.4 Mass transfer at the interface between two phases as described by the two-resistance concept. Note the interfacial concentration discontinuity, whereas, in the equivalent fugacity profile to its left, there is no discontinuity.

and

$$\left(C_A^I - C_A\right) = \frac{r^{\text{Diff}}}{k_A^M A}$$

Dividing the latter expression by the partition ratio K_{AW} and recognizing that $C_W^I = C_A^I / K_{AW}$, we may write

$$\left(\frac{C_A^I}{K_{AW}} - \frac{C_A}{K_{AW}}\right) = \frac{r^{\text{Diff}}}{k_A^M A K_{AW}}$$

or

$$\left(C_W^I - \frac{C_A}{K_{AW}}\right) = \frac{r^{\text{Diff}}}{k_A^M A K_{AW}}$$

Now, if we add this last equation to the first expression above for $\left(C_W - C_W^I\right)$, we can eliminate the undeterminable C_W^I and obtain an expression that only contains measurable quantities:

$$\left(C_W - C_W^I\right) + \left(C_W^I - \frac{C_A}{K_{AW}}\right) = \frac{r^{\text{Diff}}}{k_W^M A} + \frac{r^{\text{Diff}}}{k_A^M A K_{AW}}$$

or

$$\left(C_W - \frac{C_A}{K_{AW}}\right) = \frac{r^{\text{Diff}}}{A}\left(\frac{1}{k_W^M} + \frac{1}{k_A^M K_{AW}}\right)$$

We can define an "overall" MTC k_W^{Ov} (k-Overall-Water) that expresses the bracketed term on the right-hand side as

$$\frac{1}{k_W^{\text{Ov}}} = \left(\frac{1}{k_W^M} + \frac{1}{k_A^M K_{AW}}\right)$$

Note that, as expanded upon below, the individual MTCs add in a reciprocal manner for series processes. With this simplification, we have

$$\left(C_W - \frac{C_A}{K_{AW}}\right) = \frac{r^{\text{Diff}}}{k_W^{\text{Ov}} A}$$

Isolating the net diffusive flow rate, we have

$$r^{\text{Diff}} = k_W^{\text{Ov}} A\left(C_W - \frac{C_A}{K_{AW}}\right)$$

Often, the more conveniently available measure of air–water partitioning is the Henry's law constant H. Recasting in terms of H, we can note that, since $C_A / K_{AW} = C_W$, and since $P/C_W = H$ by Henry's law, we can also write

$$r^{\text{Diff}} = k_W^{\text{Ov}} A \left(C_W - \frac{P}{H} \right)$$

where

$$\frac{1}{k_W^{\text{Ov}}} = \left(\frac{1}{k_W^M} + \frac{RT}{H k_A^M} \right)$$

The significance of the addition of reciprocal k terms is perhaps best understood by viewing the process in terms of resistances, where the resistance, R, is $1/k$ in the same sense that the electrical resistance (ohms) is the reciprocal of conductivity (siemens or mhos). The net or overall resistance, R^{Ov}, is then the sum of the water-phase resistance R_W and the air-phase resistance R_A:

$$R_W = \frac{1}{k_W^M A}$$

$$R_A = \frac{1}{k_A^M A K_{AW}} = \frac{RT}{H k_A^M A}$$

Taking the sum of resistances, we obtain

$$R_W + R_A = \frac{1}{k_W^M A} + \frac{1}{k_A^M A K_{AW}} = \frac{1}{k_W^{\text{Ov}} A} = R^{\text{Ov}}$$

which is equivalent to the equation for $\dfrac{1}{k_W^{\text{Ov}}}$ above.

Because the resistances are in series, they add, and the total reciprocal conductivity is the sum of the individual reciprocal conductivities. The reason that K_{AW} enters the summation of resistances is that it controls the relative values of the concentrations in air and water. If K_{AW} is large, C_W^I is small compared to C_A^I, thus the concentration difference $\left(C_W - C_W^I \right)$ will be constrained to be small compared to $\left(C_A^I - C_A \right)$, and the flow rate will be constrained by the small value of $k_W^M \left(C_W - C_W^I \right)$. In general, diffusive resistances tend to be largest in phases where the concentrations are lowest, and thus the concentration gradients are lowest.

Typical environmental values of k_A^M and k_W^M are, respectively, 10 and 0.1 m h^{-1}; thus, the resistances become equal when K_{AW} is 0.01 or H is approximately 25 Pa m^{3}mol^{-1}. If H exceeds 250 Pa m^{3}mol^{-1}, the concentration in the air is relatively large, and the air resistance, R_A, is probably less than one-tenth of R_W and may be ignored. Conversely, if H is less than 2.5 Pa m^{3}mol^{-1}, the water resistance R_W is less than one-tenth of R_A, and it can be ignored.

Interestingly, when H is large, k_W^M tends to equal k_W^{Ov}, and if C_A or P/H is small, the flow rate r becomes simply $k_W^M A C_W$. This group does not contain H, thus the evaporation rate becomes independent of H or of vapor pressure. At first sight, this is puzzling. The reason is that, if H or vapor pressure is high enough, its value ceases to matter, because the overall rate is limited only by the diffusion resistance in the water phase.

An overall "air-based" MTC k_A^{Ov} (k-Overall-Air) may also be defined as

$$\frac{1}{k_A^{\text{Ov}}} = \left(\frac{1}{k_A^M} + \frac{K_{AW}}{k_W^M} \right) = \left(\frac{1}{k_A^M} + \frac{H}{RT k_W^M} \right)$$

where

$$r^{\text{Diff}} = k_A^{\text{Ov}} A \left(C_W K_{AW} - C_A \right) = k_A^{\text{Ov}} A \left(\frac{C_W H - P}{RT} \right)$$

If H is low, k_A^{Ov} approaches k_A^M, and when P/H is small, the flow rate approaches $k_A^M C_W K_{AW}$ or $k_A^M C_W H/RT$. In such cases, volatilization becomes proportional to H and may be negligible if H is very small. In the limit, when H is zero (as with sodium chloride), volatilization does not occur at all.

It may be seen that k_W^{Ov} and k_A^{Ov} are related in a simple manner:

$$k_W^{Ov} = K_{AW} k_A^{Ov} = \left(\frac{H}{RT}\right) k_A^{Ov}$$

or

$$K_{AW} = \frac{k_W^{Ov}}{k_A^{Ov}} = \frac{1/k_A^{Ov}}{1/k_W^{Ov}}$$

Figure 4.5 is a plot of log vapor pressure, P^{Sat}, versus log (solubility) in water on which the location of certain solutes is indicated. Recalling that H or $K_{AW}RT$ is the ratio of these solubilities, compounds of equal H or K_{AW} will lie on the same 45° diagonal. Compounds of $H > 250$ Pa m³mol⁻¹ lie to the upper left, are volatile, and are water-phase ("water-side") diffusion controlled. Those of

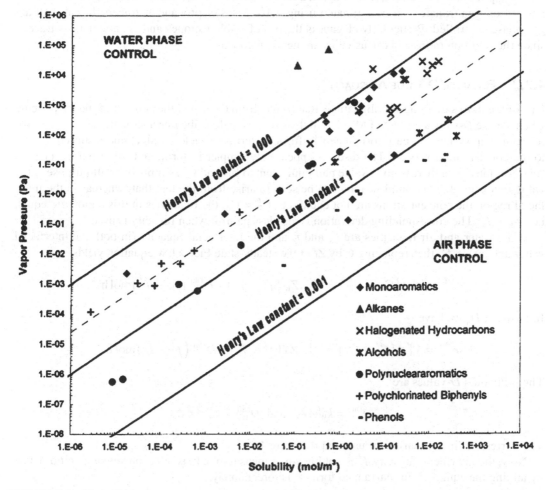

FIGURE 4.5 Plot of log vapor pressure versus log solubility in water for selected chemicals. The diagonals are lines of various Henry's law constant values. The dashed line corresponds to a Henry's law constant of 25 Pa m³mol⁻¹ at which there are approximately equal resistances in the water and air phases.

$H < 2.5$ or $K_{AW} < 0.001$ lie to the lower right, are relatively involatile, and are air-phase ("air-side") diffusion controlled. There is an intermediate band in which both resistances are important.

It is interesting to note that a homologous series of chemicals, such as the chlorobenzenes or PCBs, tends to lie along a 45° diagonal of constant K_{AW}. Substituting methyl groups or chlorines for hydrogen tends to reduce both vapor pressure and solubility by a factor of 4 to 6, thus K_{AW} tends to remain relatively constant, and the series retains a similar ratio of air and water resistances. Paradoxically, reducing vapor pressure as one ascends such a series does not reduce evaporation rate from solution, since it is K_{AW} that controls the rate of evaporation, not vapor pressure alone.

It is noteworthy that oxygen and most low-molecular-weight hydrocarbons lie in the water-phase resistant region, whereas most oxygenated organics lie in the air-phase resistant region. The H for water can be deduced from its vapor pressure of 2000 Pa at 20°C and its concentration in the water phase of 55,000 mol m^{-3} to be 0.04 Pa m^{3}mol^{-1}. If a solute has a lower H than this, it may *concentrate* in water as a result of faster water evaporation but, of course, humidity in the air alters the water evaporation rate. Water evaporation is entirely air-phase resistant, because the water need not, of course, diffuse through the water phase to reach the interface since it is already there.

Certain inferences can be made concerning the volatilization rate of one solute from another, provided that (1) their H-values are comparable, i.e., the same resistance or distribution of resistances applies, and (2) corrections are applied for differences in molecular diffusivity. For example, rates of oxygen transfer can be estimated using noble gases or propane as tracers, because all are gas-phase controlled. Particularly elegant is the use of stable isotopes and enantiomers as tracers, since the partition ratios and diffusivities are nearly identical.

4.2.2 FUGACITY/D-VALUE APPROACH

One of the most satisfying simplifications that results from the use of the concept of fugacity occurs when we use D-values instead of MTCs and diffusivities to describe intermedia transport. The two-resistance approach can be reformulated in fugacity terms to yield an algebraic result equivalent to the concentration version. The result is appealingly simpler in form, and free from the need to "choose sides", i.e., there is no D-value pair equivalent to k_W^{Ov} and k_A^{Ov}, simply one "all-purpose" D_{AW}^{Ov} value that is used. This simplification can be seen to arise from the fact that, whereas in the interfacial region the concentrations are not equal, i.e., $C_W^I \neq C_A^I$, the fugacities in this region *are* equal, i.e., $f_W^I = f_A^I$. The corresponding derivation is also less painful when fugacity is used.

If the water and air fugacities are f_W and f_A and the interfacial fugacity (in both the interfacial water and air) is f^I, then replacing C by Zf in the steady-state Fick's law equation yields

$$r^{Diff} = k_W^M A \left(C_W - C_W^I \right) = k_W^M A Z_W \left(f_W - f^I \right) = D_W^{Diff} \left(f_W - f^I \right) \text{mol h}^{-1}$$

In terms of D_A we have

$$r^{Diff} = k_A^M A \left(C_A^I - C_A \right) = k_W^M A Z_A \left(f^I - f_A \right) = D_A^{Diff} \left(f^I - f_A \right) \text{mol h}^{-1}$$

The individual D-values are

$$D_W^{Diff} = k_W^M A Z_W \text{ and } D_A^{Diff} = k_A^M A Z_A$$

The corresponding scenario is illustrated in Figure 4.4.

Now, the interfacial fugacity f^I is not known or measurable; thus, it is convenient to eliminate it by adding the equations in rearranged form as before, namely,

$$f_W - f^I = \frac{r}{D_W^{Diff}}$$

and

$$f^I - f_A = \frac{r}{D_A^{\text{Diff}}}$$

Addition eliminates f^I yielding

$$f_W - f_A = r\left(\frac{1}{D_W^{\text{Diff}}} + \frac{1}{D_A^{\text{Diff}}}\right)$$

If we define an overall D-value D_{AW}^{Ov} (D-overall air-water) in a manner analogous to the procedure with MTCs employed earlier, we can write

$$\frac{1}{D_{AW}^{\text{Ov}}} = \left(\frac{1}{D_W^{\text{Diff}}} + \frac{1}{D_A^{\text{Diff}}}\right)$$

With this simplification we can write

$$\left(f_W - f_A\right) = \frac{r}{D_{AW}^{\text{Ov}}}$$

or isolating the flow rate, we obtain

$$r = D_{AW}^{\text{Ov}}\left(f_W - f_A\right)$$

where the overall D-value is related in a reciprocal manner to the phase-specific values, analogous to MTCs discussed above:

$$\frac{1}{D_{AW}^{\text{Ov}}} = \left(\frac{1}{D_W^{\text{Diff}}} + \frac{1}{D_A^{\text{Diff}}}\right) = \left(\frac{1}{k_W^M A Z_W} + \frac{1}{k_A^M A Z_A}\right)$$

Here the terms $\frac{1}{D_W^{\text{Diff}}}$ and $\frac{1}{D_A^{\text{Diff}}}$ are effectively resistances to flow that add to give the total resistance $1/D_{AW}^{\text{Ov}}$. It can be shown that

$$D_{AW}^{\text{Ov}} = k_W^{\text{Ov}} A Z_W = k_A^{\text{Ov}} A Z_A$$

Therefore, as before, we have

$$\frac{k_W^{\text{Ov}}}{k_A^{\text{Ov}}} = \frac{Z_A}{Z_W} = K_{AW}$$

The net volatilization rate, $D_{AW}^{\text{Ov}}\left(f_W - f_A\right)$, can be viewed as the algebraic sum of an upward volatilization rate, $D_{AW}^{\text{Ov}} f_W$, and a downward absorption rate, $D_{AW}^{\text{Ov}} f_A$.

Expressions for intermedia diffusion become very simple and transparent when written in fugacity form. The necessity to choose one of two possible overall MTCs is avoided, and each conductivity is expressed in identical units containing its own Z-value. The conductivities add reciprocally, as do electrical conductivities in series.

4.2.3 Measuring Transport D-Values

Measuring non-diffusive D-values is, in principle, simply a matter of measuring the fugacity capacity Z and the flow rate G, the latter usually being the more challenging. Flows of air, water, particulate matter, rain, and food can be estimated directly. The more difficult situations involve estimations of the rates of deposition of aerosols and sedimentation of particles in the water column. The obvious approach is to place a bucket, tray, or a sticky surface at the depositing surface and measure the amount collected. This method can be criticized, because the presence of the bucket alters the hydrodynamic regime and thus the settling rate. This problem is acute when estimating aerosol deposition rates on foliage in a field or forest where the boundary layer is highly disturbed. Measurements of resuspension rates are particularly difficult, because the resuspension event may be triggered periodically by a storm or flood or by an especially energetic fish chasing prey at the bottom of the lake. Regrettably, the sediment–water interface is not easily accessible; thus, measurements are few, difficult, and expensive.

Measurement of diffusion D-values usually involves setting up a system in which there is a known fugacity driving force $(f_1 - f_2)$ and the capacity to measure r^{Diff}, leaving the overall transport D-value as the only unknown in the flow rate equation. A difficulty arises because, for a two-resistance in series system, it is impossible to measure the concentrations or fugacity at the interface; therefore, it is not possible to deduce the individual D-values that combine to give the overall D-value. The subterfuge adopted is to select systems in which one of the resistances dominates and that individual resistance can be equated to the total resistance. Guidance on chemical selection can be obtained from the location of the substance on Figure 4.5.

Air-phase (or "air-side") MTCs can be determined directly by measuring the evaporation rate of a pool of pure liquid, or even the sublimation rate of a volatile solid. The interfacial partial pressure, fugacity, or concentration of the solute can be found from vapor pressure tables. The concentration some distance from the surface can be zero if an adequate air circulation is arranged, thus DC or Df is known. The pool can be weighed periodically to determine r, and area A can be measured directly, thus the MTC or evaporation D-value is the only unknown.

Worked Example 4.4: Determination of the D-value and MTC k^M for evaporation of a liquid from a shallow pool, with efficient air removal.

A tray ($50 \times 30\,$cm in area) contains benzene at 25°C (vapor pressure 12,700 Pa). The benzene is observed to evaporate into a brisk air stream at a rate of $585\,$g h^{-1}. What are the air-phase D_A^{Diff} and k_A^M values?

Given that the molar mass of benzene is $78\,$g mol^{-1}, the molar rate of removal is

$$r = \frac{585\,\text{g h}^{-1}}{78.11\,\text{g mol}^{-1}} = 7.49\,\text{mol h}^{-1}$$

The fugacity in the air is assumed to be zero, since it is briskly removed. The fugacity at the liquid surface is benzene's vapor pressure, 12,700 Pa.

In this particular situation, the resistance is essentially completely governed by the air-side resistance, so that the overall D-value $D_{ABz}^{\text{Ov}} \approx D_A^{\text{Diff}}$. The flow rate r is related to D_A^{Diff} via $r^{\text{Diff}} = D_A^{\text{Diff}}\left(f_W - f_A\right)$. Isolating D_A^{Diff}, we obtain

$$D_A^{\text{Diff}} = \frac{r^{\text{Diff}}}{\left(f_W - f_A\right)} \simeq \frac{7.49\,\text{mol h}^{-1}}{(12,700 - 0)\,\text{Pa}} = 5.9 \times 10^{-4}\ \text{mol Pa}^{-1}\text{h}^{-1}$$

Now D_A^{Diff} is also equal to $k_A^M A Z_A$. Here $A = 0.5 \times 0.3$ m $= 0.15$ m^2 and $Z_A = 1/RT = 4.04 \times 10^{-4}$ mol m^{-3} Pa^{-1}. Solving for the MTC, we obtain

$$k_A^M = \frac{D_A^{Diff}}{AZ_A} = \frac{5.9 \times 10^{-4} \text{ mol h}^{-1} \text{ Pa}^{-1}}{0.15 \text{ m}^2 \times 4.04 \times 10^{-4} \text{ mol m}^{-3} \text{ Pa}^{-1}} = 9.7 \text{ m h}^{-1}$$

Using the concentration-based approach, ΔC is $12,700/RT$ or 5.13 mol m^{-3}. Now, given that $r^{Diff} = k_A^M A \Delta C$, we can solve for the MTC, obtaining the same value as above:

$$k_A^M = \frac{r^{Diff}}{A \Delta C} = \frac{7.5 \text{ mol h}^{-1}}{0.15 \text{ m}^2 \times 5.13 \text{ mol m}^{-3}} = 9.7 \text{ m h}^{-1}$$

The two approaches are easily seen to be algebraically equivalent. Using an experimental system of this type, the dependence of k_M on wind speed can also be measured.

Measurement of overall intermedia D-values or MTCs is similar in principle, with Df applying between two bulk phases. A convenient method of measuring water-to-air transfer is to dissolve the solute in a tank of water, blow air across the surface to simulate wind, and measure the evaporation rate indirectly by following the decrease in concentration in the water with time. If the water volume is V m^3, area is A m^2, and depth is Y m, then we can write

$$r^{Diff} = V\left(\frac{dC_W}{dt}\right) = -k_W^{Ov} A\left(C_W - \frac{C_A}{K_{AW}}\right)$$

where C_A and C_W are concentrations in air and water and k_W^{Ov} the overall MTC. The first term is merely a kinetic statement of the loss rate in mol h^{-1} and the latter was derived above. Recognizing that the depth $Y = V/A$, and assuming C_A to be zero, we can rearrange and simplify this relation as

$$\left(\frac{dC_W}{dt}\right) = \left(-\frac{k_W^{Ov}}{Y}\right)C_W$$

This is a simple first-order decay equation, and can be integrated by inspection to give

$$C_W = C_W^{t=0} \exp\left(-\left(k_W^{Ov}/Y\right)t\right)$$

Since $C = Zf$, in fugacity terms this equation becomes

$$f_W = f_W^{t=0} \exp\left(-\frac{D_{AW}^{Ov}}{VZ_W}t\right).$$

Plotting log C_W versus time gives a measurable slope of $-k_W^{Ov}/Y$, from which k_W^{Ov} can be estimated. A system of this type has been described by Mackay and Yuen (1983) and is illustrated in Figure 4.6.

A very useful quantity is the evaporation half-life, which is $0.693Y/k_W^{Ov}$ and $0.693VZ_W/D_{AW}^{Ov}$. Often, an order of magnitude estimate of this time is sufficient to show that volatilization is unimportant or that it dominates other processes, such as reaction.

As noted earlier, measurement of the individual contributing air and water D-values or MTCs is impossible, because the interfacial concentrations cannot be measured. If, however, the evaporation rates of a series of chemicals of different K_{AW} are measured, it is possible to deduce k_W^M and k_A^M or D_W^{Diff} and D_A^{Diff}. The relationship

$$\frac{1}{k_W^{Ov}} = \left(\frac{1}{k_W^M} + \frac{1}{k_A^M K_{AW}}\right)$$

suggests plotting, as in Figure 4.6, $\dfrac{1}{k_W^{Ov}}$ versus $\dfrac{1}{K_{AW}}$ for a series of chemicals. The intercept will be $\dfrac{1}{k_W^M}$ and the slope $\dfrac{1}{k_A^M}$.

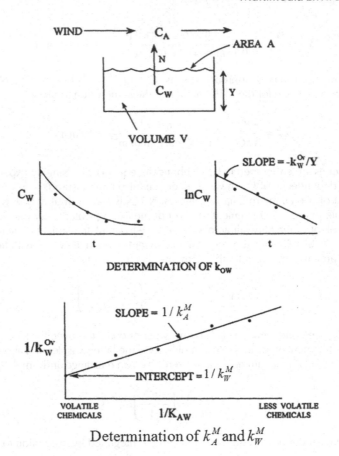

Determination of k_A^M and k_W^M

FIGURE 4.6 Determination of air phase $\left(k_A^M\right)$, water phase $\left(k_W^M\right)$, and the overall water-based $\left(k_W^{Ov}\right)$ MTCs by following the volatilization of substances with different air–water partition ratios, K_{AW}.

A correction may be necessary for molecular diffusivity differences. k_W^M or D_W^{Diff} are measured by selecting chemicals of high K_{AW} for which the term $\dfrac{1}{k_A^M K_{AW}}$ or $1/D_A^{Diff}$ is negligible. Alkanes, oxygen, or inert gases are convenient. k_A^M or D_A^{Diff} is measured by choosing chemicals of low K_{AW} such that $\dfrac{1}{k_A^M K_{AW}}$ or $1/D_A^{Diff}$ is large compared to $\dfrac{1}{k_W^M}$ or $1/D_W^{Diff}$. Alcohols are convenient for this purpose.

Worked Example 4.5: MTC-based determination of individual and overall MTCs and D-values from extent of benzene or naphthalene evaporation water after a finite time period.

A 50.0 cm deep tank contains 2.00 m³ of water at 25°C. The water contains dissolved benzene ($K_{AW} = 0.22$) and naphthalene ($K_{AW} = 0.017$), each at a concentration of 0.100 mol m⁻³. After 2.00 h, these concentrations have dropped to 47.1% and 63.9% of their initial value, respectively. What are the overall and individual MTCs and D-values?

In each case, we anticipate first-order kinetics for evaporation, and can then write for both:

$$C_W = C_W^{t=0} \exp\left(-\frac{k_W^{Ov} t}{Y}\right)$$

Taking the natural logarithm and isolating k_W^{Ov}, we can write

$$k_W^{Ov} = \left(-\frac{Y}{t}\right) \ln\left(\frac{C_W}{C_W^{t=0}}\right)$$

Substituting the various known quantities for $t = 2\,h$ gives $k_W^{Ov} = 0.188\,m\,h^{-1}$ (benzene) and $0.112\,m\,h^{-1}$ (naphthalene). For example, with benzene:

$$k_W^{Ov} = \left(-\frac{0.500\,m}{2.00\,h}\right) \ln(0.471) = 0.188\,m\,h^{-1}$$

Assuming each k_W^{Ov} to be made up of identical k_W^M and k_A^M values, i.e., that for both benzene and naphthalene:

$$\frac{1}{k_W^{Ov}} = \frac{1}{k_W^M} + \frac{1}{K_{AW} k_A^M}$$

we can express this equation twice with two pairs of K_{AW} and k_W^{Ov} values:

Benzene: $\quad \dfrac{1}{0.188\,m\,h^{-1}} = \dfrac{1}{k_W^M} + \dfrac{1}{0.22 k_A^M}$

Naphthalene: $\quad \dfrac{1}{0.112\,m\,h^{-1}} = \dfrac{1}{k_W^M} + \dfrac{1}{0.017 k_A^M}$

Here we now have two equations with two unknowns. Solving by substitution, we obtain the following values of k_W^M and k_A^M:

$$k_W^M = 0.20\,m\,h^{-1}, \quad k_A^M = 15\,m\,h^{-1}$$

Note that as a rule-of-thumb, the MTC for water is generally about 10% of that in air, so the relative magnitude of these numbers is reasonable.

Worked Example 4.6: Fugacity-based determination of individual and overall MTCs and D-values from extent of benzene or naphthalene evaporation from water after a finite time period.

The fugacity approach to the same question as in the previous Worked Example 4.5 starts by obtaining the necessary fugacity capacity Z-values: here Z_A is $4.04 \times 10^{-4}\,mol\,Pa^{-1}\,m^{-3}$ for both substances, and Z_W is $1.836 \times 10^{-3}\,mol\,Pa^{-1}\,m^{-3}$ for benzene and $23.7 \times 10^{-3}\,mol\,Pa^{-1}\,m^{-3}$ for naphthalene. Z_W values may be obtained from the appropriate Henry's law constants, as introduced earlier, e.g., for benzene:

$$Z_W = \frac{1}{H} = \frac{1}{544.7\,Pa\,mol^{-1}\,m^3} = 1.836 \times 10^{-3}\,mol\,Pa^{-1}\,m^{-3}$$

The initial fugacities in the water are

$$f_{Bz}^0 = \frac{C_W}{Z_W} = \frac{0.100\,mol\,m^{-3}}{1.836 \times 10^{-3}\,mol\,m^{-3}\,Pa^{-1}} = 54.5\,Pa$$

$$f_{Np}^0 = \frac{C_W}{Z_W} = \frac{0.100 \text{ mol m}^{-3}}{23.7 \times 10^{-3} \text{ mol m}^{-3} \text{ Pa}^{-1}} = 4.22 \text{ Pa}$$

These can be shown in an analogous manner to drop to 25 and 2.72 Pa for benzene and naphthalene, respectively, after 2 h.

Given the fugacity-based equation for change in fugacity with time for a first-order process,

$$f_W = f_W^0 \exp\left(-\frac{D_{AW}^{Ov} t}{V Z_W}\right)$$

we again take natural logs and solve for D_{AW}^{Ov} at $t = 2$ h:

$$-\frac{D_{AW}^{Ov} t}{V Z_W} = \ln\left(\frac{f_W}{f_W^0}\right)$$

or

$$D_{AW}^{Ov} = -\frac{V Z_W}{t} \ln\left(\frac{f_W}{f_W^0}\right)$$

The values thereby obtained are $D_{AW}^{Ov} = 1.38 \times 10^{-3}$ mol Pa^{-1}h^{-1} (benzene) and 10.6×10^{-3} (naphthalene).

Now, the overall D-value is related to the individual phase-specific values by

$$\frac{1}{D_{AW}^{Ov}} = \frac{1}{D_W^{Diff}} + \frac{1}{D_A^{Diff}}$$

Since the $D_A^{Diff} = k_A A_{AW} Z_A$ value is common to both chemicals, the chemical-specific water phase D_W^{Diff} values can be obtained by rearrangement of this equation as demonstrated here for benzene:

$$D_W^{Diff}(Bz) = \left[\frac{1}{\left(\dfrac{1}{D_{AW}^{Ov}(Bz)} - \dfrac{1}{\left(k_A A_{AW} Z_A^{Bz}\right)}\right)}\right]$$

$$= \left[\frac{1}{\left(\dfrac{1}{1.38 \times 10^{-3} \text{ mol Pa}^{-1}\text{h}^{-1}} - \dfrac{1}{\left(15\,\text{m h}^{-1} \times 4\,\text{m}^2 \times 4.04 \times 10^{-4} \text{ mol Pa}^{-1}\text{m}^{-3}\right)}\right)}\right]$$

$$= \left[\frac{1}{(724.63 - 41.322)} \text{ mol Pa}^{-1}\text{h}^{-1}\right] = 1.46 \times 10^{-3} \text{ mol Pa}^{-1}\text{h}^{-1}$$

Solving this equation for the corresponding value of D_{AW}^{Ov} for naphthalene yields the water D-value of 1.89×10^{-2} mol Pa^{-1}h^{-1}.

Since D_W^{Diff} contains the variable Z_W, the ratio of the two D_W^{Diff} values naphthalene and benzene in water is also given by the ratio of Z_W values. That is

$$\frac{1.89 \times 10^{-2}}{1.46 \times 10^{-3}} = \frac{D_W^{\text{Diff}}(Np)}{D_W^{\text{Diff}}(Bz)} = \frac{Z_W^{Np}}{Z_W^{Bz}} = \frac{23.7 \times 10^{-3}\,\text{mol m}^{-3}\,\text{Pa}^{-1}}{1.836 \times 10^{-3}\,\text{mol m}^{-3}\,\text{Pa}^{-1}} = 12.9$$

Therefore, D_W^{Diff} for naphthalene is equal to D_W^{Diff} for benzene times $23.7 \times 10^{-3}/1.836 \times 10^{-3}$ or 12.9. Note that $D_{AW}^{Ov} = k_W^{Ov} A Z_W$.

In practice, it is unwise to rely on only two chemicals, it being preferable to use at least five, covering a wide range of K_{AW} values. The air-phase resistance, when viewed as $1/D_A^{\text{Diff}}$, is 41.3 units in both cases, but the water-phase resistance for benzene is 685, while for naphthalene it is 52.9. Benzene experiences 5.7% of the transfer resistance in the air, while naphthalene experiences 44% resistance in the air, because it has a much lower K_{AW}.

4.2.4 COMBINING SERIES AND PARALLEL PROCESSES

Having introduced these transport D-values and shown how they combine when describing resistances in series, it is useful to set out the general flow rate equation for any combination of transport processes in series or parallel. Each transport process is quantified by a D-value (GZ, kAZ, or BAZ/Y as appropriate) that applies between two points in space such as a bulk phase and an interface, or between two bulk phases. It is helpful to prepare an arrow diagram of the processes showing the connections, as illustrated in Figure 4.7. Diffusive processes are reversible, so they actually consist of two arrows in opposing directions with the same D-value but driven by different source fugacities.

In air–water exchange, there can be deposition by the parallel processes of (1) dry particle deposition, (2) wet particle deposition, (3) rain dissolution, and (4) diffusive absorption volatilization. When processes apply in parallel between common points, the D-values add. An example is wet and dry deposition from bulk air to bulk water.

$$D_{\text{Total}} = D_1 + D_2 + D_3 + \cdots$$

When processes apply in series, the resistances add or, correspondingly, the reciprocal D-values add to give a reciprocal total.

$$\frac{1}{D_{\text{Total}}} = \frac{1}{D_1} + \frac{1}{D_2} + \frac{1}{D_3} + \cdots$$

An example is the addition of air and water boundary layer resistances, which in total control the rate of volatilization from water.

In some systems, such as with soil-based degradation and diffusion (See *Soil* model later), one arrives at a process with value D_1, in series with a pair of parallel processes with values D_2 and D_3. In this case, the parallel process D-values D_2 and D_3 are summed, and their total is then added in a reciprocal manner with the value D_1 for the process for which they are in series:

$$\frac{1}{D_{\text{Total}}} = \frac{1}{D_1} + \frac{1}{(D_2 + D_3)}$$

It is possible to assemble numerous combinations of series and parallel processes linking bulk phases and interfaces. For those familiar with electrical theory, these situations can be viewed as electrical analogs, with voltage being equivalent to fugacity, resistance equivalent to $1/D$, and current equivalent to flow rate (mol h^{-1}). Figure 4.7 gives some examples.

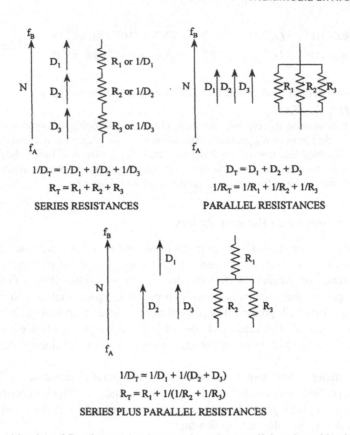

$$1/D_T = 1/D_1 + 1/D_2 + 1/D_3$$
$$R_T = R_1 + R_2 + R_3$$
SERIES RESISTANCES

$$D_T = D_1 + D_2 + D_3$$
$$1/R_T = 1/R_1 + 1/R_2 + 1/R_3$$
PARALLEL RESISTANCES

$$1/D_T = 1/D_1 + 1/(D_2 + D_3)$$
$$R_T = R_1 + 1/(1/R_2 + 1/R_3)$$
SERIES PLUS PARALLEL RESISTANCES

FIGURE 4.7 Combination of D-values and resistances in series, parallel, and combined configurations.

4.3 D-VALUES IN MULTIMEDIA CALCULATIONS

4.3.1 INTRODUCTION

In this chapter, we have examined the nature of molecular diffusivity, introduced the concept of MTCs (k), and treated the problem of resistances occurring in series and parallel as material diffuses from one phase to another. Two new D-values have been introduced, a kAZ product and a $BAZ/\Delta y$ product. We can treat situations in which various D-values apply in series and in parallel.

In some situations, diffusion D-values may be assisted or countered by advective transfer D-values. For example, PCB may be evaporating from a water surface into the atmosphere only to return by association with aerosol particles that fall by wet or dry deposition. We can add D-values when the fugacities with which they are multiplied are identical, i.e., the source is the same phase. This is convenient, because it makes the equations algebraically simple and enables us to compare the rates at which materials move by various mechanisms between phases.

We thus have at our disposal an impressive set of tools for calculating transport rates between phases. We need Z-values, MTCs, diffusivities, path lengths, and advective flow rates. Quite complicated models can be assembled describing transfer of a chemical between several media by a number of routes. In general, the total D-value for movement from phase A to B will not be the same as that from B to A. The reason is that there may be an advective process moving in only one direction. Diffusive processes always have identical D-values applying in each direction. D-values for loss by reaction can also be included in the mass-balance expression. We are now able to use these concepts to perform a Level III calculation, one in which the fugacities in two or more phases or compartments are not equal at steady state.

4.3.2 *D*-VALUES IN INTERMEDIA FUGACITY-BASED MODELS

A schematic diagram of a simple four-compartment evaluative environment is depicted in Figure 4.8, in which the intermedia transport processes are indicated by arrows. In addition to the reaction and advection *D*-values, which were introduced in Level II calculations, there are seven intermedia transfer *D*-values. Note that since the various compartments are not at the same fugacity, the emission rates of chemicals must now be specified on a medium-by-medium basis. We shall refer to this sort of situation as a Level III calculation. In contrast, Level II calculations are predicated on constant fugacity throughout the system and therefore only the total emission rate was needed.

Table 4.1 lists the intermedia D values and gives the equations in terms of transport rate parameters. Subscripts are used to designate air, 1; water, 2; soil, 3; and sediment, 4.

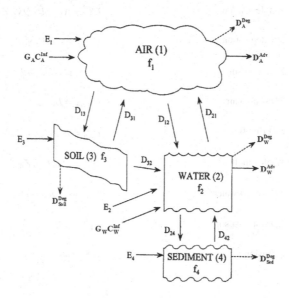

FIGURE 4.8 Four-compartment Level III diagram.

TABLE 4.1
D-Value Expressions for Various Intermedia Transport Processes

Compartments	Process	D-Values
Air (1)—water (2)	Diffusion	$D_{AW}^{Ov} = \dfrac{1}{\left(\dfrac{1}{k_A^{AW} A_{AW} Z_A} + \dfrac{1}{k_W^{AW} A_{AW} Z_W} \right)}$
	Rain dissolution (Q = aerosol)	$D_W^{QR} = A_W U_Q Z_W$
	Wet deposition	$D_W^{QW} = A_{AW} U_R Q v_Q Z_Q$
	Dry deposition	$D_W^{QD} = A_{AW} U_Q v_Q Z_Q$
	ΣD air $\to$ water	$D_{AW}^{Tot} = D_{AW}^{Ov} + D_W^{QR} + D_W^{QW} + D_W^{QD}$
	ΣD water $\to$ air	$D_{WA}^{Tot} = D_{AW}^{Ov}$

(Continued)

TABLE 4.1 (*Continued*)
D-Value Expressions for Various Intermedia Transport Processes

Compartments	Process	D-Values
Air (1)—soil (3)	Diffusion	$D_{AS}^{Ov} = \dfrac{1}{\left(\dfrac{1}{k_{A-Soil}^{M} A_{A-Soil} Z_{Soil}} + \dfrac{Y_{Soil}}{A_{A-Soil}(B_A Z_A + B_W Z_W)} \right)}$
	Rain dissolution	$D_{Soil}^{QR} = A_{A-Soil} U_R v_Q Z_W$
	Wet deposition	$D_{Soil}^{QW} = A_{A-Soil} U_R Q v_Q Z_Q$
	Dry deposition	$D_{Soil}^{QD} = A_{A-Soil} U_Q v_Q Z_Q$
	ΣD air $\to$ soil	$D_{A-Soil}^{Tot} = D_{A-Soil}^{Ov} + D_{Soil}^{QR} + D_{Soil}^{QW} + D_{Soil}^{QD}$
	ΣD soil $\to$ air	$D_{A-Soil}^{Tot} = D_{A-Soil}^{Ov}$
Soil (3)—water (2)	Soil runoff	$D_{Soil}^{SW} = A_{AW} U_{Soil-W} Z_{Soil}$
	Water runoff	$D_{Soil}^{WW} = A_{AW} U_{WW} Z_W$
	ΣD soil $\to$ water	$D_{Soil-W}^{Tot} = D_{Soil}^{SW} + D_{Soil}^{WW}$
	ΣD water $\to$ soil	$D_{W-Soil}^{Tot} = 0$
Sediment (4)—water (2)	Diffusion	$D_{Sed-W}^{Ov} = 1 \Big/ \left(\dfrac{1}{k_{Sed-W}^{M} A_{Sed-W} Z_W} + \dfrac{Y_{Sed}}{B_{Sed-W} A_{Sed-W} Z_W} \right)$
	Deposition (P = particles in water)	$D_{Sed}^{QS} = U_{Sed}^{Dep} A_{Sed-W} Z_P$
	Resuspension	$D_{Sed-W}^{QR} = U_{Sed}^{Resusp} A_{Sed-W} Z_{Sed}$
	ΣD sed $\to$ water	$D_{Sed-W}^{Tot} = D_{Sed-W}^{Ov} + D_{Sed}^{QS}$
	ΣD water $\to$ sed	$D_{W-Sed}^{Tot} = D_{Sed-W}^{Ov} + D_{Sed-W}^{QR}$
Any phase "i"	Reaction	$D_i^{Deg} = k_i^{Deg} V_i Z_i$
Total reactive loss over "n" phases	Reaction	$D_{Tot}^{Deg} = \displaystyle\sum_{i=1}^{n} k_i^{Deg} V_i Z_i$
Bulk advective loss	Advection	$D_i^{Adv} = G_i Z_i = U_i A_i Z_i$

Table 5.3 in Chapter 5 gives order-of-magnitude values for parameters used to calculate intermedia transport D-values. These values depend on the environmental conditions and to some extent on chemical transport properties such as diffusivities. The variation in diffusivity and MTC is usually small compared with the variation in Z-values, thus the use of chemical-specific diffusivities is justified only for the most accurate simulation. Most of these transport parameters can be expressed as velocities.

The values given vary considerably from place-to-place and time-to-time. If the aim is to simulate conditions in a specific region, appropriate transport rate parameters for that region can be sought.

The air-side and water-side MTCs k_A^M and k_W^M have been measured in wind-wave tanks and in lakes as a function of wind speed. Schwarzenbach et al. (1993) have reviewed these correlations. The following correlations are suggested by Mackay and Yuen (1983). Note that units are m s^{-1}.

$$k_A^M = 10^{-3} + 0.0462U^* \times \left(Sc_A\right)^{-0.67} \text{ ms}^{-1}$$

$$k_W^M = 10^{-6} + 0.0034U^* \times \left(Sc_W\right)^{-0.5} \text{ ms}^{-1}$$

$$U^* = 0.01\left(6.1 + 0.63U_{10}\right)^{0.5} U_{10} \text{ ms}^{-1}$$

where U^* is the friction velocity, which characterizes the drag of the wind on the water surface. Sc_A is the Schmidt number in air and ranges from 0.6 for water to about 2.5, and Sc_W applies to the water phase and is generally about 1000. U_{10} is the wind velocity at 10 m height.

Changing the velocity units to m h^{-1}, substituting typical values for the Schmidt number, and taking into account other studies, the following correlations are suggested, with U_{10} in units of m s^{-1}.

$$k_A^M = \left(3.6 + 5U_{10}^{1.2}\right) \text{m h}^{-1}$$

$$k_W^M = \left(0.0036 + 0.01U_{10}^{1.2}\right) \text{m h}^{-1}$$

These correlations will underestimate the MTCs under turbulent conditions of breaking waves or in rivers where there is "white water."

Sedimentation rates can be estimated by assuming a deposition velocity of about 1 m day^{-1}. Therefore, a lake containing 15 g m^{-3} of suspended solids is probably depositing 15 g m^{-2}day of solids, which corresponds to about 10 cm^{3}m^{-2}day if the density is 1.5 g cm^{-3}. This corresponds to roughly 0.4 cm^{3}m^{-2}h^{-1} or 40×10^{-8}m^{3}m^{-2}h^{-1} or 40×10^{-8}m h^{-1}. Of this, a fraction is buried, and the remainder is resuspended. In waters of lower solid concentration, the deposition rate is correspondingly slower.

The air-to-water D-value (D_{12}) consists of diffusive absorption (D_V) and non-diffusive wet and dry aerosol deposition. Each D-value can be estimated and summed to give D_{12}.

The half-life for loss from a phase of volume V and Z-value Z by process D is clearly 0.693 VZ/D. If a half-life $t_{1/2}$ is suggested, D is 0.693 $VZ/t_{1/2}$. Short half-lives represent large D-values and fast, important processes. It is always useful to calculate a half-life or a characteristic time VZ/D to ensure that it is reasonable.

4.3.3 EQUATIONS FOR MASS BALANCE IN MULTIMEDIA FUGACITY-BASED MODELS

Once the necessary D-values for advection, deposition, reaction, inter-compartment diffusion, volatilization, etc. are defined, it is then possible to write mass-balance equations for each medium. As an example, we consider a four-compartment system of air, water, soil, and sediment

Air

$$E_A + G_A^{Adv}C_A^{Inf} + f_W D_{WA} + f_{Soil} D_{Soil-A} = f_A\left(D_{AW} + D_{A-Soil} + D_A^{Deg} + D_A^{Adv}\right) = f_A D_A^{Tot}$$

Water

$$E_W + G_W^{Adv}C_W^{Inf} + f_A D_{AW} + f_{Soil} D_{Soil-W} + f_{Sed} D_{Sed-W} = f_W\left(D_{WA} + D_{W-Sed} + D_W^{Deg} + D_W^{Adv}\right) = f_W D_W^{Tot}$$

Soil

$$E_{Soil} + f_A D_{A-Soil} = f_{Soil}\left(D_{Soil-A} + D_{Soil-W} + D_{Soil}^{Deg}\right) = f_{Soil}D_{Soil}^{Tot}$$

Sediment

$$E_{Sed} + f_W D_{W-Sed} = f_{Sed}\left(D_{Sed-W} + D_{Sed}^{Deg} + D_{Sed}^{Adv}\right) = f_{Sed}D_{Sed}^{Tot}$$

For each compartment "i," E_i is the emission rate (mol h^{-1}), G_i^{Adv} is the advective inflow rate (m^{3}h^{-1}), C_i^{Inf} is the background inflow concentration (mol m^{-3}), D_i^{Deg} is the decomposition reaction rate D-value, and D_i^{Adv} is the advection rate D-value. D_i^{Tot} is the sum of all loss D-values from medium i. Sediment burial and air-to-stratospheric transfer can be included as an advection process or as a pseudo reaction.

These four equations contain four unknowns (the fugacities), thus solution is possible, although it does require a fair bit of algebra! First solving for the compartment fugacities, and expressing the sum of emission and advection inputs into each medium as I $\left(\text{that is: } I_i = E_i + G_i^{Adv}C_i^{Inf}\right)$, we can write

$$f_A = \frac{I_A + f_W D_{WA} + f_{Soil}D_{Soil-A}}{D_A^{Tot}}$$

$$f_W = \frac{I_W + f_A D_{AW} + f_{Soil}D_{Soil-W} + f_{Sed}D_{Sed-W}}{D_W^{Tot}}$$

$$f_{Soil} = \frac{I_{Soil} + f_A D_{A-Soil}}{D_{Soil}^{Tot}}$$

$$f_{Sed} = \frac{I_{Sed} + f_W D_{W-Sed}}{D_{Sed}^{Tot}}$$

Note that for the soil and sediment expressions, there is no advective term, so "I" equals "E" and simply replaces it for consistency.

Working with the air fugacity expression first and substituting the soil fugacity expression to eliminate f_{Soil}, we may write

$$f_A = \frac{I_A + f_W D_{WA} + f_{Soil}D_{Soil-A}}{D_A^{Tot}} = \frac{I_A + f_W D_{WA} + \left(\dfrac{I_{Soil} + f_A D_{A-Soil}}{D_{Soil}^{Tot}}\right)D_{Soil-A}}{D_A^{Tot}}$$

Fully expanding we have

$$f_A\left(1 - \frac{D_{A-Soil}D_{Soil-A}}{D_{Soil}^{Tot}D_A^{Tot}}\right) = f_W\left(\frac{D_{WA}}{D_A^{Tot}}\right) + \left(\frac{I_A}{D_A^{Tot}} + \frac{I_{Soil}D_{Soil-A}}{D_{Soil}^{Tot}D_A^{Tot}}\right)$$

This and the following expression for the other fugacities are simplified (!) by defining the following symbols for terms that reoccurs in more than one equation:

$$J_1 = \left(\frac{I_A}{D_A^{Tot}} + \frac{I_{Soil}D_{Soil-A}}{D_{Soil}^{Tot}D_A^{Tot}}\right)$$

$$J_2 = \left(\frac{D_{WA}}{D_A^{Tot}} \right)$$

$$J_3 = \left(1 - \frac{D_{Soil-A}D_{A-Soil}}{D_{Soil}^{Tot}D_A^{Tot}} \right)$$

$$J_4 = \left(D_{AW} + \frac{D_{A-Soil}D_{Soil-W}}{D_{Soil}^{Tot}} \right)$$

Upon substitution of these expressions, the air fugacity expression now becomes

$$f_A = f_W \left(\frac{J_2}{J_3} \right) + \left(\frac{J_1}{J_3} \right)$$

Taking the same approach with the water fugacity expression, we can eliminate the soil and sediment fugacities first by substitution:

$$f_W = \frac{I_W + f_A D_{AW} + f_{Soil} D_{Soil-W} + f_{Sed} D_{Sed-W}}{D_W^{Tot}}$$

$$= \frac{I_W + f_A D_{AW} + \left(\dfrac{I_{Soil} + f_A D_{A-Soil}}{D_{Soil}^{Tot}} \right) D_{Soil-W} + \left(\dfrac{I_{Sed} + f_W D_{W-Sed}}{D_{Sed}^{Tot}} \right) D_{Sed-W}}{D_W^{Tot}}$$

Expanding all terms, and isolating all terms in water fugacity, we obtain

$$f_W \left\{ 1 - \left(\frac{D_{W-Sed}D_{Sed-W}}{D_{Sed}^{Tot}D_W^{Tot}} \right) \right\} = \frac{I_W + f_A \left\{ D_{AW} + \dfrac{D_{A-Soil}D_{Soil-W}}{D_{Soil}^{Tot}} \right\} + \left[\dfrac{I_{Soil}D_{Soil-W}}{D_{Soil}^{Tot}} \right] + \left[\dfrac{I_{Sed}D_{Sed-W}}{D_{Sed}^{Tot}} \right]}{D_W^{Tot}}$$

or

$$f_W = \frac{I_W + f_A J_4 + \left[\dfrac{I_{Soil}D_{Soil-W}}{D_{Soil}^{Tot}} \right] + \left[\dfrac{I_{Sed}D_{Sed-W}}{D_{Sed}^{Tot}} \right]}{\left\{ D_W^{Tot} - \left(\dfrac{D_{W-Sed}D_{Sed-W}}{D_{Sed}^{Tot}} \right) \right\}}$$

Combining these expressions for the air and water fugacity, we obtain

$$f_W = \frac{I_W + \left(f_W \left(\dfrac{J_2}{J_3} \right) + \left(\dfrac{J_1}{J_3} \right) \right) J_4 + \left[\dfrac{I_{Soil}D_{Soil-W}}{D_{Soil}^{Tot}} \right] + \left[\dfrac{I_{Sed}D_{Sed-W}}{D_{Sed}^{Tot}} \right]}{\left\{ D_W^{Tot} - \left(\dfrac{D_{W-Sed}D_{Sed-W}}{D_{Sed}^{Tot}} \right) \right\}}$$

Solving for the water fugacity, we arrive at (finally!)

$$f_W = \frac{\left(I_W + \left[\dfrac{J_1 J_4}{J_3} \right] + \left[\dfrac{I_{Soil} D_{Soil-W}}{D_{Soil}^{Tot}} \right] + \left[\dfrac{I_{Sed} D_{Sed-W}}{D_{Sed}^{Tot}} \right] \right)}{\left(D_W^{Tot} - \left[\dfrac{J_2 J_4}{J_3} \right] - \left(\dfrac{D_{W-Sed} D_{Sed-W}}{D_{Sed}^{Tot}} \right) \right)}$$

With this solution, we may now go back and back-substitute for the water fugacity in our first derived expression for the air fugacity to obtain f_A and then solve for f_{Soil} and f_{Sed}.

When the fugacities in the various compartments are not equal, i.e., equilibrium is not assumed, it is necessary to specify the emissions into each compartment separately. Different mass distributions, concentrations, and residence times result if $100\,mol\ h^{-1}$ is emitted to air, water, or soil; thus, "mode of entry" is an important determinant of environmental fate and persistence.

Having obtained the fugacities, all process rates can be deduced as Df, and a steady-state mass balance may be then stated in which the total inputs to each medium equal the outputs. The amounts and concentrations can be calculated from such expressions. An overall residence time can be calculated as the sum of the amounts present divided by the total input (or output) rate. A reaction residence time can be calculated as the amount divided by the total reaction rate, and a corresponding advection residence time can also be deduced. Doubling emissions simply doubles fugacities, masses, and concentrations, but the residence times are unchanged.

An important property of this model is its *linear additivity*. This is also called the *principle of superposition*. Because all the equations are linear, the fugacity in, for example, water, deduced as a result of emissions to air, water, and soil, is simply the sum of fugacities in water deduced from each emission separately. It is thus possible to attribute the fugacity to sources, e.g., 50% is from emission to water, 30% is from emission to soil, and 20% from emission to air. The masses and flow rates are also linearly additive. We will revisit this concept in the next chapter, where we look at the details of multimedia modeling in the environment.

SYMBOLS AND DEFINITIONS INTRODUCED IN CHAPTER 4

Symbol	Common Units	Description
B_i	–	Molecular diffusivity in phase "i"
C_i^{Inf}	mol m^{-3}	Inflow concentration of chemical in phase "i"
D_i^j	mol Pa^{-1}h^{-1}	"D-value" for process "j" in phase "i"
D_{ij}^{Ov}	mol Pa^{-1}h^{-1}	Overall D-value for diffusion between phases "i" and "j"
D_{Tot}	mol Pa^{-1}h^{-1}	Total D-value for two or more processes in series or parallel
E	mol h^{-1}	Emission rate into system
F_A	–	Area factor
F_Y	–	Tortuosity factor
f_i^I	Pa	Unmeasured fugacity of interfacial zone in medium "i"
f_{Sys}^t	Pa	Fugacity of chemical in entire system at time t
I	mol h^{-1}	Total chemical influx or inflow rate
k_i^j	h^{-1}	Rate constant for process "j" in phase "i"
k^M	m h^{-1}	MTC, mass transfer coefficient
k_{ij}^{Ov}	m h^{-1}	Overall MTC for transfer from phase "i" into phase "j", defined with respect to phase "i"
m	mol	Total amount of chemical in system
Q	–	Rain scavenging ratio for aerosols

(Continued)

Symbol	Common Units	Description
r_i^j	mol h^{-1}	Molar rate of reaction for process "j" in phase "i"
R_i	h	Resistance at an interface associated with phase "i"
R^{Ov}	h	Overall resistance for an interface
U_R	m h^{-1}	Velocity of a flowing medium
τ	h	Residence time
τ_i	h	Residence time of chemical in phase "i"
v_{Void}^f	–	Void fraction

PRACTICE PROBLEM FOR CHAPTER 4

Ten kilograms each of benzene, 1,4 dichlorobenzene, and p-cresol are spilled into a pond 5 m deep with an area of 1 km^2. If k_W is 0.1 m h^{-1}, and k_A is 10 m h^{-1}, what will be the times necessary for half of each chemical to be evaporated? Use the property data from Appendix A, and ignore other loss processes.

Answer
Benzene, 36 h; dichlorobenzene, 38 h; p-cresol, 12,400 h.

5 Evaluative Fugacity Models and Calculations

THE ESSENTIALS

Type	State	Compartment Fugacities	Key Processes (Cumulative List)	Key Parameters (Cumulative List)
Level I	Equilibrium	Constant and equal	Equilibration	Concentration: C Fugacity: Z
Level II	Equilibrium. steady-state, well-mixed	Constant and equal	Advection and/or reaction	Concentration: G, k Fugacity: $D = GZ$ or kVZ
Level III	Steady-state, well-mixed	Constant and unequal	Interphase diffusion	Concentration: k_M Fugacity: $D_{12}^{Oy} = 1/(1/D_1 + 1/D_2)$
Level IV	Non-steady-state, dynamic	Time-dependent	Time-dependent flow rates	Concentration: C_t Fugacity: f_t

Key Equations:

Level I	Level II	Level III	Level IV
$f_{Sys} = \dfrac{M}{\sum\limits_{i=1}^{n} V_i Z_i}$	$f_{Sys} = \dfrac{E}{\sum\limits_{i=1}^{n} D_i}$	—	—
$C_i = Z_i f_{Sys}$	$C_i = Z_i f_{Sys}$	$C_i = Z_i f_i$	$C_i^t = Z_i f_i^t$
$m_i = V_i C_i = V_i Z_i f_{Sys}$	$m_i = V_i C_i = V_i Z_i f_{Sys}$	$m_i = V_i C_i = V_i Z_i f_i$	$m_i^t = V_i C_i^t = V_i Z_i f_i^t$
$Z_T = \sum\limits_{i=1}^{n} v_i^f Z_i$	$Z_T = \sum\limits_{i=1}^{n} v_i^f Z_i$	$Z_T = \sum\limits_{i=1}^{n} v_i^f Z_i$	$Z_T = \sum\limits_{i=1}^{n} v_i^f Z_i$
$m_{Sys} = \sum\limits_{i=1}^{n} V_i C_i$	$m_{Sys} = \sum\limits_{i=1}^{n} V_i C_i$	$m_{Sys} = \sum\limits_{i=1}^{n} V_i C_i$	$m_{Sys}^t = \sum\limits_{i=1}^{n} V_i C_i^t$

5.1 LEVEL I MODELS

A significant advance in environmental science was made in 1978, when Burns et al. (1982) proposed that chemicals may be assessed in "evaluative environments" that have fictitious but realistic properties such as volume, composition, and temperature. Evaluative environments can be decreed to consist of a few homogeneous phases of specified dimensions with constant temperature and composition. Essentially, the environmental scientist designs a "world" with desired specifications and then explores mathematically the likely behavior of chemicals in that world. No claim is made that the evaluative world is identical to any real environment, although broad similarities in chemical behavior are expected. There are good precedents for this approach. In 1824, Carnot devised an evaluative steam engine, now termed the *Carnot cycle*, which leads to a satisfying explanation of entropy and the second law of thermodynamics. The kinetic theory of gases uses an evaluative assumption of the behavior of gas molecules.

Multimedia calculations involve the determination of concentrations and flow rates of a given chemical into and out of a system, as well as between compartments within a system. These may be time-dependent and therefore need to be evaluated at certain key times, or at long enough times to ensure that steady-state conditions have been reached. Such calculations are classified in broad terms as one of four possible types, as noted at various points in the previous chapters. In what follows we will outline and demonstrate aspects of these computations when done in the fugacity framework. All can be done using the equivalent concentration-based logic, but generally take more effort. Table 5.1 summarizes the characteristics of the four fugacity-based model levels.

Programs to accomplish generic versions of these types of calculations in Excel spreadsheet format are available from the CEMC website (www.trentu.ca/cemc/resources-and-models). This site also hosts other non-generic models developed for particular environmental and geographical conditions, either in spreadsheet or older Visual Basic formats. Those interested in using the currently supported versions of the models should go to the CEMC site for information and download links.

"Level I" calculations are the simplest multimedia environmental calculations possible. All compartments are assumed to be at equilibrium and there are no inputs or losses. Such a calculation is limited to the equilibrium distribution of a fixed amount of a chemical among two or more compartments. The principal variability comes from (1) the number of compartments, (2) the relative volumes of the compartments, and (3) the partition ratios and corresponding Z-values associated with the chemicals in question, within each compartments.

The basic steps involved in a Level I calculation are outlined in the flowchart given in Figure 5.1. The most significant part of a Level I calculation is the calculation of appropriate fugacity capacities or Z-values. Usually this begins with the calculation of Z_A the Z-value for air. Since the ideal gas law is generally assumed, this generally results in a value of about 4×10^{-4} mol m^{-3} Pa^{-1} at 298 K. From this, or the value for some other appropriate temperature, the remaining Z-values are estimated, usually using one of the relations outlined in Table 5.2.

TABLE 5.1
Key Characteristics of Generic "Mackay-Type" Fugacity-Based Model Levels

Type	State	Compartment Fugacities	Key Processes (Cumulative)	Key Parameters (Cumulative)
Level I	Equilibrium	Constant and equal	Equilibration	Concentration: K Fugacity: Z
Level II	Equilibrium, steady-state, well-mixed	Constant and equal	Advection and/or reaction	Concentration: G, k Fugacity: $D = GZ$ or kVZ
Level III	Steady-state, well-mixed	Constant and unequal	Interphase diffusion	Concentration: k_M Fugacity: $1/$ $D_{12} = (1/D_1 + 1/D_2)$
Level IV	Non-steady-state, dynamic	Time-dependent and unequal	Time-dependent flow rates	Concentration: C_i Fugacity: f_i

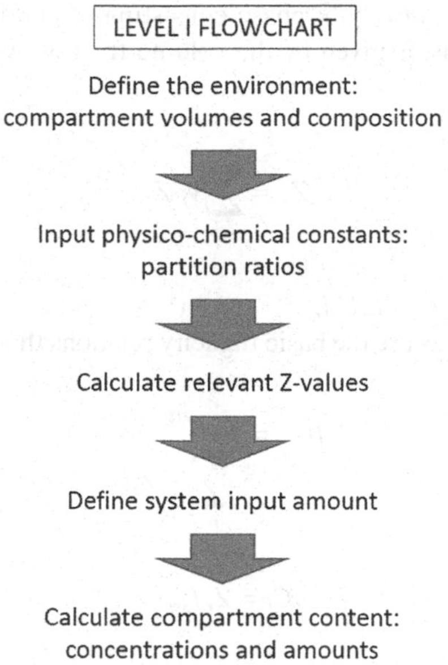

LEVEL I FLOWCHART

Define the environment:
compartment volumes and composition

Input physico-chemical constants:
partition ratios

Calculate relevant Z-values

Define system input amount

Calculate compartment content:
concentrations and amounts

FIGURE 5.1 Flowchart for a generic Level I fugacity-based calculation.

TABLE 5.2
Definitions of Z-Values and Equations Used in Fugacity Model Calculations

Air	$Z_A = \dfrac{1}{RT}$
Water	$Z_W = \dfrac{1}{H} = \dfrac{C^{Sat}}{P^{Sat}} = \dfrac{Z_A}{K_{AW}}$
Octanol	$Z_O = Z_W K_{OW}$
Pure chemical phase	$Z_P = \dfrac{1}{v_P P^{Sat}}$
Soil, sediment	$Z_{Soil} = m_{OC}^f K_{OC} Z_W \left(\dfrac{\left(\rho_{Soil} \ (\mathrm{kg\,m^{-3}}) \right)}{1000} \right)$
Aerosols	$Z_Q = Z_A K_{QA}$ and $K_{QA} = \dfrac{6 \times 10^6}{P^{Sat}/\mathrm{Pa}} = m_{OM}^f K_{OA} \left(\dfrac{\left(\rho_Q \ (\mathrm{kg\,m^{-3}}) \right)}{1000} \right)$
Biota	$Z_{Biota} = v_L^f Z_O$

Here R is the gas constant (8.314 Pa m³mol⁻¹ K⁻¹), T is the absolute temperature (K), H is the Henry's law constant (Pa m³mol⁻¹), C^{Sat} is the contaminant solubility in water (mol m⁻³), P^{Sat} is the contaminant (solute) liquid or super-cooled liquid vapor pressure (Pa), K_{AW} is the air–water partition ratio, K_{OW} is the octanol–water partition ratio, K_{OC} is the organic carbon–water partition ratio, v_P is the molar volume of pure chemical (m³mol⁻¹), m_{OC}^f is the mass fraction organic carbon, m_{OM}^f is the mass fraction organic matter, ρ_S is the density of soil (kg m⁻³), ρ_Q is the density of aerosols (kg m⁻³), K_{QA} is the aerosol–air partition ratio, and v_L^f is the lipid content (volume fraction).

As noted earlier, the Z-value of a bulk phase consisting of continuous and dispersed material, e.g., water plus suspended solids, is given by the volume fraction-weighted Z-values summed over all constituents in the bulk phase:

$$Z_T = \sum_{i=1} v_i^f Z_i$$

where v_i^f is the volume fraction of phase i.

Associated with these Z-values are the basic fugacity relations that apply for a Level I calculation:

$$f_{Sys} = \frac{m}{\sum_{i=1}^{n} V_i Z_i}$$

$$C_i = Z_i f_{Sys}$$

$$m_i = V_i C_i = V_i Z_i f_{Sys}$$

Here f_{Sys} is the fugacity (Pa), M is the total amount of chemical (mol), V_i is the volume (m^3), and m_i is the molar amount in phase i.

The steps and details of a Level I calculation are best grasped by working through examples. A chemical can be selected from those listed in Appendix A and the properties used with assumed media volumes to deduce the distribution of a defined quantity of the substance between these media. Worked Example 5.1 demonstrates such a computation. For pedagogical purposes, it has been found to be effective to allow students to select a chemical of their choice and proceed through Level I, II, III and IV calculations, using their results as a basis for an environmental fate assessment report. Here we use the standard evaluative "EQuilibrium Criterion (EQC)" model environment parameters for this and the subsequent Worked Examples for Levels II and III. The EQC environment is introduced formally later in the book.

Worked Example 5.1: Level I calculation of a four-phase model environment (air, water, soil, and sediment) for a given mass of chemical added by emission to the system.

A fictitious chemical called "hypothene" is released into a pond. The total quantity released is 20 kg. The water is assumed to be in equilibrium contact with the surrounding air, soil, and pond bottom sediment phases. Determine the equilibrium concentration of hypothene in each phase.

Assume the following physico-chemical properties for hypothene: molar mass of 200 g mol^{-1}, melting point 0°C, Henry's law constant = 10 Pa mol^{-1} m^3, log $K_{OW} = 4$, water solubility 20 g m^{-3}, and vapor pressure of 1 Pa. The temperature at which this data was collected was 27.5°C.

Assume the following environmental data: soil organic mass fraction = 0.02, sediment organic mass fraction = 0.04, air density = 1.1756 kg m^{-3}, water density 1.0 × 10^3 kg m^{-3}, bulk soil and sediment density = 1500 kg m^{-3}, and compartment volumes of 6 × 10^9 m^3 (air), 7 × 10^6 m^3 (water), 4.5 × 10^4 m^3 (soil), and 2.1 × 10^4 m^3 (sediment).

In a Level I calculation, the fugacity is constant throughout the specified system, and there are no inputs or outputs. Therefore, the essence of the calculation here is to (i) estimate Z-values for the chemical in all phases, (ii) determine the prevailing system fugacity, and (iii) use these to determine the concentrations in the air, soil, and sediment. Here we use subscripts W for water, A for air, Soil for soil and Sed for sediment.

i. Begin by calculating the necessary Z-values:

$$\text{Air:}\quad Z_A = \frac{1}{RT} = \frac{1}{8.3145\,\text{Pa mol}^{-1}\,\text{m}^3\,\text{K}^{-1} \times 300.6_5\,\text{K}}$$

$$= 4.000 \times 10^{-4}\,\text{mol Pa}^{-1}\,\text{m}^{-3}$$

$$\text{Water:}\quad Z_W = \frac{1}{H} = \frac{1}{10\,\text{Pa mol}^{-1}\,\text{m}^3}$$

$$= 0.10\,\text{mol Pa}^{-1}\,\text{m}^{-3}$$

$$\text{Soil:}\quad Z_{\text{Soil}} = y_{OC}K_{OC}Z_W\left(\frac{\rho_{\text{Soil}}\ \text{kg}^{-1}\,\text{m}^3}{1000}\right)$$

$$= 0.02\left(0.41 \times 10^4\right)0.1\left(\frac{1500}{1000}\right) = 12.3\,\text{mol Pa}^{-1}\,\text{m}^{-3}$$

$$\text{Sediment:}\quad Z_{\text{Sed}} = y_{OC}(K_{OC})Z_W\left(\frac{\rho_{\text{Sed}}\ \text{kg}^{-1}\,\text{m}^3}{1000}\right)$$

$$= 0.04\left(0.41 \times 10^4\right)0.1\left(\frac{1500}{1000}\right) = 24.6\,\text{mol Pa}^{-1}\,\text{m}^{-3}$$

ii. Next, determine the prevailing system fugacity:

$$f_{\text{Sys}} = \frac{m_{\text{Sys}}}{\displaystyle\sum_{i=1}^{n} Z_i V_1}$$

The numerator is the number of moles released into the system:

$$m_{\text{Sys}} = \frac{20\,\text{kg} \times 1000\,\text{g kg}^{-1}}{200\,\text{g mol}^{-1}} = 100\,\text{moles}$$

The denominator is:

$$\sum_{i=1}^{n} Z_i V_i = Z_A V_A + Z_W V_W + Z_{\text{Soil}} V_{\text{Soil}} + Z_{\text{Sed}} V_{\text{Sed}}$$

$$= 4.00 \times 10^{-4} \times 6 \times 10^9 + 0.10 \times 7 \times 10^6 + 12.3 \times 4.5 \times 10^4 + 24.6 \times 2.1 \times 10^4$$

$$= 2.40 \times 10^6 + 7.00 \times 10^5 + 5.54 \times 10^5 + 5.17 \times 10^5 = 4.17 \times 10^6\,\text{mol Pa}^{-1}$$

Therefore, the fugacity is

$$\therefore f_{\text{Sys}} = \frac{100\,\text{mol}}{4.17 \times 10^6\,\text{mol Pa}^{-1}} = 2.40 \times 10^{-5}\,\text{Pa}$$

iii. Finally, use the fugacity and fugacity capacities to determine the concentrations in the various phases at equilibrium with the water:

$$C_A = Z_A f_{\text{Sys}} = 4.00 \times 10^{-4}\,\text{mol Pa}^{-1}\,\text{m}^{-3} \times 2.40 \times 10^{-5}\,\text{Pa} = 9.59 \times 10^{-9}\,\text{mol m}^{-3}$$

$$C_W = Z_W f_{Sys} = 1.0 \times 10^{-1}\,\text{mol Pa}^{-1}\,\text{m}^{-3} \times 2.40 \times 10^{-5}\,\text{Pa} = 2.40 \times 10^{-6}\,\text{mol m}^{-3}$$

$$C_{Soil} = Z_{Soil} f_{Soil} = 12.3\,\text{mol Pa}^{-1}\,\text{m}^{-3} \times 2.40 \times 10^{-5}\,\text{Pa} = 2.95 \times 10^{-4}\,\text{mol m}^{-3}$$

$$C_{Sed} = Z_{Sed} f_{Sys} = 24.6\,\text{mol Pa}^{-1}\,\text{m}^{-3} \times 2.40 \times 10^{-5}\,\text{Pa} = 5.90 \times 10^{-4}\,\text{mol m}^{-3}$$

From this point, the volumes of the various compartments may be used to calculate the molar amounts of benzene in each, by way of $m = CV$ (moles):

$$m_A = C_A V_A = 9.59 \times 10^{-9}\,\text{mol m}^{-3} \times 6 \times 10^9\,\text{m}^3 = 57.54\,\text{mol}$$

$$m_W = C_W V_W = 2.40 \times 10^{-6}\,\text{mol m}^{-3} \times 7 \times 10^6 = 16.8\,\text{mol}$$

$$m_{Soil} = C_{Soil} V_{Soil} = 2.95 \times 10^{-4}\,\text{mol m}^{-3} \times 4.5 \times 10^4 = 13.27\,\text{mol}$$

$$m_{Sed} = C_{Sed} V_{Sed} = 5.90 \times 10^{-4}\,\text{mol m}^{-3} \times 2.1 \times 10^4 = 12.39\,\text{mol}$$

As a check, we can do a mass balance to make sure we can account for all the quantity of chemical introduced, since we know that the following must be true:

$$m_{sys} = \sum_{i=1}^{n} m_i = 57.54 + 16.8 + 13.27 + 12.39 = 100\ \text{mol}$$

Such a calculation as outlined in Worked Example 5.1 may be done for an arbitrarily chosen evaluative environment with the Level I spreadsheet model. Here the calculation may be made more complex, involving several more compartments at equilibrium (e.g., soil, suspended particles, aerosols, and fish).

The chemical properties input page for such a "full-blown" Level I calculation is shown in the following figure for hypothene, with various chemical properties being user-entered or auto-filled from the chemical database, if the data have been previously stored there:

The corresponding environmental properties input page of the Level I spreadsheet program is shown in the following figure for the specified environment. Some additional density values are included to ensure that the program is able to calculate "zero" without an error for compartments that are not included in the system under consideration:

ENVIRONMENTAL PROPERTIES

Please complete all fields or select an environment from the database: [Hypothene ▼]   [Input Environment]

Environment Name	Hypothene

[Clear Form]

Fractions

Aerosols in Air	0
Susp. Particles in Water	0
Fish in Water	0

Organic Carbon

Suspended Particles (g/g)	0.2
Soil (g/g)	0.02
Sediment (g/g)	0.04

Volume

Air (m³)	6.00E+09
Water (m³)	7.00E+06
Soil (m³)	4.50E+04
Sediment (m³)	2.10E+04

Density

Air (kg/m³)	1.175
Aerosol (kg/m³)	2000
Water (kg/m³)	1000
Suspended Particles (kg/m³)	1500
Fish (kg/m³)	1000
Soil (kg/m³)	1500
Sediment (kg/m³)	1500

Lipid Fraction

Fish (g/g)	0.05

Additional Comments

Click **Add to DB** to add this environment to the Environment Database

[Add to DB]

The system chemical quantity entry for the Level I spreadsheet program for a calculation in which a direct emission is present is shown in the following figure:

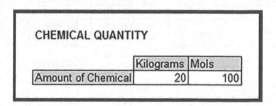

CHEMICAL QUANTITY

	Kilograms	Mols
Amount of Chemical	20	100

Running the program with this set of input data results in the following output:

Level I RESULTS

Chemical Name	Hypothene
Environment Name	Hypothene

Fugacity Ratio	1

Subcooled Liq. Vap. Press. (Pa)	1

Fugacity (Pa)	2.40E-05

Partition Ratio	dimensionless
Air-Water	4.00E-03
Soil-Water	1.23E+02
Sediment-Water	2.46E+02
Suspended Particles-Water	1.23E+03
Fish-Water	5.00E+02
Aerosol-Water	-
Aerosol-Air	6.00E+06

	Z Value	VZ
	mol/m³*Pa	mol/Pa
Air	4.0004E-04	2.4002E+06
Aerosol	2.4002E+03	0.0000E+00
Water	1.0000E-01	7.0000E+05
Suspended Particles	1.2300E+02	0.0000E+00
Fish	5.0000E+01	0.0000E+00
Soil	1.2300E+01	5.5350E+05
Sediment	2.4600E+01	5.1660E+05

Total		4.17E+06

	Concentration		
	mol/m³	g/m³	ng/L
Air	9.5925E-09	1.9185E-06	1.6328E+00
Aerosol	5.7555E-02	1.1511E+01	5.7555E+03
Water	2.3979E-06	4.7958E-04	4.7958E-01
Suspended Particles	2.9494E-03	5.8988E-01	3.9325E+02
Fish	1.1989E-03	2.3979E-01	2.3979E+02
Soil	2.9494E-04	5.8988E-02	3.9325E+01
Sediment	5.8988E-04	1.1798E-01	7.8651E+01

	Quantity		
	mol	kg	%
Air	57.5551	1.1511E+01	57.56
Aerosol	0.0000	0.0000E+00	0.00
Water	16.7852	3.3570E+00	16.79
Suspended Particles	0.0000	0.0000E+00	0.00
Fish	0.0000	0.0000E+00	0.00
Soil	13.2723	2.6545E+00	13.27
Sediment	12.3875	2.4775E+00	12.39

Total	1.00E+02	2.00E+01	100

The same information is presented in compact diagrammatic form as follows, with concentrations (mol m^{-3}), total quantity (kg), and percentage of chemical in each compartment:

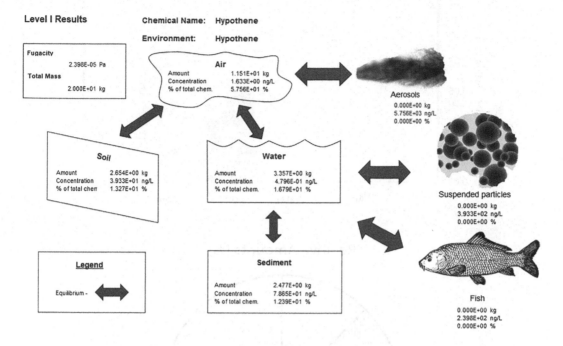

Level I Results

| Chemical Name: | Hypothene |
| Environment: | Hypothene |

Fugacity 2.398E-05 Pa

Total Mass 2.000E+01 kg

Air
Amount   1.151E+01 kg
Concentration   1.633E+00 ng/L
% of total chem. 5.756E+01 %

Aerosols
0.000E+00 kg
5.756E+03 ng/L
0.000E+00 %

Soil
Amount   2.654E+00 kg
Concentration   3.933E+01 ng/L
% of total chem 1.327E+01 %

Water
Amount   3.357E+00 kg
Concentration   4.796E-01 ng/L
% of total chem. 1.679E+01 %

Suspended particles
0.000E+00 kg
3.933E+02 ng/L
0.000E+00 %

Legend

Equilibrium - ⬌

Sediment
Amount   2.477E+00 kg
Concentration   7.865E+01 ng/L
% of total chem. 1.239E+01 %

Fish
0.000E+00 kg
2.398E+02 ng/L
0.000E+00 %

Finally, the mass distribution of the chemical amongst the compartments, the concentrations in each compartment, and the relative amounts in each compartment are outputted graphically as follows:

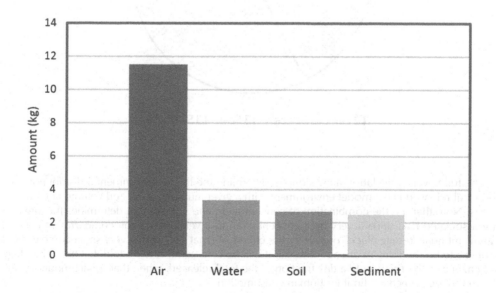

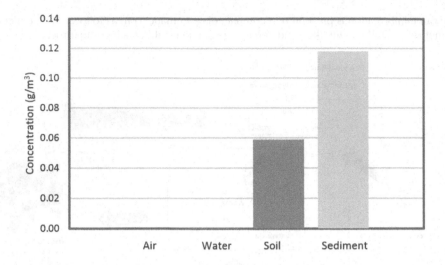

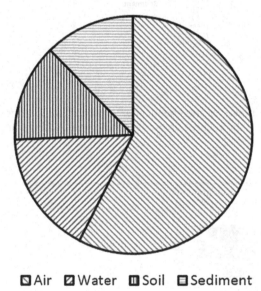

Relative Amounts (%)

▨Air ▨Water ▥Soil ▤Sediment

The full Level I calculation allows one to determine the likely environmental distribution of a chemical released into a model environment with a given number, type, and volume of compartments. Note that it is the combination of concentration *and* volume that determines the fate and any associated risk, since a very high concentration in a very, very small compartment is not a significant final "resting place" or sink for the chemical. That said, if that very small compartment happens to be the fish you are about to eat, it will be cold comfort to be told that the percentage of benzene in the fish is only a tiny fraction of the total released, when that small amount is well above the recommended limit for human consumption.

Note that stickler agrochemists may contend that the approximation of K_{OC} as $0.41K_{OW}$ is inappropriate. Such users (and others) are free to input a more accurate experimentally determined value. As well, some biologists may feel that the assumption of a 5% lipid fraction implied in $K_{FW} = 0.05K_{OW}$ is not appropriate for a particular species of fish, and are free to use a different lipid fraction via direct input of a chosen K_{FW} value.

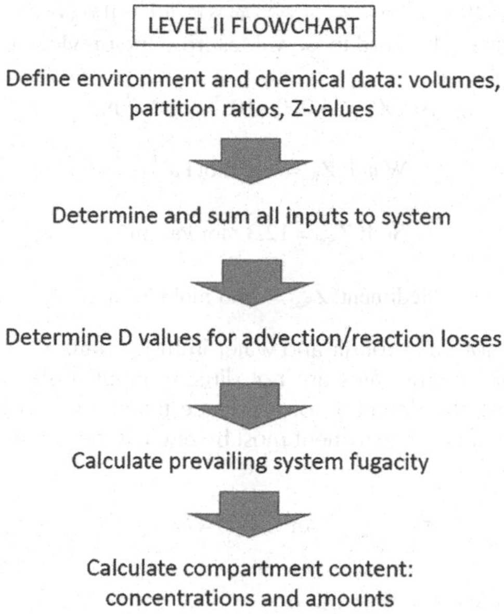

LEVEL II FLOWCHART

Define environment and chemical data: volumes,
partition ratios, Z-values

Determine and sum all inputs to system

Determine D values for advection/reaction losses

Calculate prevailing system fugacity

Calculate compartment content:
concentrations and amounts

FIGURE 5.2 Flowchart for a generic Level II fugacity-based calculation.

5.2 LEVEL II MODELS AND CALCULATIONS

Level II calculations are more complex than Level I, since they include constant flow input and output rates in a system assumed to be at steady state. Equilibrium is assumed between all compartments, and the essential computation is one of mass balance of input and output flows. The standard input data include the properties of the environment, the properties of the chemical in question, chemical input rates by emission and advection, and information on reaction and advection rates. The fugacity is calculated, followed by a complete mass balance. Since equilibrium is assumed to apply within the environment, it is immaterial into which phase the chemical is introduced.

The essential features of a Level II calculation are illustrated in Figure 5.2. As was done for Level I, we illustrate the essential steps and details with a Worked Example.

Worked Example 5.2: Level II calculation of a four-phase model environment (air, water, soil, sediment) with a direct emission and advective flow into air and water.

Calculate the partitioning of the hypothetical chemical "hypothene" in the same system as in Worked Example 5.1, but assume that the chemical is directly emitted into the system at a rate of 20 kg h^{-1}, and that the half-lives for removal by reaction are 1×10^{11} (air), 693 h (water), 69.3 (soil), and 6930 h (sediment). Assume residence times of 600 and 7000 h for the air and water bodies, respectively, and hypothene inflow concentrations of 2×10^5 ng m^{-3} for air and 2×10^6 ng L^{-1} for water.

In a Level II calculation, as with Level I, the fugacity is constant throughout the system. However, in contrast to Level I, we now have inputs and losses in any of the compartments. The system is assumed to be at steady state and all compartments are "well-mixed" such that the concentrations in each compartment are constant in time and space. Moreover, the entire system is assumed to be at equilibrium, such that the determination of system fugacity is all that is needed as long as the necessary Z-values are available or at least estimable. Thus, the calculation steps are (i) estimate Z-values for all phases, (ii) determine the total chemical input flow rate, (iii) determine the sum of D-values for all advection and reaction loss paths or pathways, (iv) determine the prevailing system fugacity, and (v) determine the concentrations and then, if needed, amounts in all compartments.

i. Calculate or estimate the various Z-values, as we did in the Level I example above. Here we simply carry forward these values from the previous Level I example:

$$\text{Air: } Z_A = 4.000 \times 10^{-4} \text{ mol Pa}^{-1} \text{ m}^{-3}$$

$$\text{Water: } Z_W = 0.10 \text{ mol Pa}^{-1} \text{ m}^{-3}$$

$$\text{Soil: } Z_{\text{Soil}} = 12.3 \text{ mol Pa}^{-1} \text{ m}^{-3}$$

$$\text{Sediment: } Z_{\text{Sed}} = 24.6 \text{ mol Pa}^{-1} \text{ m}^{-3}$$

ii. Calculate the advection rates for air and water in $m^3 h^{-1}$ units:
   The air and water inflow rates are not directly input. Instead, they may be determined by rearranging the definition of residence time and recognizing that the inflow and outflow rates for any compartment must be equal if the volume is not changing with time. Thus, we have

$$\tau_i = \frac{V_i}{G_i^{\text{Outf}}} \quad \text{or} \quad G_i^{\text{Inf}} = G_i^{\text{Outf}} = \frac{V_i}{\tau_i}$$

Using this approach, we can determine

$$\text{Air inflow rate} \left(G_A^{\text{Inf}} \right) = 6 \times 10^9 \text{ m}^3 / 600 \text{ h} = 1 \times 10^7 \text{ m}^3 \text{ h}^{-1}$$

$$\text{Water inflow rate} \left(G_W^{\text{Inf}} \right) = 7 \times 10^6 \text{ m}^3 / 7000 \text{ h} = 1 \times 10^3 \text{ m}^3 \text{ h}^{-1}$$

iii. Calculate the concentration of the air and water inflow in $mol\ m^{-3}$ units:
   The inflow concentrations for air and water are inputted in the convenient units of ng m^{-3} and ng L^{-1}, respectively. The conversion is straightforward:

$$\text{Air inflow concentration} \left(C_A^{\text{Inf}} \right) = 2 \times 10^5 \text{ ng m}^{-3} \times 10^{-9} \text{ g ng}^{-1} / 200 \text{ g mol}^{-1} = 1.0 \times 10^{-6} \text{ mol m}^{-3}$$

$$\text{Water inflow concentration} \left(C_W^{\text{Inf}} \right) = 2 \times 10^6 \text{ ng L}^{-1} \times 10^{-9} \text{ g ng}^{-1} \times 10^3 \text{ L m}^{-3} / 200 \text{ g mol}^{-1} = 1.0 \times 10^{-2} \text{ mol m}^{-3}$$

iv. Calculate the molar inflow rates and the total input flow rate E (mol h^{-1}):
   The molar inflow rates are given by the product of molar inflow concentration and flow rate:

$$\text{Input flow rate into air} \left(C_A^{\text{Inf}} G_A^{\text{Inf}} \right) = 10^{-6} \text{ mol m}^{-3} \times 10^7 \text{ m}^3 \text{ h}^{-1} = 10 \text{ mol h}^{-1}$$

$$\text{Input flow rate into water} \left(C_A^{\text{Inf}} G_A^{\text{Inf}} \right) = 10^{-2} \text{ mol m}^{-3} \times 10^3 \text{ m}^3 \text{ h}^{-1} = 10 \text{ mol h}^{-1}$$

$$\text{Direct emission rate} = 20 \text{ kg h}^{-1} \times 1000 \text{ g kg}^{-1} / 200 \text{ g mol} = 100 \text{ mol h}^{-1}$$

The total input flow rate E is the sum of all inputs, here 120 mol h^{-1}

v. Determine and sum the D-values for all advection losses:
   For advective loss, we have $D_i^{\text{Adv}} = G_i \left(\text{m}^3 \text{ h}^{-1} \right) Z_i \left(\text{mol Pa}^{-1} \text{ m}^{-3} \right)$. Therefore, we find

$$\text{Air: } D_A^{\text{Adv}} = 10^7 \text{ m}^3 \text{ h}^{-1} \times 4.0 \times 10^{-4} \text{ mol Pa}^{-1} \text{ m}^{-3} = 4.0 \times 10^3 \text{ mol Pa}^{-1} \text{ h}^{-1}$$

$$\text{Water: } D_W^{\text{Adv}} = 10^3 \text{ m}^3 \text{ h}^{-1} \times 0.1 \text{ mol Pa}^{-1} \text{ m}^{-3} = 1 \times 10^2 \text{ mol Pa}^{-1} \text{ h}^{-1}$$

The advective losses are parallel processes, so their D-values sum to make the corresponding value for total loss due to advection:

Total advective loss D-value: $D_{\text{Total}}^{\text{Adv}} = D_A^{\text{Adv}} + D_W^{\text{Adv}} = \left(4\times10^3 + 1\times10^2\right) = 4.1\times10^3 \text{ mol Pa}^{-1}\text{ h}^{-1}$

vi. Determine and sum the D-values for all reaction losses:

For reaction losses, we have $D_i^{\text{Deg}} = V\left(\text{m}^3\right)Z\left(\text{mol Pa}^{-1}\text{ m}^{-3}\right)k\left(\text{h}^{-1}\right)$. To determine these, we express the first-order rate constants k in terms of the more easily recalled half-lives for loss by reaction, recalling the definition of half-life:

$$t_{1/2} = \frac{0.693}{k^{\text{Loss}}} \quad \text{or} \quad k^{\text{Loss}} = \frac{0.693}{t_{1/2}}$$

Thus, we have $D_i^{\text{Deg}} = V\left(\text{m}^3\right)Z\left(\text{mol Pa}^{-1}\text{m}^{-3}\right)\dfrac{0.693}{t_{1/2}\left(\text{h}\right)}$, and therefore

Air: $D_A^{\text{Deg}} = 6\times10^9 \text{ m}^3 \times 4.00\times10^{-4} \text{ mol Pa}^{-1}\text{ m}^{-3} \times \left(0.693/1\times10^{11}\text{ h}^{-1}\right) = 1.66\times10^{-5} \text{ mol Pa}^{-1}\text{ h}^{-1}$

Water: $D_W^{\text{Deg}} = 7\times10^6 \text{ m}^3 \times 0.1 \text{ mol Pa}^{-1}\text{ m}^{-3} \times \left(0.693/693\text{ h}^{-1}\right) = 7\times10^2 \text{ mol Pa}^{-1}\text{ h}^{-1}$

Soil: $D_{\text{Soil}}^{\text{Deg}} = 4.5\times10^4 \text{ m}^3 \times 12.3 \text{ mol Pa}^{-1}\text{ m}^{-3} \times \left(0.693/69.3\text{ h}^{-1}\right) = 5.54\times10^3 \text{ mol h}^{-1}\text{ Pa}^{-1}$

Sediment: $D_{\text{Sed}}^{\text{Deg}} = 2.1\times10^4 \text{ m}^3 \times 24.6 \text{ mol Pa}^{-1}\text{ m}^{-3} \times \left(0.693/6930\text{ h}^{-1}\right) = 51.7 \text{ mol h}^{-1}\text{ Pa}^{-1}$

Again, as these reactive losses are parallel processes, their D-values sum to the total D-value for losses due to reaction:

Total reaction losses: $D_{\text{Total}}^{\text{Deg}} = D_A^{\text{Deg}} + D_W^{\text{Deg}} + D_{\text{Soil}}^{\text{Deg}} + D_{\text{Sed}}^{\text{Deg}} = 6.29\times10^3 \text{ mol Pa}^{-1}\text{ h}^{-1}$

vi. Determine and sum the D-values for total losses:

The sum of these advection and reaction D-values is $= 1.04 \times 10^4 \text{mol Pa}^{-1}\text{h}^{-1}$

viii. Calculate the prevailing system fugacity:

$$f_{\text{Sys}} = \frac{E}{\displaystyle\sum_{i=1}^{n} D_i} = \frac{120 \text{ mol h}^{-1}}{1.04\times10^4 \text{ mol Pa}^{-1}\text{ h}^{-1}} = 1.16\times10^{-2} \text{ Pa}$$

ix. Calculate the concentrations in the various compartments as $C = Zf$:

Air: $C_A = Z_A f_{\text{Sys}} = 4.00\times10^{-4} \text{ mol Pa}^{-1}\text{ m}^{-3} \times 1.16\times10^{-2} \text{ Pa} = 4.64\times10^{-6} \text{ mol m}^{-3}$

Water: $C_W = Z_W f_{\text{Sys}} = 0.10 \text{ mol Pa}^{-1}\text{ m}^{-3} \times 1.16\times10^{-2} \text{ Pa} = 1.16\times10^{-3} \text{ mol m}^{-3}$

Soil: $C_{\text{Soil}} = Z_{\text{Soil}} f_{\text{Sys}} = 12.3 \text{ mol Pa}^{-1}\text{ m}^{-3} \times 1.16\times10^{-2} \text{ Pa} = 1.43\times10^{-1} \text{ mol m}^{-3}$

Sediment: $C_{\text{Sed}} = Z_{\text{Sed}} f_{\text{Sys}} = 24.6 \text{ mol Pa}^{-1}\text{ m}^{-3} \times 1.16\times10^{-2} \text{ Pa} = 2.85\times10^{-1} \text{ mol m}^{-3}$

Such a Level II calculation as outlined in Worked Example 5.2 may be done for this same set of conditions using the Level II spreadsheet model.

The chemical properties input page for such a Level II calculation for hypothene following user entry of the necessary input data is shown in the following figure:

Chemical Name CAS	Hypothene

Molar Mass (g/mol)	200
Data Temperature (°C)	27.5
Melting Point (°C)	0
Vapor Pressure (Pa)	1.00E+00
Solubility in Water (g/m³)	20
Henry's Law Constant (Pa·m³/mol)	10

Reaction Half-Lives (hours)

In Air	1.00E+11
In Aerosol	1.00E+11
In Water	6.93E+02
In Suspended Particles	1.00E+11
In Fish	1.00E+11
In Soil	6.93E+01
In Sediment	6.93E+03

Partition Ratios

$\log K_{OW}$	4
K_{OC}	4.10E+03

The environmental properties input page of the Level II spreadsheet program is illustrated below:

Environment Name	Hypothene

Fractions

Aerosol in Air	0.00E+00
Susp. Particles in Water	0.00E+00
Fish in Water	0.00E+00

Volumes (m³)

Air	6.00E+09
Water	7.00E+06
Soil	4.50E+04
Sediment	2.10E+04

Density (kg/m³)

Air	1.18
Aerosol	2000
Water	1000
Susp. Particles	1500
Fish	1000
Soil	1500
Sediment	1500

Advective Flow Residence Times (h)

Air	600
Water	7000
Sediment (burial)	1.00E+11

Organic Carbon (g/g)

Susp. Particles	0.2
Fish Lipid Fraction	0.05
Soil	0.02
Sediment	0.04

The system emission rate entry for the Level II spreadsheet program appears as follows for a calculation in which a direct emission is present:

Emission Rate	20	kg/h	100.00	mol/h

Advective Inflow Concentrations

Concentration in Inflow Water	2.00E+06	ng/L	1.000E-02	mol/m³
Concentration in Air	2.00E+05	ng/m³	1.000E-06	mol/m³

The model output for this set of input data is

Level II Results

Chemical Name	Hypothene
Environment Name	Hypothene

Fugacity Ratio	1.000E+00

Subcooled Liq. Vap. Press. (Pa)	1.000E+00

Fugacity (Pa)	1.155E-02

Aquivalence (mol/m³)	-

Mass Balance

	Rate		Concentration			
	kg/h	mol/h	ng/m³	ng/L	kg/m³	mol/m³
Emission	20.0	100.0	-	-	-	-
Inflow Air	2.0	10.0	2.000E+05	-	2.000E-07	1.000E-06
Inflow Water	2.0	10.0	-	2.000E+06	2.000E-03	1.000E-02

Total	24.0	120.0

	Loss rate		D Value	Res. Time	
	kg/h	mol/h	m³/h	hours	days
Advection	9.474E+00	4.737E+01	4.100E+03	1.017E+03	4.238E+01
Reaction	1.453E+01	7.263E+01	6.287E+03	6.634E+02	2.764E+01
Overall	2.400E+01	1.200E+02	1.039E+04	4.015E+02	1.673E+01

Phase Properties

	Z Value	VZ
	mol/m³*Pa	mol/Pa
Air	4.000E-04	2.400E+06
Aerosol	2.400E+03	0.000E+00
Water	1.000E-01	7.000E+05
Suspended Particles	1.230E+02	0.000E+00
Fish	5.000E+01	0.000E+00
Soil	1.230E+01	5.535E+05
Sediment	2.460E+01	5.166E+05

Total		4.170E+06

Phase Properties

	Concentration		
	mol/m³	g/m³	Alternative units
Air	4.622E-06	9.243E-04	9.243E+06 ng/m³
Aerosol	27.72963128	5.546E+03	2.773E+07 ng/g
Water	1.155E-03	2.311E-01	2.311E+06 ng/L
Suspended Particles	1.421E+00	2.842E+02	1.895E+06 ng/g
Fish	5.776E-01	1.155E+02	1.155E+06 ng/g
Soil	1.421E-01	2.842E+01	1.895E+05 ng/g
Sediment	2.842E-01	5.684E+01	3.789E+05 ng/g

	Quantity		
	mol	kg	%
Air	57.5551	1.1511E+01	57.56
Aerosol	0.0000	0.0000E+00	0.00
Water	16.7852	3.3570E+00	16.79
Suspended Particles	0.0000	0.0000E+00	0.00
Fish	0.0000	0.0000E+00	0.00
Soil	13.2723	2.6545E+00	13.27
Sediment	12.3875	2.4775E+00	12.39

1.00E+02	2.00E+01	100

Advection

	Res. Time		Flow Rate	D Value	Rate		% of Total
	hours	days	m³/h	mol/Pa*h	kg/h	mol/h	Losses
Air	6.000E+02	2.500E+01	1.000E+07	4.000E+03	9.243E+00	4.622E+01	3.851E+01
Aerosol	6.000E+02	2.500E+01	0.000E+00	0.000E+00	0	0	0.000E+00
Water	7.000E+03	2.917E+02	1.000E+03	1.000E+02	2.311E-01	1.155E+00	9.627E-01
Suspended Particles	7.000E+03	2.917E+02	0.000E+00	0.000E+00	0.000E+00	0.000E+00	0.000E+00
Fish	7.000E+03	2.917E+02	0.000E+00	0.000E+00	0.000E+00	0.000E+00	0.000E+00
Soil	-	-	-	-	-	-	-
Sediment	1.000E+11	4.167E+09	2.100E-07	5.166E-06	1.194E-08	5.968E-08	4.973E-08

Total		4.100E+03	9.474E+00	4.737E+01	3.948E+01

Reaction

	Half-Life		Rate Constant	D Value	Rate		% of Total
	hours	days	1/h	mol/Pa*h	kg/h	mol/h	Losses
Air	1.000E+11	4.167E+09	6.930E-12	1.663E-05	3.843E-08	1.922E-07	1.601E-07
Aerosol	1.000E+11	4.167E+09	6.930E-12	0.000E+00	0.000E+00	0.000E+00	0.000E+00
Water	6.930E+02	2.888E+01	1.000E-03	7.000E+02	1.617E+00	8.087E+00	6.739E+00
Suspended Particles	1.000E+11	4.167E+09	6.930E-12	0.000E+00	0.000E+00	0.000E+00	0.000E+00
Fish	1.000E+11	4.167E+09	6.930E-12	0.000E+00	0.000E+00	0.000E+00	0.000E+00
Soil	6.930E+01	2.888E+00	1.000E-02	5.535E+03	1.279E+01	6.394E+01	5.329E+01
Sediment	6.930E+03	2.888E+02	1.000E-04	5.166E+01	1.194E-01	5.968E-01	4.973E-01

Total			Total	6.287E+03	1.453E+01	7.263E+01	6.052E+01

The same information is presented in compact diagrammatic form as follows, with

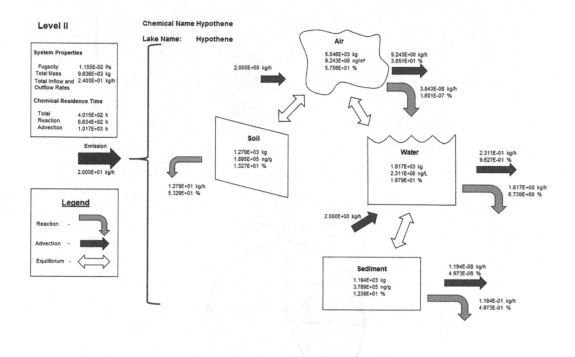

concentrations, total load in the environment (either in moles or in kilograms), and percentage of chemical in each compartment:

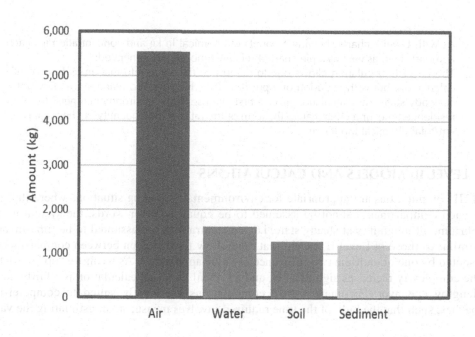

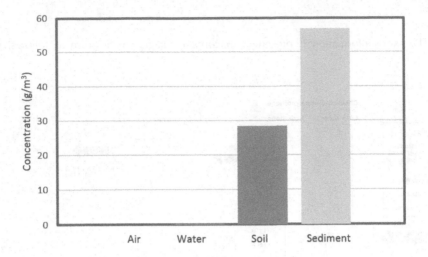

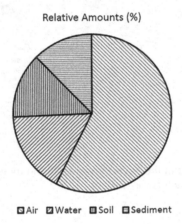

Relative Amounts (%)

☐ Air ☑ Water ▥ Soil ⊟ Sediment

As with Level I, charts showing amounts of chemical in kg and concentration for each compartment, as well as a pie chart of relative amount, are generated:

The Level II calculation allows one to determine the same information as in a Level I calculation, but with inclusion of input flow by advection and reaction loss flow rates at steady state. The calculation gives a first glimpse of the environmental persistence or residence time for a chemical, in the form of the ratio of total quantity of chemical present/total chemical input rate.

5.3 LEVEL III MODELS AND CALCULATIONS

Level III computations are appropriate for environmental modeling situations where the fugacity in each compartment cannot be assumed to be equal. In other words, the system is not at equilibrium, although it is at steady state, i.e., concentrations are assumed to be time-invariant. It is similar to those of Level II, except that they allow for diffusion between compartments, as represented by one or multiple two-resistance D-value approaches. As we move "up" in such levels, the complexity increases significantly, and a typical Level III calculation is a fairly detailed and lengthy endeavor. Fortunately, such computations are ideally suited to computer-based approaches, such that the bulk of the time required involves measuring or estimating the various transport parameter values and inputting the necessary chemical and environmental data. As with the previous Levels I and II, a generic fugacity-based spreadsheet model for Level III is available for download from the CEMC website (www.trentu.ca/cemc/resources-and-models). This spreadsheet allows for air containing aerosol particles; water containing fish and suspended solids; soil containing air, water, and solids; and sediment containing water and solids. A variety of flow vectors are possible, including dry and wet atmospheric deposition, soil runoff, sedimentation, etc., and multiple direct emissions to individual compartments are also possible. The user must provide physical chemical (partitioning) properties, reaction half-lives, and sufficient information to deduce intermedia transport D-values. To assist in this task, Table 5.3 gives suggested order-of-magnitude values for the various parameters. Such values are included as defaults in some programs, but they can be modified as desired.

The water-to-air D-value for diffusive volatilization is the same as for absorption. The air-to-soil D-value is similar, but the areas differ, and the absorption-volatilization D-value is also different. The soil-to-air D-value is for volatilization. The water-to-sediment D-value represents diffusive transfer plus non-diffusive sediment deposition. The sediment-to-water D-value represents diffusive transfer plus non-diffusive resuspension. Finally, the soil-to-water D-value consists of non-diffusive water and particle runoff.

Assembly of an entire Level III model for a chemical is a fairly demanding task, since there are numerous areas, flows, mass-transfer coefficients, and diffusivities to be estimated. Quite complicated models can be assembled describing transfer of a chemical between several media by a number of routes. In general, the total D-value for movement from phase A to B will not be the same as that from B to A. The reason is that there may be an advective process moving in only one direction. Diffusive processes always have identical D-values applying in each direction. D-values for loss by reaction can also be included in the mass-balance expression.

TABLE 5.3

Order-or-Magnitude Values of Transport Parameters Used in Multimedia Modeling

Parameter	Symbol	Suggested Typical Value
Air-side MTC over water	k_A^M	$3\,\text{m}\,\text{h}^{-1}$
Water-side MTC	k_W^M	$0.03\,\text{m}\,\text{h}^{-1}$
Transfer rate to higher altitude	U_S	$0.01\,\text{m}\,\text{h}^{-1}$ ($90\,\text{m}\,\text{year}^{-1}$)
Rain rate (m³ rain/m² area h)	U_{Rain}	$9.7 \times 10^{-5}\,\text{m}\,\text{h}^{-1}$ ($0.85\,\text{m}\,\text{year}^{-1}$)
Scavenging ratio	Q	$200{,}000$
Vol. fraction aerosols	v_Q^f	30×10^{-12}
Dry deposition velocity	U_{Dry}	$10.8\,\text{m}\,\text{h}^{-1}$ ($0.003\,\text{m}\,\text{s}^{-1}$)
Air-side MTC over soil	k_A^M	$1\,\text{m}\,\text{h}^{-1}$
Diffusion path length in soil	Y_{Soil}	$0.05\,\text{m}$
Molecular diffusivity in air	B_{MA}	$0.04\,\text{m}^2\text{h}^{-1}$
Molecular diffusivity in water	B_{MW}	$4.0 \times 10^{-6}\,\text{m}^2\text{h}^{-1}$
Water runoff rate from soil	U_{WW}	$3.9 \times 10^{-5}\,\text{m}\,\text{h}^{-1}$ ($0.34\,\text{m}\,\text{year}^{-1}$)
Solids runoff rate from soil	U_{EW}	$2.3 \times 10^{-8}\,\text{m}^3\text{m}^{-2}\text{h}^{-1}$ ($0.0002\,\text{m}\,\text{year}^{-1}$)
Water-side MTC over sediment	k_{Sed-W}^M	$0.01\,\text{m}\,\text{h}^{-1}$
Diffusion path length in sediment	Y_{Sed}	$0.005\,\text{m}$
Sediment deposition rate	U_{Dep}	$4.6 \times 10^{-8}\,\text{m}^3\text{m}^{-2}\text{h}^{-1}$ ($0.0004\,\text{m}\,\text{year}^{-1}$)
Sediment resuspension rate	U_{Resusp}	$1.1 \times 10^{-}\,\text{m}^3\text{m}^{-2}\text{h}^{-1}$ ($0.0001\,\text{m}\,\text{year}^{-1}$)
Sediment burial rate	U_{Burial}	$3.4 \times 10^{-8}\,\text{m}^3\text{m}^{-2}\text{h}^{-1}$ ($0.0003\,\text{m}\,\text{year}^{-1}$)
Leaching rate of water from soil to ground water	U_L	$3.9 \times 10^{-5}\,\text{m}^3\text{m}^{-2}\text{h}^{-1}$ ($0.34\,\text{m}\,\text{year}^{-1}$)

Level III-type calculations were first suggested and illustrated in a series of papers on fugacity models (Mackay, 1979, Mackay and Paterson, 1981, 1982, Mackay et al., 1985). It is important to emphasize that these models will give the same results as other concentration-based models, provided that the intermedia transport expressions are equivalent. A major advantage of the fugacity approach is that an enormous amount of detail can be contained in one D-value, which can be readily compared with other D-values for different processes. It is quite difficult, on the other hand, to compare a reaction rate constant, a mass-transfer coefficient, and a sedimentation rate and identify their relative importance.

An important feature of the Level III results is that a complete quantitative picture of the various chemical pathways is now apparent. Important processes and compartments can now be identified. A Level III calculation gives a first, approximate picture of the complexities of multimedia environmental fate of chemicals.

Figure 5.3 gives a schematic illustration of a Level III calculation flowchart. The computation again shares some common features to Levels I and II, and the three Level "types" may be seen as reflecting differing levels of approximation for a true system.

A manual Level III calculation is a long and detailed task. We illustrate aspects of such a calculation for the system summarized in Figure 5.4. It is a comprehensive multimedia picture of chemical emission, advection, reaction, intermedia transport, and residence time or persistence. The important processes are now clear, and it is possible to focus on them when seeking more accurate rate data and a better simulation of a real system. Figure 5.4 contains information about 21 processes, some of which, such as air–water transfer, consist of several contributing processes. The human mind is incapable of making sense of the vast quantity of physical chemical and environmental data without the aid of a conceptual tool such as a Level III program.

Note that it is possible for the modeler to add more compartments and to subdivide the existing compartments. For example, it may be advantageous to add vegetation as a separate compartment. The atmosphere or water column could be segmented vertically. The soil can be treated as several layers. If information is available to justify these changes, they can be implemented, albeit at the expense of greater algebraic complexity. If the number of compartments becomes large and highly connected, it is preferable to solve the equations by matrix algebra.

LEVEL III FLOWCHART

Define environment and chemical data: volumes, areas, depths densities, residence times, transport velocities, partition ratios, Z-values, reaction half-lives

Determine D values for all relevant processes

Construct mass balances of input and output fluxes for each compartment

Calculate fugacity in each compartment

Calculate compartment contents: concentrations and amounts

FIGURE 5.3   Flowchart for a generic Level III fugacity-based calculation.

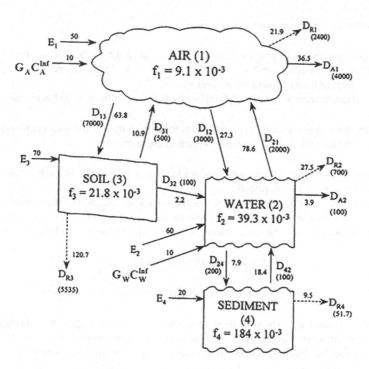

FIGURE 5.4 Schematic representation of the results corresponding to a sample computed output. Here the subscripts A and R refer to losses due to advection and reaction (degradation), respectively.

Worked Example 5.3: Level III calculation for a four-compartment system (air, water, soil, sediment).

Determine the steady-state concentrations of hypothene in the same four-compartment system as in Worked Examples 5.1 and 5.2, but as a Level III system, that is, without the assumption of equal fugacity in the four compartments. Assume the total direct emission rate is the same, but that it is apportioned to each compartment in the following manner:

Air: $10\,kg\,h^{-1} = 50\,mol\,h^{-1}$
Water: $12\,kg\,h^{-1} = 60\,mol\,h^{-1}$
Soil: $14\,kg\,h^{-1} = 70\,mol\,h^{-1}$
Sediment: $4\,kg\,h^{-1} = 20\,mol\,h^{-1}$

For simplicity, assume only diffusive transfer between compartments. Assume the same half-lives for degradation as in the Level II example, with only a change to the air half-life:

Air: 693 h; Water: 693 h; Soil: 69.3 h; Sediment: 6930 h

The determination of overall D-values for intermedia transport is rather detailed. In order to avoid unnecessary complexity in this Level III illustration, we will use predetermined total D-values for the various intermedia transport phenomena and delay the derivation of these until Chapter 6. For this exercise, use the following overall D-values $\left(D_{ij}^{Ov}\right)$:

Air–water diffusion: $500\,mol\,Pa^{-1}h^{-1}$
Soil–air diffusion: $150\,mol\,Pa^{-1}h^{-1}$
Sediment–water diffusion: $3.50\,mol\,Pa^{-1}h^{-1}$

Assume the following environmental properties:

Depth of compartments in metres: 2000 (air), 20 (water), 0.1 (soil), 0.06 (sediment)
Areas of compartments in m^2: 3×10^6 (air), 3.5×10^5 (water and sediment), 4.5×10^5 (soil).
Volume fraction of water in sediments and soils: 0.377
Organic carbon mass fractions of soil and sediment, respectively: 0.02 and 0.04

Write down the four mass-balance equations and solve by hand for the four corresponding fugacities and then calculate the concentrations in each compartment.

i. Calculate or estimate the various Z-values for the components of each phase, and then the Z-value for each bulk phase.
   The various bulk phase Z-values are given by the volume fraction-weighted sum of the Z-values for each component of the phase. For air and water, we assume "pure" phases, so we can use the same values as we did in the Level I and II cases above:

$$\text{Air: } Z_A = 4.000\times10^{-4} \, \text{mol Pa}^{-1}\,\text{m}^{-3}$$

$$\text{Water: } Z_W = 0.10 \, \text{mol Pa}^{-1}\,\text{m}^{-3}$$

For the soil and the sediment, these are assumed to be comprised of both water and solids, the latter having the mass fraction of organic carbon noted above. We first need to determine the Z-values for the solid component of the soil and sediment:

$$\text{Soil solids: } Z_{\text{Soil}}^{\text{Solids}} = y_{OC} K_{OC} Z_W \left(\frac{\rho_{\text{Solids}}(\text{kg}^{-1}\,\text{m}^3)}{1000} \right)$$

$$= 0.02\left(0.41\times10^4\right)0.1\left(\frac{2400}{1000}\right) = 19.68 \, \text{mol Pa}^{-1}\,\text{m}^{-3}$$

$$\text{Sediment solids: } Z_{\text{Sed}}^{\text{Solids}} = y_{OC} K_{OC} Z_W \left(\frac{\rho_{\text{Solids}}(\text{kg}^{-1}\,\text{m}^3)}{1000} \right)$$

$$= 0.04\left(0.41\times10^4\right)0.1\left(\frac{2400}{1000}\right) = 39.36 \, \text{mol Pa}^{-1}\text{m}^{-3}$$

With these values in hand, we can determine the Z-values for the bulk phases as volume fraction-weighted sums:

$$\text{Bulk soil: } Z_{\text{Soil}}^{\text{Bulk}} = v_{\text{Soil}}^{f-W} Z_W + v_{\text{Soil}}^{f-\text{Solids}} Z_{\text{Soil}}^{\text{Solids}}$$

$$= (0.377\times0.1+0.623\times19.68)\,\text{Pa mol}^{-1}\,\text{m}^{-3}$$

$$= 12.3\,\text{Pa mol}^{-1}\,\text{m}^{-3}$$

$$\text{Bulk sediment: } Z_{\text{Sed}}^{\text{Bulk}} = v_{\text{Sed}}^{f-W} Z_W + v_{\text{Sed}}^{f-\text{Solids}} Z_{\text{Sed}}^{\text{Solids}}$$

$$= (0.377\times0.1+0.623\times39.36)\,\text{Pa mol}^{-1}\,\text{m}^{-3}$$

$$= 24.6\,\text{Pa mol}^{-1}\,\text{m}^{-3}$$

Note that we arrive at the same Z-values, but in a more detailed manner in a Level III context, since we acknowledge the fact that the soil and sediment have water content, whereas in Levels I and II, it is treated as a single bulk material from the start.

ii. Calculate the total background and direct emission input rate in each compartment where relevant (mol h^{-1}):
Inflow rate for air: 10 mol h^{-1}
Inflow rate for water: 10 mol h^{-1}
Direct emission rate into air: 50 mol h^{-1}
Direct emission rate into water: 60 mol h^{-1}
Direct emission rate into soil: 70 mol h^{-1}
Direct emission rate into sediment: 20 mol h^{-1}
  The total input flow rate E is the sum of all inputs, here 220 mol h^{-1}.

iii. Determine and sum the D-values for all advection and reaction losses:
For advective loss, we have from Level II:

$$\text{Air: } D_A^{Adv} = 4.0 \times 10^3 \, \text{mol Pa}^{-1} \text{h}^{-1}$$

$$\text{Water: } D_W^{Adv} = 1 \times 10^2 \, \text{mol Pa}^{-1} \text{h}^{-1}$$

For reaction losses, we have similar results to Level II, except we have assumed a shorter half-life for atmospheric degradation of 693 h for this Level III example. Thus, again we have $D_i^{Deg} = V(\text{m}^3) Z(\text{mol Pa}^{-1} \text{m}^{-3}) \times \dfrac{0.693}{t_{1/2}(\text{h})}$ and therefore:

$$\text{Air: } D_A^{Deg} = 6 \times 10^9 \, \text{m}^3 \times 4.00 \times 10^{-4} \, \text{mol Pa}^{-1} \text{m}^{-3} \times (0.693 / 693 \text{h}^{-1}) = 2.40 \times 10^3 \, \text{mol Pa}^{-1} \text{h}^{-1}$$

$$\text{Water: } D_W^{Deg} = 7 \times 10^6 \, \text{m}^3 \times 0.1 \, \text{mol Pa}^{-1} \text{m}^{-3} \times (0.693 / 693 \text{h}^{-1}) = 7 \times 10^2 \, \text{mol Pa}^{-1} \text{h}^{-1}$$

$$\text{Soil: } D_{Soil}^{Deg} = 4.5 \times 10^4 \, \text{m}^3 \times 12.3 \, \text{mol Pa}^{-1} \text{m}^{-3} \times (0.693 / 69.3 \text{h}^{-1}) = 5.54 \times 10^3 \, \text{mol h}^{-1} \text{Pa}^{-1}$$

$$\text{Sediment: } D_{Sed}^{Deg} = 2.1 \times 10^4 \, \text{m}^3 \times 24.6 \, \text{mol Pa}^{-1} \text{m}^{-3} \times (0.693 / 6930 \text{h}^{-1}) = 51.7 \, \text{mol h}^{-1} \text{Pa}^{-1}$$

iv. Write out balanced mass-transfer equations for each compartment (input rates = output rates):

$$\text{Air: } E_A + G_A^{Adv} C_A^{Inf} + f_W D_{AW}^{Ov} + f_{Soil} D_{Soil-A}^{Ov} = f_A \left(D_{AW}^{Ov} + D_{Soil-A}^{Ov} + D_A^{Adv} + D_A^{Deg} \right)$$

$$50 \, \text{mol h}^{-1} + 10 \, \text{mol h}^{-1} + f_W \times 500 \, \text{mol Pa}^{-1} \text{h}^{-1} + f_{Soil} \times 150 \, \text{mol Pa}^{-1} \text{h}^{-1}$$

$$= f_A \times (500 + 150 + 4000 + 2400) \, \text{mol Pa}^{-1} \text{h}^{-1} = 7050 f_A \, \text{mol Pa}^{-1} \text{h}^{-1}$$

$$\text{Water: } E_W + G_W^{Adv} C_W^{Inf} + f_A D_{AW}^{Ov} + f_{Sed} D_{Sed-W}^{Ov}$$

$$= f_W \left(D_{AW}^{Ov} + D_{Sed-W}^{Ov} + D_W^{Adv} + D_W^{Deg} \right) = f_W D_W^{Tot}$$

$$60 \, \text{mol h}^{-1} + 10 \, \text{mol h}^{-1} + f_A \times 500 \, \text{mol Pa}^{-1} \text{h}^{-1} + f_{Sed} \times 3.50 \, \text{mol Pa}^{-1} \text{h}^{-1}$$

$$= f_W (500 + 3.50 + 100 + 700) \, \text{mol Pa}^{-1} \text{h}^{-1}$$

$$= f_W \times 1303.5 \, \text{mol Pa}^{-1}$$

$$\text{Soil: } E_{Soil} + f_A D_{Soil-A}^{Ov} = f_{Soil} \left(D_{Soil-A}^{Ov} + D_{Soil}^{Deg} \right)$$

$$70 \, \text{mol h}^{-1} + f_A \times 150 \, \text{mol Pa}^{-1} \text{h}^{-1} = f_{Soil} (150 + 5540) \, \text{mol Pa}^{-1} \text{h}^{-1}$$

$$= 5690 f_{Soil} \, \text{mol Pa}^{-1} \text{h}^{-1}$$

Sediment: $E_{\text{Sed}} + f_W D_{\text{Sed}-W}^{\text{Ov}} = f_{\text{Sed}} \left(D_{\text{Sed}-W}^{\text{Ov}} + D_{\text{Sed}}^{\text{Deg}} \right)$

$$20 \, \text{mol} \, \text{h}^{-1} + f_W \times 3.5 \, \text{mol} \, \text{Pa}^{-1} \, \text{h}^{-1} = f_{\text{Sed}} \left(3.50 + 51.7 \right) \text{mol} \, \text{Pa}^{-1} \, \text{h}^{-1}$$

$$= 55.2 f_{\text{Sed}} \, \text{mol} \, \text{Pa}^{-1} \, \text{h}^{-1}$$

v. Solve for the fugacity in each compartment:
Using the algebraic approach introduced in Chapter 4, we have the following:

Air: $E_A + G_A^{\text{Adv}} C_A^{\text{Inf}} + f_W D_{AW}^{\text{Ov}} + f_{\text{Soil}} D_{\text{Soil}-A}^{\text{Ov}} = f_A \left(D_{AW}^{\text{Ov}} + D_{\text{Soil}-A}^{\text{Ov}} + D_A^{\text{Deg}} + D_A^{\text{Adv}} \right) = f_A D_A^{\text{Tot}}$

$$60 \, \text{mol} \, \text{h}^{-1} + 500 f_W \, \text{mol} \, \text{Pa}^{-1} \, \text{h}^{-1} + 150 f_{\text{Soil}} \, \text{mol} \, \text{Pa}^{-1} \, \text{h}^{-1} = 7050 f_A \, \text{mol} \, \text{Pa}^{-1} \, \text{h}^{-1}$$

Water: $E_W + G_W^{\text{Adv}} C_W^{\text{Inf}} + f_A D_{AW}^{\text{Ov}} + f_{\text{Soil}} D_{\text{Soil}-W}^{\text{Ov}} + f_{\text{Sed}} D_{\text{Sed}-W}^{\text{Ov}} = f_W \left(D_{AW}^{\text{Ov}} + D_{\text{Sed}-W}^{\text{Ov}} + D_W^{\text{Deg}} + D_W^{\text{Adv}} \right) = f_W D_W^{\text{Tot}}$

$$70 \, \text{mol} \, \text{h}^{-1} + 500 f_A \, \text{mol} \, \text{Pa}^{-1} \, \text{h}^{-1} + 3.5 f_{\text{Sed}} \, \text{mol} \, \text{Pa}^{-1} \, \text{h}^{-1} = 1303.5 f_W \, \text{mol} \, \text{Pa}^{-1} \, \text{h}^{-1}$$

Soil: $E_{\text{Soil}} + f_A D_{\text{Soil}-A}^{\text{Ov}} = f_{\text{Soil}} \left(D_{\text{Soil}-A}^{\text{Ov}} + D_{\text{Soil}-W}^{\text{Ov}} + D_{\text{Soil}}^{\text{Deg}} \right) = f_{\text{Soil}} D_{\text{Soil}}^{\text{Tot}}$

$$70 \, \text{mol} \, \text{h}^{-1} + 150 f_A \, \text{mol} \, \text{Pa}^{-1} \, \text{h}^{-1} = 5690 f_{\text{Soil}} \, \text{mol} \, \text{Pa}^{-1} \, \text{h}^{-1}$$

Sediment: $E_{\text{Sed}} + f_W D_{\text{Sed}-W}^{\text{Ov}} = f_{\text{Sed}} \left(D_{\text{Sed}-W}^{\text{Ov}} + D_{\text{Sed}}^{\text{Deg}} + D_{\text{Sed}}^{\text{Adv}} \right) = f_{\text{Sed}} D_{\text{Sed}}^{\text{Tot}}$

$$20 \, \text{mol} \, \text{h}^{-1} + 3.5 f_W \, \text{mol} \, \text{Pa}^{-1} \, \text{h}^{-1} = 55.2 f_{\text{Sed}} \, \text{mol} \, \text{Pa}^{-1} \, \text{h}^{-1}$$

Using the same simplification definitions as before, we have

$$J_1 = \left(\frac{I_A}{D_A^{\text{Tot}}} + \frac{I_{\text{Soil}} D_{\text{Soil}-A}^{\text{Ov}}}{D_{\text{Soil}}^{\text{Tot}} D_A^{\text{Tot}}} \right) = \left(\frac{60}{7050} + \frac{70 \times 150}{5690 \times 7050} \right) = 8.772 \times 10^{-2}$$

$$J_2 = \left(\frac{D_{AW}^{\text{Ov}}}{D_A^{\text{Tot}}} \right) = \left(\frac{500}{7050} \right) = 7.092 \times 10^{-2}$$

$$J_3 = \left(1 - \frac{D_{\text{Soil}-A}^{\text{Ov}} D_{\text{Soil}-A}^{\text{Ov}}}{D_A^{\text{Tot}} D_{\text{Soil}}^{\text{Tot}}} \right) = \left(1 - \frac{150 \times 150}{7050 \times 5690} \right) = 9.994 \times 10^{-1}$$

$$J_4 = \left(D_{AW}^{\text{Ov}} + \frac{D_{\text{Soil}-W}^{\text{Ov}} D_{\text{Soil}-A}^{\text{Ov}}}{D_{\text{Soil}}^{\text{Tot}}} \right) = \left(500 + \frac{0 \times 150}{5690} \right) = 500$$

$$I_A = E_A + G_A^{\text{Adv}} C_A^{\text{Inf}} = 50 + 10 = 60$$

$$I_W = E_W + G_W^{\text{Adv}} C_W^{\text{Inf}} = 60 + 10 = 70$$

$$I_{\text{Soil}} = E_{\text{Soil}} + G_{\text{Soil}}^{\text{Adv}} C_{\text{Soil}}^{\text{Inf}} = 70$$

$$I_{\text{Sed}} = E_{\text{Sed}} + G_{\text{Sed}}^{\text{Adv}} C_{\text{Sed}}^{\text{Inf}} = 20$$

We can then solve for the water fugacity, according to the expression derived earlier:

$$f_W = \frac{\left(I_W + \left[\dfrac{J_1 J_4}{J_3}\right] + \left[\dfrac{I_{Soil} D_{Soil-W}^{Ov}}{D_{Soil}^{Tot}}\right] + \left[\dfrac{I_{Sed} D_{Sed-W}^{Ov}}{D_{Sed}^{Tot}}\right]\right)}{\left(D_W^{Tot} - \left[\dfrac{J_2 J_4}{J_3}\right] - \left[\dfrac{D_{Sed-W}^{Ov} D_{Sed-W}^{Ov}}{D_{Sed}^{Tot}}\right]\right)} = 5.97 \times 10^{-2} \, Pa$$

With the water fugacity in hand, we can determine the other remaining fugacities as follows:

$$f_A = f_W \left(\frac{J_2}{J_3}\right) + \left(\frac{J_1}{J_3}\right) = 1.30 \times 10^{-2} \, Pa$$

$$f_{Soil} = \frac{I_{Soil} + f_A D_{Soil-A}^{Ov}}{D_{Soil}^{Tot}} = 1.26 \times 10^{-2} \, Pa$$

$$f_{Sed} = \frac{I_{Sed} + f_W D_{Sed-W}^{Ov}}{D_{Sed}^{Tot}} = 3.66 \times 10^{-1} \, Pa$$

Note that the fugacities in the various compartments are no longer equal, due to the non-equilibrium nature of a Level III calculation. Here, the fugacity is greatest in the sediment, due mainly to the significant direct emission into this relatively low volume compartment.

vi. Calculate the concentrations in the various compartments as $C = Zf$:

Air: $C_A = Z_A f_A = 4.00 \times 10^{-4} \, mol \, Pa^{-1} \, m^{-3} \times 1.30 \times 10^{-2} \, Pa = 5.20 \times 10^{-6} \, mol \, m^{-3}$

Water: $C_W = Z_W f_W = 0.10 \, mol \, Pa^{-1} \, m^{-3} \times 5.97 \times 10^{-2} \, Pa = 5.97 \times 10^{-3} \, mol \, m^{-3}$

Soil: $C_{Soil} = Z_{Soil} f_{Soil} = 12.3 \, mol \, Pa^{-1} \, m^{-3} \times 1.26 \times 10^{-2} \, Pa = 1.55 \times 10^{-1} \, mol \, m^{-3}$

Sediment: $C_{Sed} = Z_{Sed} f_{Sed} = 24.6 \, mol \, Pa^{-1} \, m^{-3} \times 3.66 \times 10^{-1} \, Pa = 9.00 \, mol \, m^{-3}$

Note that the concentration is very large in the sediment, which is a reflection of the relatively high fugacity and fugacity capacity of the sediment, compared with the air or water. Such an outcome is also evident if we compute the partition ratios:

Air–water: $\log K_{AW} = \log\left(\dfrac{Z_A}{Z_W}\right) = \log\left(\dfrac{4.00 \times 10^{-4} \, mol \, Pa^{-1} \, m^{-3}}{1.00 \times 10^{-1} \, mol \, Pa^{-1} \, m^{-3}}\right) = \log 4.00 \times 10^{-3} = -2.40$

Soil–air: $\log K_{Soil-A} = \log\left(\dfrac{Z_{Soil}}{Z_A}\right) = \log\left(\dfrac{12.6 \, mol \, Pa^{-1} \, m^{-3}}{4.00 \times 10^{-4} \, mol \, Pa^{-1} \, m^{-3}}\right) = \log 3.15 \times 10^4 = 4.50$

Sediment–water: $\log K_{Sed-W} = \log\left(\dfrac{Z_{Sed}}{Z_W}\right) = \log\left(\dfrac{24.6 \, mol \, Pa^{-1} \, m^{-3}}{0.10 \, mol \, Pa^{-1} \, m^{-3}}\right) = \log 246 = 2.39$

"Eyeballing" these partition ratios immediately shows their predictive value, even without any further computation. The high values for soil–air and sediment–water indicate that it is the organic carbon-rich media that are most "attractive" to a relatively hydrophobic, low-vapor pressure chemical such as hypothene. Given enough time, it will concentrate in the soil and sediment.

For this calculation, the "Chemical Properties" input panel of the Level III spreadsheet program available from the CEMC website should be as follows:

Chemical Name	Hypothene
CAS	

Molar Mass (g/mol)	200
Data Temperature (°C)	27.5
Melting Point (°C)	0
Vapor Pressure (Pa)	1.00E+00
Solubility in Water (g/m³)	20
Henry's Law Constant (Pa·m³/mol)	10

Reaction Half-Lives (hours)

In Air (gaseous)	693.0
In Aerosol	1.0E+11
In Water (no susp. particles)	693.0
In Suspended Particles	1.0E+11
In Fish	1.0E+11
In Soil	69.3
In Sediment	6930.0

Partition Ratios

$\log K_{OW}$	4
K_{OC}	4100

The "Environmental Properties," which contains additional detailed mass-transport data needed for a full analysis, is given in the following figures:

Environment Name	Hypothene

Volume Fractions

Aerosol in Air	0.000
Susp. Particles in Water	0.000
Fish in Water	0.000
Air in Soil	0.000
Water in Soil	0.377
Solids in Soil	0.623
Water in Sediment	0.377
Solids in Sediment	0.623

Area (m²)

Air	3.00E+06
Water	3.50E+05
Soil	4.50E+05
Sediment	3.50E+05

Depth (m)

Air	2000
Water	20
Soil	0.1
Sediment	0.06

Densities for subcompartments (kg/m³)

Air in Air	1.175556
Aerosol in Air	2400
Water in Water	1000
Sup. Particles in Water	1000
Fish in Water	1000
Air in Soil	1.175556
Water in Soil	1000
Solid in Soil	2400
Water in Sediment	1000
Solid in Sediment	2400

Advective Flow Residence Times (h)

Air	600
Water	7000
Sediment (burial)	1.00E+11

Organic Carbon (g/g)

Susp. Particles	0.02
Fish Lipid	0.05
Soil	0.02
Sediment	0.04

Transport velocities (m/h)

Air-side air-water MTC	5.00
Water-side air-water MTC	0.0500
Rain rate	0.0000000
Aerosol dry deposition	0.0000000
Soil-air phase diffusion MTC	1.00
Soil-water phase diffusion MTC	0.0000100
Soil-air boundary layer MTC	5.0000000
Water-side sed.-water diffusion MTC	0.0001000
Sediment deposition	0.0000000
Sediment resuspension	0.0000000
Soil-water runoff rate	0.0000000
Soil-solids runoff rate	0.0000000

Scavenging ratio (unitless)	200000

The emission rate and advective inflow parameter data entry are shown in this figure:

Emission Rate

Into Air	10	kg/h
Into Water	12	kg/h
Into Soil	14	kg/h
Into Sediment	4	kg/h

Advective Inflow Concentrations

Concentration in Air	2.00E+05	ng/m^3
Concentration in Water	2.00E+06	ng/L

The results of the calculation as outputted by the Level III program are given in the following figures:

Mass Balance

	Emission Rate	
	kg/h	mol/h
Air	10.0	50.0
Water	12.0	60.0
Soil	14.0	70.0
Sediment	4.0	20.0
	Total:	200

	Inflow Rate		Concentration			
	kg/h	mol/h	ng/m³	ng/L	kg/m³	mol/m³
Air	2.0	10.0	2.000E+05	-	2.000E-07	1.000E-06
Water	2.0	10.0	-	2.000E+06	2.000E-03	1.000E-02
	Total:	20.0				
Total	44.0	220.0				

	kg	mol
Total Amount of Chemical in System	5.385E+04	2.693E+05

	Loss rate		D Value	Res. Time	
	kg/h	mol/h	m³/h	hours	days
Advection	1.160E+01	5.802E+01	8.201E+03	4640.85	193.37
Reaction	3.240E+01	1.620E+02	1.179E+04	1662.26	69.26
Overall	4.400E+01	2.200E+02	1.999E+04	1223.89	51.00

Fugacity

	Fugacity	VZ
	Pa	mol/Pa
Air	1.30E-02	2.400E+06
Water	5.97E-02	7.000E+05
Soil	1.27E-02	5.534E+05
Sediment	3.67E-01	5.157E+05
	Total:	4.169E+06
Fugacity Ratio	1.000E+00	

Phase Properties

	Z Value	Concentration			Quantity		
	mol/m³·Pa	mol/m³	g/m³	µg/g	mol	kg	%
Air							
Bulk	4.00E-04	5.205E-06	1.041E-03	8.856E-01	3.123E+04	6.246E+03	11.6
Air Vapour	4.00E-04	5.205E-06	1.041E-03	8.856E-01	3.123E+04	6.246E+03	11.6
Aerosol	2.40E+03	3.123E+01	6.246E+03	2.603E+03	0.000E+00	0.000E+00	0.0
Water							
Bulk	1.00E-01	5.968E-03	1.194E+00	1.194E+00	4.177E+04	8.355E+03	15.5
Water	1.00E-01	5.968E-03	1.194E+00	1.194E+00	4.177E+04	8.355E+03	15.5
Susp. Particles	8.20E+00	4.893E-01	9.787E+01	9.787E+01	0.000E+00	0.000E+00	0.0
Fish	5.00E+01	2.984E+00	5.968E+02	5.968E+02	0.000E+00	0.000E+00	0.0
Soil							
Bulk	1.23E+01	1.557E-01	3.113E+01	1.663E+01	7.005E+03	1.401E+03	2.6
Air	4.00E-04	5.064E-06	1.013E-03	8.615E-01	0.000E+00	0.000E+00	0.0
Water	1.00E-01	1.266E-03	2.532E-01	2.532E-01	2.147E+01	4.295E+00	0.0
Solids	1.97E+01	2.491E-01	4.982E+01	2.076E+01	6.984E+03	1.397E+03	2.6
Sediment							
Bulk	2.46E+01	9.012E+00	1.802E+03	9.627E+02	1.892E+05	3.785E+04	70.3
Water	1.00E-01	3.669E-02	7.339E+00	7.339E+00	2.905E+02	5.810E+01	0.1
Solids	3.94E+01	1.444E+01	2.889E+03	1.204E+03	1.890E+05	3.779E+04	70.2

Advection

	Res. Time		Flow Rate	D Value	Removal Rate		% of Total
	hours	days	m³/h	mol/Pa*h	kg/h	mol/h	Losses
Air							
Bulk	600.00	25.00	1.000E+07	4000.4	10.41015795	52.05079	23.7
Air Vapour	-	-	1.000E+07	4000.4	1.041E+01	5.205E+01	23.7
Aerosol	-	-	0.000E+00	0.0	0.000E+00	0.000E+00	0.0
Water							
Bulk	7000.00	291.67	1.000E+03	100.0	1.194E+00	5.968E+00	2.7
Water	-	-	1.000E+03	100.0	1.194E+00	5.968E+00	2.7
Susp. Particles	-	-	0.000E+00	0.0	0.000E+00	0.000E+00	0.0
Fish	-	-	0.000E+00	0.0	0.000E+00	0.000E+00	0.0
Soil							
Bulk	1.00E+11	4.17E+09	-	-	-	-	-
Sediment							
Bulk	1.00E+11	4.17E+09	2.100E-07	0.0	3.785E-07	1.892E-06	0.0

Total				4.100E+03	1.160E+01	5.802E+01	26.37

Reaction

	Half-Life		Rate Constant	D Value	Removal Rate		% of Total
	hours	days	1/h	mol/Pa*h	kg/h	mol/h	Losses
Air							
Bulk	-	-	-	2400.2	6.246E+00	3.123E+01	1.42E+01
Air Vapour	693.0	2.89E+01	1.000E-03	2400.2	6.246E+00	3.123E+01	1.42E+01
Aerosol	1.00E+11	4.17E+09	6.930E-12	0.0	0.000E+00	0.000E+00	0.00E+00
Water							
Bulk	-	-	-	700.0	8.355E+00	4.177E+01	1.90E+01
Water	693.00	28.88	1.000E-03	700.0	8.355E+00	4.177E+01	1.90E+01
Susp. Particles	1.00E+11	4.17E+09	6.930E-12	0.0	0.000E+00	0.000E+00	0.00E+00
Fish	1.00E+11	4.17E+09	6.930E-12	0.0	0.000E+00	0.000E+00	0.00E+00
Soil							
Bulk	69.30	2.89	1.000E-02	5534.3	1.401E+01	7.005E+01	3.18E+01
Sediment							
Bulk	6930.00	288.75	1.000E-04	51.6	3.785E+00	1.892E+01	8.60E+00

Total				8.686E+03	3.240E+01	1.620E+02	73.63

Intermedia Transport

	Half Times		Equiv. Flows	D Value	Transport Rate	
	hours	days	m³/h	mol/Pa*h	kg/h	mol/h
Air to water	3.33E+03	1.39E+02	1.250E+06	500.0	1.30	6.5061571
Air to soil	1.11E+04	4.62E+02	3.750E+05	150.0	0.39	1.95
Water to air	9.70E+02	4.04E+01	5.000E+03	500.0	5.97	29.84
Water to sediment	1.39E+05	5.78E+03	3.500E+01	3.5	0.04	0.21
Soil to air	2556.89	106.54	1.220E+01	150.0	0.38	1.90
Soil to water	0.00	0.00	0.000E+00	0.0	0.00E+00	0.00E+00
Sediment to water	102137.93	4255.75	1.425E-01	3.5	2.57E-01	1.28E+00

Process D Values

All D values are in units of mol/Pa*h

Air-water diffusion (air-side)	700.07
Air-water diffusion (water-side)	1750.00
Air-water diffusion (total)	500.04

Rain dissolution to water	0.00
Rain dissolution to soil	0.00

Aerosol deposition to water (dry)	0.00
Aerosol deposition to water (wet)	0.00
Aerosol deposition to water (total)	0.00

Aerosol deposition to soil (dry)	0.00
Aerosol deposition to soil (wet)	0.00
Aerosol deposition to soil (total)	0.00

Soil to water runoff (water)	0.00E+00
Soil to water runoff (solids)	0.00E+00

Soil-air diffusion (air-phase)	1.80E+02
Soil-air diffusion (water-phase)	4.50E-01
Soil-air diffusion (boundary layer)	9.00E+02
Soil-air diffusion (total)	150.03

Water-sediment diffusion	3.50

Water to sediment deposition	0.00E+00
Sediment to water resuspension	0.00E+00

Finally, these results are summarized in the following graphical representations:

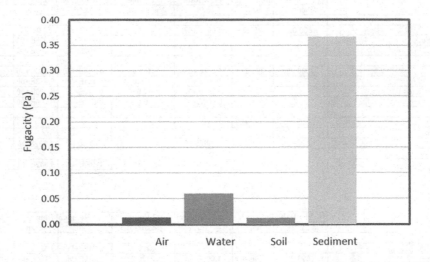

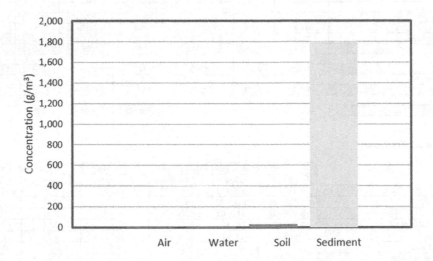

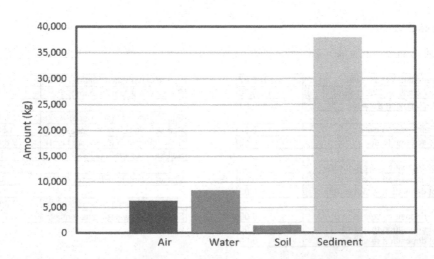

Relative Amounts (%)

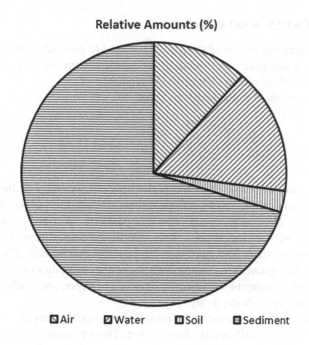

☒ Air ☒ Water ☐ Soil ☱ Sediment

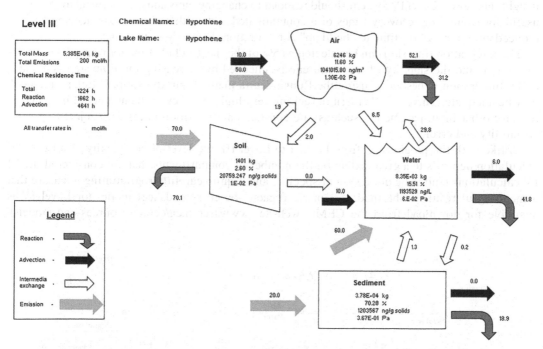

FIGURE 5.5 Schematic representation of the results of Worked Example 5.3 corresponding to computed output.

Figure 5.5 summarizes the various transport rates in mol h^{-1}.

Consideration of these figures clearly shows that the sediment is the main sink for hypothene. This result is due to a combination of a significant direct emission rate and low compartment volume, as well as the only modest diffusional transfer to water, the only compartment that is in direct contact with the sediment. Nearly all loss in the sediment is due to decomposition.

5.4 LEVEL IV MODELS AND CALCULATIONS

It is relatively straightforward to extend the Level III model to unsteady-state conditions. Instead of writing the steady-state mass-balance equations for each medium, we write differential equations reflecting the time-dependent rate of change in the fugacity.

In general, for compartment i, this takes the form:

$$\frac{df_i}{dt} = \left(\frac{1}{V_i Z_i}\right)\left[I_i + \sum_{j \neq i}^{n}\left(D_{ji} f_j\right) - D_i^{\mathrm{Tot}} f_i \right]$$

Here V_i is the compartment volume, Z_i is the bulk chemical Z-value, I_i is the chemical input rate (which may be a function of time), all for compartment "i." Each term of the form D_{ji} represents an intermedia input transfer D-value from any other compartment "j" to the compartment in question, and D_i^{Tot} is the total of D-values for output or loss specific to compartment "i". If an initial fugacity is defined for each medium, these four equations can be integrated numerically to give the fugacities as a function of time, thus quantifying the time response characteristics of the system. Figure 5.6 is a schematic diagram of a three-compartment system of air, water, and sediment, which illustrates these differential equations more explicitly.

It is noteworthy that the characteristic response time of a compartment is $\tau_i = V_i Z_i / D_i^{\mathrm{Tot}}$, which can be deduced from the Level III steady-state version. These characteristic times provide advance insight into how a Level IV system should respond to changing emissions. This calculation is most useful for estimating recovery times of a contaminated system that is now experiencing zero or reduced emissions, or the time to "build up" concentrations to steady-state values.

The integration time step can be selected as 5% of the shortest half-time for transport or transformation, and the stability of the result can be checked by increasing the time step systematically. Integration is best done using the Runge–Kutta method, but the simple "Euler's method" may be adequate. There is little merit in seeking very high accuracy in the numerical integration, because other input parameters such as rate constants and partition ratios are subject to greater variability and error.

Unlike with the progression from Level I to Level III, the level of complexity of a Level IV calculation depends to a great extent on the number of compartments that are considered. Level IV calculations tend to be quite system-specific, and require careful programming to ensure that a meaningful result is obtained. A generic fugacity-based spreadsheet model for Level IV is available for download from the CEMC website (www.trentu.ca/cemc/resources-and-models).

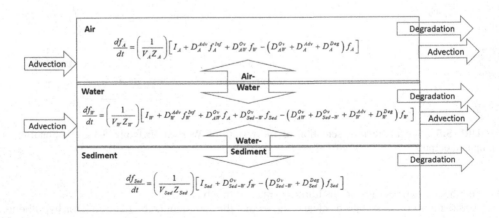

FIGURE 5.6 Schematic representation of a three-compartment Level IV system, with the corresponding differential equations for the rate of change in fugacity with time for each compartment.

LEVEL IV FLOWCHART

Define environment and chemical data

Determine D values for all relevant processes

Construct differential equations describing rate of change in fugacity with time for each phase

Define initial or boundary conditions

Determine minimum time step needed via shortest characteristic response time

Calculate change in fugacity over time step for each phase, and integrate over time until equilibrium.

Calculate concentrations in each phase from fugacity at any desired time

FIGURE 5.7   Flowchart for a generic Level IV fugacity-based calculation.

This spreadsheet allows for time-dependent distribution of a chemical between two compartments, which may be air–water, water–sediment, etc. As with the Level III program, the user must provide physical chemical (partitioning) properties, reaction half-lives, and sufficient information to deduce intermedia transport D-values.

The essence of a Level IV calculation is detailed in Worked Example 5.4 and the essential steps are outlined in Figure 5.7.

Worked Example 5.4: A simple Level IV calculation involving equilibration in a two-level system (air–water).

A fish tank with an air–water interface area of $0.15\,m^2$ and a depth of $0.3\,m$ is set up in an $18\,m^3$ room. The air in the room has been contaminated at time $t = 0$ with $10^{-6}\,mol\,m^{-3}$ of a chemical with a Henry's law constant $H = 1\,Pa\,m^3mol$. The water is initially pure. Calculate and plot the concentration of the chemical in the air and water as a function of time. Assume the room is closed so that no air can enter or leave, and the chemical does not leave the air by any other route than transfer into the water.
  Data needed:

  Air-side mass-transfer coefficient (MTC): $k_{AW}^{m-A} = 3.0\,m\,h^{-1}$
  Water-side MTC: $k_{AW}^{m-W} = 3.0\times10^{-2}\,m\,h^{-1}$

  i. First determine the fugacity capacities or Z-values:
     Air: $Z_A = 1/RT = 4.00 \times 10^{-4}\,mol\,Pa^{-1}m^{-3}$
     Water: $Z_W = 1/H = 1\,mol\,Pa^{-1}m^{-3}$

ii. Determine the necessary D-values for all processes:

In this case, we have only to consider intermedia diffusional transfer by the two-resistance model.

$$D_A^{\text{Diff}} = k_{AW}^{M-A} A_{AW} Z_A$$

$$= 3\,\text{m}\,\text{h}^{-1} \times 1.5 \times 10^{-1}\,\text{m}^2 \times 4.00 \times 10^{-4}\,\text{mol}\,\text{Pa}^{-1}\,\text{m}^{-3}$$

$$= 1.8 \times 10^{-4}\,\text{mol}\,\text{Pa}^{-1}\,\text{h}^{-1}$$

$$D_W^{\text{Diff}} = k_{AW}^{M-W} A_{AW} Z_W$$

$$= 3 \times 10^{-2}\,\text{m}\,\text{h}^{-1} \times 1.5 \times 10^{-1}\,\text{m}^2 \times 1.00\,\text{mol}\,\text{Pa}^{-1}\,\text{m}^{-3}$$

$$= 4.5 \times 10^{-3}\,\text{mol}\,\text{Pa}^{-1}\,\text{h}^{-1}$$

$$D_{AW}^{\text{Ov}} = \frac{1}{\left(\dfrac{1}{D_A^{\text{Diff}}} + \dfrac{1}{D_W^{\text{Diff}}}\right)} = \frac{1}{\left(\dfrac{1}{1.8 \times 10^{-4}} + \dfrac{1}{4.5 \times 10^{-3}}\right)} = 1.73 \times 10^{-4}\,\text{mol}\,\text{Pa}^{-1}\,\text{h}^{-1}$$

Recall that D-values are analogous to electrical conductivities and are inversely proportional to "resistances," in a manner that is directly analogous to electrical conductivity and resistance. Series resistances add, and thus series conductivities add reciprocally. It is useful to determine where most of the resistance lies, i.e., in the air or water. In this case, the smaller D-value and large resistance is D_A^{Diff} such that the "air-side" dominates the transfer rate, and D_{AW}^{Ov} is approximately equal to D_A^{Diff}.

iii. Write out differential equations reflecting the changes in fugacity with time:

Air: $\dfrac{df_A}{dt} = \left(\dfrac{1}{V_A Z_A}\right)\left[D_{AW}^{\text{Ov}}\left(f_W - f_A\right)\right]$

$$\frac{df_A}{dt} = \left(\frac{1}{18\,\text{m}^3 \times 4 \times 10^{-4}\,\text{mol}\,\text{Pa}^{-1}\,\text{m}^{-3}}\right)\left[1.73 \times 10^{-4}\,\text{mol}\,\text{Pa}^{-1}\,\text{h}^{-1}\left(f_W - f_A\right)\right]$$

Water: $\dfrac{df_W}{dt} = \left(\dfrac{1}{V_W Z_W}\right)\left[D_{AW}^{\text{Ov}}\left(f_A - f_W\right)\right]$

$$\frac{df_W}{dt} = \left(\frac{1}{4.5 \times 10^{-2}\,\text{m}^3 \times 0.1\,\text{mol}\,\text{Pa}^{-1}\,\text{m}^{-3}}\right)\left[1.73 \times 10^{-4}\,\text{mol}\,\text{Pa}^{-1}\,\text{h}^{-1}\left(f_A - f_W\right)\right]$$

iv. Determine the integration time step, starting with 5% of the smallest characteristic response time:

Characteristic time for air: $\tau_A = \dfrac{V_A Z_A}{D_{AW}^{\text{Ov}}} = \dfrac{18\,\text{m}^3 \times 4 \times 10^{-4}\,\text{mol}\,\text{Pa}^{-1}\,\text{m}^{-3}}{1.73 \times 10^{-4}\,\text{mol}\,\text{Pa}^{-1}\,\text{h}^{-1}} = 18.5\,\text{h}$

Characteristic time for water: $\tau_W = \dfrac{V_W Z_W}{D_{AW}^{\text{Ov}}} = \dfrac{4.5 \times 10^{-2}\,\text{m}^3 \times 1.0\,\text{mol}\,\text{Pa}^{-1}\,\text{m}^{-3}}{1.73 \times 10^{-4}\,\text{mol}\,\text{Pa}^{-1}\,\text{h}^{-1}} = 260\,\text{h}$

The shortest characteristic response time is associated with air, for which a time step Δt of $0.05 \times 18.5\,\text{h} = 0.93\,\text{h}$ is conservative, and a time step of 1 h is used in the following numerical integration.

v. Determine the initial conditions for the integration:

The initial fugacity in the air is $C_A/Z_A = 10^{-6}\,\text{mol}\,\text{m}^{-3}/4 \times 10^{-4}\,\text{mol}\,\text{Pa}^{-1}\,\text{m}^{-3} = 2.5 \times 10^{-3}$ Pa, and that of the water is initially 0.

vi. Use a spreadsheet to numerically integrate using the Euler method, by incrementally changing the fugacities in the air and water by the product of the instantaneous rate of change in the fugacities times the time step:

$$\Delta f_i = \Delta t \times \frac{df_i}{dt}$$

The first few rows of a spreadsheet calculation of this progression are illustrated in the following figure.

T /h	f(air)	f(water)	C(air)	C(water)
0	2.50E-03	0	1.00E-06	0
1	2.44E-03	9.61E-06	9.76E-07	9.61E-06
2	2.38E-03	1.90E-05	9.53E-07	1.90E-05
3	2.32E-03	2.80E-05	9.30E-07	2.80E-05
4	2.27E-03	3.69E-05	9.08E-07	3.69E-05
5	2.22E-03	4.55E-05	8.86E-07	4.55E-05...

One can see that as time progresses, the fugacity in the air drops while that of the water increases. After some time, we find that the fugacities in the air and water have nearly reached the same value, i.e., the two-level system is nearly at equilibrium or steady state:

400	3.45E-04	3.45E-04	1.38E-07	3.45E-04
401	3.45E-04	3.45E-04	1.38E-07	3.45E-04
402	3.45E-04	3.45E-04	1.38E-07	3.45E-04
403	3.45E-04	3.45E-04	1.38E-07	3.45E-04
404	3.45E-04	3.45E-04	1.38E-07	3.45E-04
405	3.45E-04	3.45E-04	1.38E-07	3.45E-04

Note that the concentrations are not equal near equilibrium, as is appropriate, but instead the ratio $C_A/C_W = K_{AW} = 4 \times 10^{-4}$ for this chemical, as required for the system at equilibrium, for which $K_{AW} = Z_A/Z_W = H/RT$.

The full progression plotted against time shows the familiar "approach to equilibrium" form (See Figure 5.8). Note that, while the concentrations have hugely different asymptotic limits (requiring a log scale to plot), the fugacities tend to the same "equifugacity" limit.

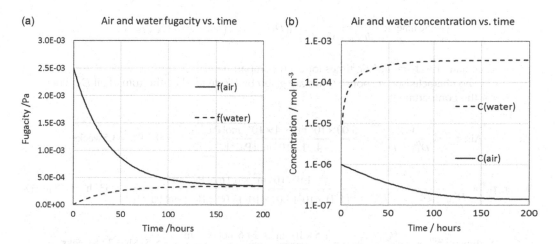

FIGURE 5.8 Plots of the time-dependence of fugacity (a) and concentration (b) for Worked Example 5.4.

Worked Example 5.5: A Level IV calculation involving steady release of a chemical into a three-level system (air–water–sediment).

Consider an environmental system similar to that used in Worked Example 5.3, but without soil. Such a scenario might be appropriate for a lake that receives no significant run-off from soils. Suppose that 1×10^4 mol h^{-1} of hypothene is released into the water starting at time $t = 0$. Assume for simplicity that all compartments are rapidly mixed relative to all other processes. Use a Level IV approach to make plots of the evolution of hypothene fugacity and concentration as a function of elapsed time. Assume that there is no other emission or background hypothene concentration in any of air, water, or sediment before the release. For simplicity, treat the system as being static from the point of view of advective flow, but that the following degradation D-values (mol Pa^{-1} h^{-1}): air: 2400, water: 1000, sediment: 369.

    i. Calculate or estimate the various Z-values, as we did in the Level I example:
       Air: $Z_A = 4.000 \times 10^{-4}$ mol Pa^{-1} m^{-3}
       Water: $Z_W = 0.10$ mol Pa^{-1} m^{-3}
       Sediment: $Z_{Sed} = 24.6$ mol Pa^{-1} m^{-3}
    ii. Determine and sum the D-values for all reaction losses:
       For reaction losses, we have:
       Air: $D_A^{Deg} = 2400$ mol Pa^{-1} h^{-1}
       Water: $D_W^{Deg} = 1000$ mol Pa^{-1} h^{-1}
       Sediment: $D_{Sed}^{Deg} = 369$ mol Pa^{-1} h^{-1}
       The sum of these reaction D-values is 3769 mol Pa^{-1}h^{-1}

    iii. Calculate overall D-values for intermedia transfer, as in Worked Example 5.3:
       Air–water: $D_{AW}^{Ov} = 1710$ mol Pa^{-1} h^{-1}
       Sediment–water: $D_{Sed-w}^{Ov} = 300$ mol Pa^{-1} h^{-1}
       Note that again we assume that the sediment–water diffusion is completely controlled by the water-side mass-transfer coefficient:
    iv. Write out differential equations for mass transfer into and out of each compartment:

$$\text{Air: } \frac{df_A}{dt} = \left(\frac{1}{V_A Z_A} \right) \left(D_{AW}^{Ov} f_W - \left(D_A^{Deg} + D_{AW}^{Ov} \right) f_A \right)$$

$$\text{Water: } \frac{df_W}{dt} = \left(\frac{1}{V_W Z_W} \right) \left(E_W + D_{AW}^{Ov} f_A + D_{Sed-w}^{Ov} f_{Sed} - \left(D_{AW}^{Ov} + D_{Sed-w}^{Ov} + D_W^{Deg} \right) f_W \right)$$

$$\text{Sediment: } \frac{df_{Sed}}{dt} = \left(\frac{1}{V_{Sed} Z_{Sed}} \right) \left(D_{Sed-w}^{Ov} f_W - \left(D_{Sed-w}^{Ov} + D_{Sed}^{Deg} \right) f_{Sed} \right)$$

    v. Calculate the characteristic times for each compartment.
       The characteristic response times are given by VZ divided by the sum of all D-values for the compartment:

$$\text{Air: } \tau_A^{CR} = \frac{V_A Z_A}{D_{AW}^{Ov} + D_A^{Deg}} = \frac{6.00 \times 10^9 \text{ m}^3 \times 4 \times 10^{-4} \text{ mol Pa}^{-1} \text{ m}^{-3}}{4.11 \times 10^3 \text{ mol Pa}^{-1} \text{ h}^{-1}} = 583.9 \text{ h} = 3.5 \text{ weeks}$$

$$\text{Water: } \tau_W^{CR} = \frac{V_W Z_W}{D_{AW}^{Ov} + D_{Sed-w}^{Ov} + D_W^{Deg}} = \frac{1.0 \times 10^7 \text{ m}^3 \times 0.1 \text{ mol Pa}^{-1} \text{ m}^{-3}}{\left(1.71 \times 10^3 \left(3.01 \times 10^3 \right) 3.00 \times 10^2 \right) \text{ mol Pa}^{-1} \text{ h}^{-1}} = 332 \text{ h} = 1.98 \text{ weeks}$$

$$\text{Sediment: } \tau_{Sed}^{CR} = \frac{V_{Sed} Z_{Sed}}{D_{Sed-w}^{Ov} + D_{Sed}^{Deg}} = \frac{1.5 \times 10^5 \text{ m}^3 \times 24.6 \text{ mol Pa}^{-1} \text{ m}^{-3}}{6.69 \times 10^2 \text{ mol Pa}^{-1} \text{ h}^{-1}} = 5,515 \text{ h} = 32.8 \text{ weeks}$$

vi. Perform a spreadsheet-based numerical integration using the Euler method, stepping the fugacity in each compartment by the product of the instantaneous rate of change in fugacity for the compartment times a time interval = 5% of the shortest response time (here $0.05 \times 332\,h = 16.6\,h$).

The first few rows of the corresponding spreadsheet are illustrated below, calculated using a time step of 16.8 h, which gives agreeable week divisions:

T /weeks	T /h	f(air)	f(water)	f(sed)	C(air)	C(water)	C(sed)
0	0.0	0.00E+00	0.00E+00	0.00E+00	0.00E+00	0.00E+00	0.00000E+00
0.1	16.8	0.00E+00	1.68E-01	0.00E+00	0.00E+00	1.68E-02	0.00E+00
0.2	33.6	2.02E-03	3.27E-01	2.29E-04	8.06E-07	3.27E-02	5.64E-03
0.3	50.4	5.89E-03	4.79E-01	6.76E-04	2.36E-06	4.79E-02	1.66E-02
0.4	67.2	1.15E-02	6.23E-01	1.33E-03	4.59E-06	6.23E-02	3.27E-02
0.5	84.0	1.86E-02	7.60E-01	2.17E-03	7.44E-06	7.60E-02	5.35E-02

The resulting plots are illustrated in Figures 5.9 and 5.10.

Worked Example 5.6: A Level IV calculation involving recovery of a three-level system (air–water–sediment).

Consider the same Level IV environmental system used in Worked Example 5.5 for which the near-steady-state concentrations in the three compartments were 7.70×10^{-4} mol m^{-3} in air, 4.62×10^{-1} mol m^{-3} in water, and 5.08×10^{1} mol m^{-3} in sediment, respectively. Suppose that all sources of chemical are cut off at this point in time. Use a Level IV approach to make plots of the evolution of hypothene fugacity and concentration as a function of elapsed time. Assume that there is no other emission or background hypothene concentration in any of air, water, or sediment before the release. For simplicity, again treat the system as being static from the point of view of advective flow, but that the same reaction degradation rates apply as in Worked Example 5.5. Assume for simplicity that all compartments are rapidly mixed relative to all other processes (Figures 5.9 and 5.10).

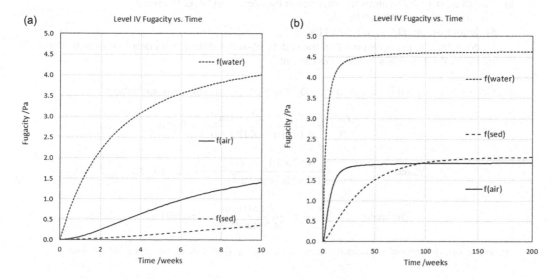

FIGURE 5.9 Plots of the time-dependence of fugacity for Worked Example 5.5 for a short (a) and long (b) time range.

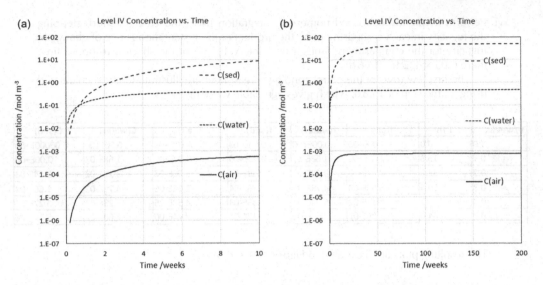

FIGURE 5.10 Plots of the time-dependence of concentration for Worked Example 5.5 for a short (a) and long (b) time range.

i. Calculate or estimate the various Z-values, as we did in the Level I example above:
   Air: $Z_A = 4.000 \times 10^{-4} \, \text{mol Pa}^{-1} \, \text{m}^{-3}$
   Water: $Z_W = 0.10 \, \text{mol Pa}^{-1} \, \text{m}^{-3}$
   Sediment: $Z_{\text{Sed}} = 24.6 \, \text{mol Pa}^{-1} \, \text{m}^{-3}$

ii. Determine and sum the D-values for all reaction losses:
   For reaction losses, we have from before:
   Air: $D_A^{\text{Deg}} = 2400 \, \text{mol Pa}^{-1} \, \text{h}^{-1}$
   Water: $D_W^{\text{Deg}} = 1000 \, \text{mol Pa}^{-1} \, \text{h}^{-1}$
   Sediment: $D_{\text{Sed}}^{\text{Deg}} = 369 \, \text{mol Pa}^{-1} \, \text{h}^{-1}$
   The sum of these reaction D-values is $3769 \, \text{mol Pa}^{-1} \text{h}^{-1}$.

iii. Calculate overall D-values for intermedia transfer, as in Level III above:
   Air–water: $D_{AW}^{\text{Ov}} = 1710 \, \text{mol Pa}^{-1} \, \text{h}^{-1}$
   Sediment–water: $D_{\text{Sed}-W}^{\text{Ov}} = 300 \, \text{mol Pa}^{-1} \, \text{h}^{-1}$
   Note that again we assume that the sediment–water diffusion is completely controlled by the water-side mass-transfer coefficient.

iv. Calculate the initial fugacities at "t = 0," the time when emissions are halted:

$$\text{Air: } f_A^{t=0} = \frac{C_A^{t=0}}{Z_A} = \frac{7.70 \times 10^{-4} \, \text{mol m}^{-3}}{4.0 \times 10^{-4} \, \text{mol Pa}^{-1} \, \text{m}^{-3}} = 1.92 \, \text{Pa}$$

$$\text{Water: } f_W^{t=0} = \frac{C_W^{t=0}}{Z_W} = \frac{4.62 \times 10^{-1} \, \text{mol m}^{-3}}{1.0 \times 10^{-1} \, \text{mol Pa}^{-1} \, \text{m}^{-3}} = 4.62 \times 10^{-1} \, \text{Pa}$$

$$\text{Sediment: } f_{\text{Sed}}^{t=0} = \frac{C_{\text{Sed}}^{t=0}}{Z_{\text{Sed}}} = \frac{50.8 \, \text{mol m}^{-3}}{24.6 \, \text{mol Pa}^{-1} \, \text{m}^{-3}} = 2.07 \, \text{Pa}$$

v. Write out differential equations for mass transfer into and out of each compartment:

$$\text{Air: } \frac{df_A}{dt} = \left(\frac{1}{V_A Z_A} \right) \left(D_{AW}^{\text{Ov}} f_W - \left(D_{AW}^{\text{Ov}} + D_A^{\text{Deg}} \right) f_A \right)$$

$$\text{Water: } \frac{df_W}{dt} = \left(\frac{1}{V_W Z_W}\right)\left(D_{AW}^{Ov} f_A + D_{Sed-W}^{Ov} f_{Sed} - \left(D_{AW}^{Ov} + D_{Sed-W}^{Ov} + D_W^{Deg}\right) f_W\right)$$

$$\text{Sediment: } \frac{df_{Sed}}{dt} = \left(\frac{1}{V_{Sed} Z_{Sed}}\right)\left(D_{Sed-W}^{Ov} f_W - \left(D_{Sed-W}^{Ov} + D_{Sed}^{Deg}\right) f_{Sed}\right)$$

vi. Calculate characteristic times for each compartment (from before).

$$\text{Air: } \tau_A = \frac{V_A Z_A}{\left(D_{AW}^{Ov} + D_A^{Deg}\right)} = 584\,\text{h} = 3.5\,\text{weeks}$$

$$\text{Water: } \tau_W = \frac{V_W Z_W}{\left(D_{AW}^{Ov} + D_{Sed-W}^{Ov} + D_W^{Deg}\right)} = 332\,\text{h} = 1.98\,\text{weeks}$$

$$\text{Sediment: } \tau_{Sed} = \frac{V_{Sed} Z_{Sed}}{D_{Sed-W}^{Ov} + D_{Sed}^{Deg}} = 5,515\,\text{h} = 32.8\,\text{weeks}$$

vii. Perform a spreadsheet-based numerical integration using the Euler method, stepping the fugacity in each compartment by the product of the instantaneous rate of change in fugacity for the compartment times a time interval = 5% of the shortest response time (here again 0.05 × 332 h = 16.6 h and we again use 16.8 h).
The first few rows of such a spreadsheet are shown here:

T /weeks	T /h	f(air)	f(water)	f(sed)	C(air)	C(water)	C(sed)
0	0.0	1.92E+00	4.62E+00	2.07E+00	7.70E-04	4.62E-01	5.08E+01
0.1	16.8	1.92376	4.44912	2.06528	7.70E-04	4.45E-01	5.08E+01
0.2	33.6	1.92175	4.28963	2.06507	7.69E-04	4.29E-01	5.08E+01
0.3	50.4	1.91788	4.13816	2.06464	7.67E-04	4.14E-01	5.08E+01
0.4	67.2	1.91230	3.99425	2.06400	7.65E-04	3.99E-01	5.08E+01
0.5	84.0	1.90516	3.85745	2.06317	7.62E-04	3.86E-01	5.08E+01
0.6	100.8	1.89658	3.72738	2.06216	7.59E-04	3.73E-01	5.07E+01

The resulting plots of fugacity versus time are illustrated in Figure 5.11.

Note that the response is quite fast for air and water, which relatively rapidly exchange the chemical until they "lock" into nearly equal fugacities that undergo a much slower degradation rate. In contrast, the sediment fugacity decreases very slowly over a much longer time. Extending the time of integration greatly, we can better see this slow approach to a common fugacity for all compartments. Viewed in terms of concentration, the fact that the system is striving to achieve a state of equifugacity is no longer evident (Figure 5.12).

Other more complicated Level IV calculations follow this same approach, with complexity increasing with the number of compartments, processes for advection, reaction, diffusion, and other contributors to the movement of the chemical through the system. However, the basic approach is the same.

5.5 THE EQC MODEL

Level I–III models were combined into a single "EQC" model by Mackay et al. (1996). This model is available from the CEMC website (www.trentu.ca/cemc/resources-and-models) and has the advantage that it leads the user through the series of levels in steps of increasing complexity, and thus, it is hoped, toward fidelity with respect to real environmental conditions.

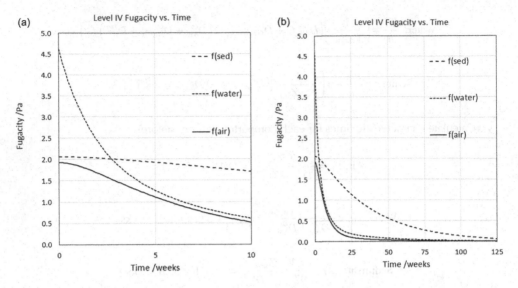

FIGURE 5.11 Plots of the time-dependence of fugacity for Worked Example 5.6 for a short (a) and long (b) time range.

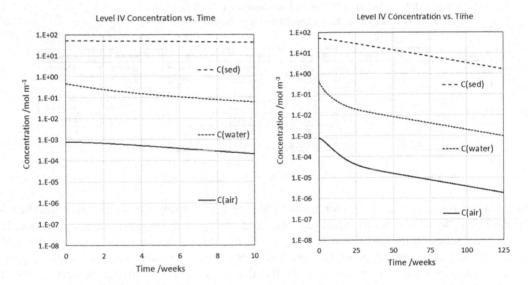

FIGURE 5.12 Plots of the time-dependence of concentration for Worked Example 5.6 for a short (a) and long (b) time range.

The figures below illustrate the final output diagrams from an EQC calculation for a typical scenario, from which one is able to note the increasing level of complexity as the program moves to higher levels:

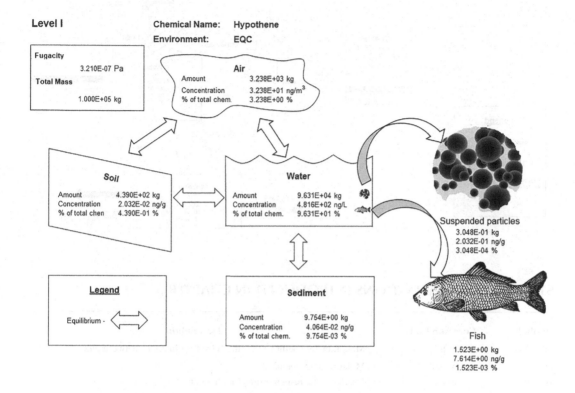

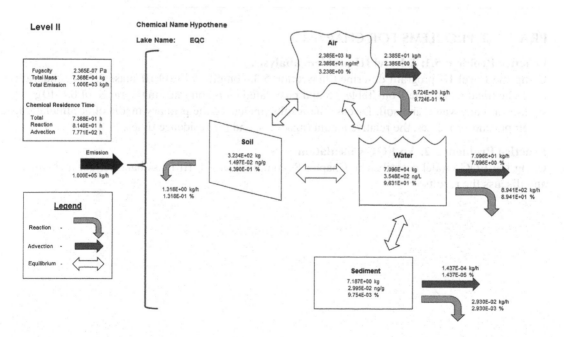

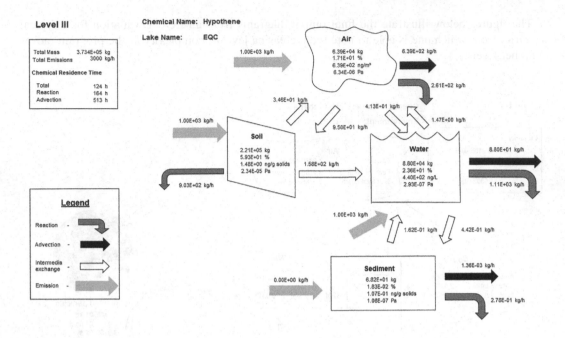

SYMBOLS AND DEFINITIONS INTRODUCED IN CHAPTER 5

Symbol	Common Units	Description
k_{ij}^{M-i}	m h^{-1}	Mass-transfer coefficient for the "i" side of interface between phases "i" and "j"
M^m	g mol^{-1}	Molar mass of chemical
τ_i^{CR}	mol h^{-1}	Characteristic response time for phase "i"

PRACTICE PROBLEMS FOR CHAPTER 5

Practice Problem 5.1: A Level III problem analysis.

Using the Level III program described in Section 5.3, compile a Level III mass-balance diagram for a chemical using data from Table 3.5 and postulated emission rates in the range of 0.1–10 g h^{-1} sq. km into air, water, and soil. Discuss the results, including the primary media of accumulation, the important processes, the relative media fugacities, and the residence times.

Practice Problem 5.2: An EQC Calculation

Using the EQC model described in Chapter 8, compile Level I–III mass balances for a chemical and discuss the results.

6 Specific Fugacity Models and Calculations

THE ESSENTIALS

$$\text{Sensitivity of a parameter: } S = \frac{\left(\dfrac{\Delta \text{Output}}{\text{Output}}\right)}{\left(\dfrac{\Delta \text{Input}}{\text{Input}}\right)}$$

Air-in-soil and water-in-soil diffusion D-values:

$$D_{\text{Soil}}^{\text{PoreAir}} = \frac{B_{EA} A Z_A}{Y} \quad \text{where} \quad B_{EA} = \frac{B_A v_A^{f(10/3)}}{\left(v_A^f + v_W^f\right)^2}$$

$$D_{\text{Soil}}^{\text{PoreWater}} = \frac{B_{EW} A Z_W}{Y} \quad \text{where} \quad B_{EA} = \frac{B_W v_W^{f(10/3)}}{\left(v_A^f + v_W^f\right)^2}$$

Conversions from mass fractions to volume fractions for a pure mixture of components A and B:

$$v_A^f = \left[\frac{1}{1 + \left(\dfrac{m_B^f/\rho_B}{m_A^f/\rho_A}\right)}\right]$$

Biota Z-values:

$$Z_{\text{Biota}} = K_{\text{Biota}-W} Z_W \simeq v_{\text{Biota}}^{f-\text{Lipid}} Z_O$$

Urban air film Z-values:

$$Z_{\text{Film}} = v_{OC}^f \left(K_{OA} Z_A\right)$$

$$Z_{\text{Film}}^Q = Z_A^Q = \left[10^{(\log K_{OA} + \log v_{OC}^f - 11.91)}\right] Z_A \rho_Q \times 10^9$$

6.1 SPECIFIC MODELS

6.1.1 INTRODUCTION

The scope of application of the strategies outlined above for Level I–IV calculations is very broad in terms of specific environmental fate questions. In this section, we provide overviews of several multimedia models that are illustrative of the variety of applications that exist. Most are available as spreadsheet implementations from the CEMC website (www.trentu.ca/cemc/resources-and-models).

Some of the most satisfying moments in environmental science come when a model is successfully fitted to experimental or observed data, and it becomes apparent that the important chemical transport and transformation processes are being represented with fidelity. Even more satisfying is the subsequent use of the model to predict chemical fate in as-yet-uninvestigated situations leading to gratifying and successful "validation." Failure of the model may be disappointing, but it is a positive demonstration that our fundamental understanding of environmental processes is flawed and further investigation is needed. For a review of the history of environmental mass-balance models, the reader is referred to Wania and Mackay (1999).

The simplest of such models is used to explore how a chemical migrates or is exchanged across the interface between two media, given the concentrations or fugacities in both. We refer to these as intermedia exchange models. No mass balance is necessarily sought—merely a knowledge of how fast, and by what mechanism, the chemical is migrating. Compartment volumes are not necessary, but they may be included for the purpose of calculating half-lives. An example is air–water exchange in which both concentrations are defined and the aim is to deduce in what direction and at what rate the chemical is moving. Often, it is not clear if a substance in a lake is experiencing net input or output as a result of exchange with the atmosphere. An important conclusion is that zero net flux does not necessarily correspond to equilibrium or equifugacity.

The simplest mass balance model is a one-compartment "box" that receives various defined inputs either as an emission term or as the product of a D-value and a fugacity from an adjoining compartment. The various D-values for output or loss processes are then calculated. The steady-state fugacity at which inputs and outputs are equal is then deduced. An unsteady-state version of the model can also be devised. Examples are a "box" of soil to which sludge or pesticide is applied, a one-compartment fish with input of chemical from respired water and food, and a mass balance for the water in a lake.

The complexity can be increased by adding more connected compartments. The Quantitative Water, Air, Sediment Interaction (*QWASI*) model includes mass balances in two compartments (water and sediment), the concentration in air being defined. A river, harbor, or estuary can be treated as a series of connected Eulerian *QWASI* boxes or using Lagrangian (follow a parcel of water as it flows) coordinates. A sewage treatment plant *(STP)* and a wastewater treatment plant *(WWTP)* model have been devised in which the compartments are the three principal vessels in the activated sludge process. This illustrates that the modeling concepts can also be applied to engineered systems. Indeed, such systems are often easier to model, because they are well defined in terms of volumes, flows, and other operating conditions such as temperature. Multi-compartment approaches can be applied to chemical fate in organisms ranging from plants to humans and whales, these being known as physiologically based pharmacokinetic (PBPK) or "toxicokinetic" models. Fairly complex models containing multiple compartments can also be assembled, an example being the *POPCYCLING–BALTIC* model of chemical fate in the Baltic region. The ultimate model is one of chemical fate in the entire global environment, *GloboPOP* (Wania et al., 1999)

6.1.2 MODEL-BUILDING STRATEGIES

The general model-building strategy is first to evaluate the system being simulated, then to decide how many compartments and thus mass balances are required. There is a compelling incentive to start with a simple model and then build up complexity only when justified. The volumes and bulk Z-values are then deduced for each compartment. All inputs and outputs are identified, and equations are written for each flux, either as an emission or a Df product. For a steady-state model, the inputs are equated to outputs for each of the n compartments, leading to n equations with n unknown fugacities. These equations are solved, either algebraically or using a matrix method.

For a dynamic system, the differential equations for each medium are written in the form

$$\frac{df_i}{dt} = \left(\frac{1}{V_i Z_i}\right)\left[\text{Inputs}(\text{mol h}^{-1}) - \text{Outputs}(\text{mol h}^{-1})\right]$$

These equations are then solved, either analytically or numerically, for a defined set of initial conditions and inputs, with the integration time step selected as 5% or less of the shortest half-time for transport or transformation, as noted above.

Results should be checked for a mass balance. For steady-state models, this is simply a comparison of inputs and outputs for each compartment. For dynamic models, the initial mass plus the cumulative inputs should equal the final mass and the cumulative outputs. To gain a pictorial appreciation of the results, a mass-balance diagram can be drawn listing all inputs and outputs beside the appropriate arrows. The dominant processes then become apparent.

It is often useful to play "sensitivity games" with the model to gain an appreciation of how variation in an input quantity, such as an emission rate, Z-value, or D-value, propagates through the calculation and affects the results. An input quantity can be increased by 1% and the effect on the desired output quantity determined. The best way to quantify this sensitivity S is to deduce

$$S = \frac{\left(\dfrac{\Delta\text{Output}}{\text{Output}}\right)}{\left(\dfrac{\Delta\text{Input}}{\text{Input}}\right)}$$

For example, if the input quantity of 100 is increased to 101 and the output quantity changes from 1000 to 1005, then S is 0.5:

$$S = \frac{\left(\dfrac{1005 - 1000}{1000}\right)}{\left(\dfrac{101 - 100}{100}\right)} = \frac{\left(\dfrac{5}{1000}\right)}{\left(\dfrac{1}{100}\right)} = \frac{(0.005)}{(0.01)} = 0.5$$

The sensitivity is actually a finite-difference estimate of the partial derivative of log output with respect to log input, and it is dimensionless. For linear systems of the types treated here, all values of S should be less than 1.0. It may be useful to list the input parameters, deduce S for each, and then rank them in order of decreasing S. The most sensitive parameters, for which the most accurate data are necessary, are then identified. Often, the sensitivity of a parameter is surprisingly small, and only a rough input estimate is needed. It may be desirable to revisit and improve the accuracy of the estimates of most sensitive parameters.

Another approach is to employ Monte Carlo analysis and run the model repeatedly, allowing the input data to vary between prescribed limits and deducing the variation in the output quantities. This gives an estimate of the likely variability in the output as a result of the anticipated uncertainty in all variables when they are acting in concert. However, the Monte Carlo approach does not necessarily reveal the individual quantitative sources of this variability.

The model results can then be compared with measured values to achieve a measure of validation. Complete validation is impossible, because chemicals or conditions can always be found for which the model fails. For a philosophical discussion of the feasibility of validation, the reader is referred to a review by Oreskes et al. (1994). A model can be useful, even if not validated, because it can give reliable results for a restricted set of conditions. The trick is to know when it is inappropriate to use the model, which is usually when one or more of the model's inherent assumptions are violated.

A final note on transparency: It is unethical for an environmental scientist to assert that a chemical experiences certain fate characteristics as a result of model calculations unless the full details of the calculations inherent in the model are made available. The scientific basis on which the conclusions are reached must be fully transparent. For various reasons, the modeler may elect to prevent the user from modifying the code, but the calculations themselves must be readable. For this reason, all model calculations described here are fully transparent, a criterion for good modeling practice (Buser et al., 2012).

To aid in the construction of models that are specific to environmental processes, a summary of Z- and D-values that occur commonly in environmental models is given in Tables 6.1 and 6.2.

Recall that Z-values for bulk phases comprised of more than one defined constituent are calculated as sums of volume fraction-weighted Z-values for each component:

$$Z_{\text{Bulk}} = \sum_{i=1}^{n} v_i^f Z_i$$

where v_i^f is the volume fraction of phase i, and Z_i is the corresponding Z-value of the chemical in question in that particular phase. For example, for bulk air containing aerosol,

$$Z_A^{\text{Bulk}} = v_A^f Z_A + v_Q^f Z_Q$$

For a solid phase containing organic carbon (OC) of mass fraction m_{OC}^f, in which sorption to mineral matter (MM) is negligible and the partitioning ratio with respect to water is

$$K_D = m_{OC}^f K_{OC} \left(\text{L kg}^{-1} \right), K_{\text{Solid}-W} = \frac{K_D \left(\rho (\text{kg}^{-1}\, \text{m}^3) \right)}{1000}, \text{ and } Z_{\text{Solid}}^{\text{Bulk}} = Z_W K_{\text{Solid}-W}$$

TABLE 6.1

Summary of Z-Value Equations

Phase	Z-Value	Comment
Air	$Z_A = \dfrac{1}{RT}$	$R = 8.314$ Pa m^3 mol^{-1} $T =$ Kelvin temperature
Water	$Z_W = \dfrac{1}{H} = \dfrac{Z_A}{K_{AW}}$	$H =$ Henry's law constant in Pa m^3 mol^{-1} $K_{AW} =$ Air–water partition ratio
Lipid	$Z_L = Z_O$	$Z_O =$ water-saturated n-octanol fugacity capacity
Aerosols	$Z_Q = K_{QA} Z_A$	$K_{QA} =$ aerosol–air partition ratio
Organic Carbon	$Z_{OC} = \dfrac{K_{OC} Z_W \rho_{OC}}{1000}$	$K_{OC} =$ OC–water partition ratio, usually estimated as $0.41 K_{OW}$. $\rho_{OC} =$ density of OC, usually about 1000 kg m^{-3}
Mineral Matter	$Z_{MM} = \dfrac{K_{MM} Z_W \rho_{MM}}{1000}$	$K_{MM} =$ MM–water partition ratio. $\rho_{MM} =$ density of MM in kg m^{-3}
Biota	$Z_{\text{Biota}} = K_{\text{Biota}-W} Z_W \simeq v_{\text{Biota}}^{f-\text{Lipid}} Z_O$	$v_{\text{Biota}}^{f-\text{Lipid}} =$ Lipid volume fraction
Heterogeneous fluid phase	$Z_{\text{Bulk}} = \sum_{i=1}^{n} v_i^f Z_i$	v_i^f is the volume fraction of constituent phase "i," and Z_i is its corresponding Z-value
Heterogeneous solid phase	$Z_{\text{Solid}}^{\text{Bulk}} = Z_W K_{\text{Solid}-W}$	$K_{\text{Solid}-W} = K_D \left(\dfrac{\rho}{1000} \right) = y_{OC} K_{OC} \left(\dfrac{\rho}{1000} \right) =$ partition ratio between solid phase and water

TABLE 6.2
Summary of D-Value Equations

Process	D-Value	Comment
Advection or flow	$D_i = G_i Z_i = U_i A Z_i$	G_i = medium flow rate (m³ h⁻¹)
		$G = U$ (flow velocity m h⁻¹) × A (flow cross-sectional area m²)
Decomposition or reaction	$D = kVZ$	V = volume (m³)
		k = rate constant (h⁻¹)
		B = molecular or effective diffusivity (m² h⁻¹)
Diffusion	$D = \dfrac{BAZ}{Y}$	A = interfacial contact area (m²)
		Y = path length (m)
		dV/dt = growth rate (m³ h⁻¹)
Growth dilution	$D = Z\left(\dfrac{dV}{dt}\right) = kVZ$	$k = (dV/dt)/V$ is the growth rate constant (h⁻¹)

As noted above, K_{OC} can be estimated as $0.41K_{OW}$ (Karickhoff, 1981) or as $0.35K_{OW}$ plus or minus a factor of 2.5 (Seth et al., 1999). Note that K_{OM} is typically $0.56K_{OC}$, i.e., organic matter (OM) is typically about 56% OC (Sposito, 1989). K_{QA} can be estimated as $6 \times 10^6/P^{Sat}$, where P^{Sat} is the liquid vapor pressure in Pa, or from other correlations using vapor pressure or K_{OA} as described above.

6.2 AIR–WATER EXCHANGE: THE *AIRWATER* MODEL

Air–water exchange process calculations are useful when estimating chemical loss from treatment lagoons, ponds, and lakes; for estimating deposition rates of atmospheric contaminants; and for interpreting observed air and water concentrations to establish the direction and rate of transfer. The complexity of the several processes and the widely varying physico-chemical properties of chemicals of interest lead to situations in which chemical behavior is not necessarily intuitively obvious. The simple model derived here provides a rational method of estimating exchange characteristics and exploring the sensitivity of the results to assumed values of the various chemical and environmental parameters. Bidleman (1988) gives an excellent review of atmospheric processes treated by the model.

An elegant application of the fugacity concept to elucidating chemical exchange in the air–water system is that of Jantunen and Bidleman (1996). Samples of air and surface water from the Bering and Chukchi Seas (between Alaska and Russia) were analyzed for α-hexachlorocyclohexane (α-HCH) over a multiyear period, and the ratios of air to water fugacities were deduced using the Henry's law constant for seawater at the appropriate temperature. Initially, in the mid-1980s, this ratio was greater than 1, indicating that α-HCH was being absorbed by the ocean. This is consistent with the source of α-HCH being evaporation of technical grade lindane following its application in South East Asia, India, and China, with subsequent atmospheric transport. Later, in the mid-1990s, after the use of technical grade lindane was greatly reduced, the fugacity ratio became less than 1 (because of the drop in air fugacity), and net volatilization of α-HCH started. Essentially, the ocean acted first as a "sponge," absorbing α-HCH, then it desorbed the α-HCH in response to changes in the concentration in air. Interpretation of data using the fugacity ratio illustrated this clearly. It is an example to be followed in cases where there is doubt about the direction of net transport of chemicals between air and water.

The situation treated here, and the resulting *AirWater* model, are largely based on the study of air–water exchange by Mackay et al. (1986), as is depicted in Figure 6.1.

The basic concept is that, if one knows the bulk air and bulk water concentrations of a chemical, the net direction of chemical movement and its relative apportioning to different modes of

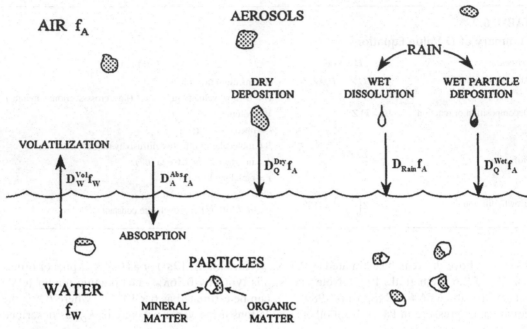

FIGURE 6.1 Air–water exchange processes.

transfer may be calculated in a "snapshot-in-time" manner. Since the model only focuses on instantaneous rates of transfer between water and air, the volumes of the two compartments are not needed. The transfer rates can be compared on a flux or "rate per 1 m² of air–water interface contact area" basis, as is illustrated below. The model assumes static, well-mixed phases of air containing aerosols and water containing suspended particles. The transport processes between air and water that are included are deposition by rain, wet and dry particle deposition, and air–water diffusional exchange. These are mediated by aerosol–air, rain–air, and suspended particle–water diffusional transfer. The overall result is a measure of the relative rates of the various contributors to an overall rate of air–water transfer.

Worked Example 6.1: Model AirWater model calculation.

We will carry out a typical *AirWater* Model calculation using a model chemical "AirWater86" and the model environment "EnvAirWater86," for which the following tabulated data are needed. These values are the same as those given in the illustrative example in Mackay et al. (1986), and are included in the *AirWater* spreadsheet program database:

"AirWater86" and "EnvAirWater86" Data Required for *AirWater* Calculations

Chemical and physical properties $T = 15°C$

Molar mass $M^m = 350$ g mol^{-1}

Vapor pressure $P_A^{Sat} = 5 \times 10^{-3}$ Pa

Water solubility $C_W^{Sat} = 3.5 \times 10^{-2}$ g m^{-3}

$H = P_A^{Sat}/C_W^{Sat} = 50$ Pa m^3 mol^{-1}

log $K_{OW} = 6.0$

Melting point = 5°C

(Continued)

	"AirWater86" and "EnvAirWater86" Data Required for *AirWater* Calculations
Air phase	Air–water surface interface area = 1 m²
	Height = 1000 m
	Density = 1.227 kg m⁻³
	Bulk concentration = 6 ng m⁻³(1.71 × 10⁻¹¹ mol m⁻³)
	Mass-transfer coefficient (air side) = 18 m h⁻¹
Aerosol particles in air	Density = 2000 kg m⁻³
	Concentration = 100 µg m⁻³
	Volume fraction = 5 × 10⁻¹¹
	Aerosol dry deposition velocity = 0.3 cm s⁻¹
	Wet aerosol scavenging coefficient or ratio = 200,000
Rain in air	Rain rate = 0.5 m year⁻¹
Water phase	Average depth = 10 m
	Density = 1000 (water, organic) or 2500 (mineral) kg m⁻³
	Bulk concentration = 1 ng L⁻¹ (2.86 × 10⁻¹² mol m⁻³)
	Mass-transfer coefficient (water side) = 7.2 × 10⁻² m h⁻¹
Suspended solids in water	Density = 1500 kg m⁻³
	Concentration = 12 g m⁻³
	Volume fraction = 8 × 10⁻⁶
	Suspended matter OC mass fraction = 0.05

The water phase area and depth (and hence volume) are defined, it being assumed that the water is well mixed. The water contains suspended particulate matter, to which the chemical can be sorbed, and which may contain mineral and organic material. The concentration (mg L⁻¹ or g m⁻³) of suspended matter is defined, as is its mass-based organic carbon (OC) content (mass OC per mass dry particulates). By assuming the OC content of organic matter (OM) to be 56% by mass, the masses of mineral matter (MM) and OM can be deduced. Densities of 1000, 1000, and 2500 kg m⁻³ are assumed for water, OM, and MM, respectively, thus enabling the volumes and volume fractions to be deduced.

The air phase is treated similarly, having the same area as the water and a defined (possibly arbitrary) height and containing a specified concentration (ng m⁻³) of aerosols or atmospheric particulates to which the chemical may sorb. By assuming an aerosol density of 1500 kg m⁻³, the volume fraction of aerosols can be deduced. No information on aerosol composition, size distribution, or area is sought or used. If the concentration of aerosols or total suspended particulates is C_A^{TSP} ng m⁻³, this corresponds to $10^{-12} C_A^{TSP}$ kg m⁻³ and to a volume fraction v_Q^f of $10^{-12} C_A^{TSP}/\rho_Q$, where ρ_Q is the aerosol density (1500 kg m⁻³). Thus, a typical C_A^{TSP} of 30,000 ng m⁻³ or 30 mg m⁻³ is equivalent to a volume fraction of $v_Q^f = 20 \times 10^{-12}$.

The steps in the calculation are closely related to a Level III calculation, since the system is assumed to be at steady state but not necessarily at equilibrium. The volume fractions are calculated based on the logic outlined above for aerosols and suspended particles. The volume fractions of air and water are then determined by difference:

i. Calculate volume fractions:

Aerosol in air	$v_Q^f = \dfrac{C_A^{TSP}}{\rho_Q} = \dfrac{100\,\mu\text{g m}^{-3} \times 10^{-9}\,\text{kg }\mu\text{g}^{-1}}{2000\,\text{kg m}^{-3}} = 5 \times 10^{-11}$
Vapor in air	$v_A^f = 1 - v_Q^f \approx 1$
Total suspended particles in water	$v_{TSP}^f = \dfrac{C_W^{TSP}}{\rho_{TSP}} = \dfrac{12\,\text{g m}^{-3} \times 1 \times 10^{-3}\,\text{kg g}^{-3}}{1500\,\text{kg m}^{-3}} = 8.0 \times 10^{-6}$
Liquid in water	$v_W^f = 1 - v_{TSP}^f \approx 1$

The Z-values for the chemical in all phases present are then calculated based on the expressions given in Table 6.1. The Z-value for water may be conveniently computed from either the reciprocal of the Henry's law constant, or from the ratio of the chemical's saturation vapor pressure to the liquid or sub-cooled-liquid solubility.

ii. Calculate Z-values for all constituents of all phases:

Constituent	Fugacity Capacity (Z-value)(mol Pa^{-1} m^{-3})
Air	$Z_A = \dfrac{1}{RT} = 4.17 \times 10^{-4}$
Aerosol	$Z_Q = K_{QA} Z_A \simeq \left(6 \times 10^6 / P^{Sat}\right) Z_A = \left(6 \times 10^6 / 5 \times 10^{-3}\right) \times 4.17 \times 10^{-4} = 5.01 \times 10^5$
Water	$Z_W = \dfrac{1}{H} = \dfrac{C_W^{Sat}}{P^{Sat}} = \dfrac{3.5 \times 10^{-2}\,\mathrm{g\,m^{-3}} / 350\,\mathrm{g\,mol^{-1}}}{5 \times 10^{-3}\,\mathrm{Pa}} = 2.00 \times 10^{-2}$
Suspended solids	$Z_{TSP} = \left(0.41 \times K_{OW}\right) Z_W \left(\dfrac{\rho_{TSP}}{1000}\right) v_{TSP}^{f-OC} = 0.41 \times 10^6 \times 2.0 \times 10^{-2} \times 1.5 \times 0.5 = 615$

Effective Z-values for the bulk water and air phases are then calculated. Recall from the discussion of phases at equilibrium that Z-values are additive as volume-weighted fractions. Total or bulk mass concentrations in the water and air phases are converted to molar concentrations and then divided by the bulk Z-values to give the prevailing water and air fugacities.

iii. Calculate effective Z-values and fugacities in bulk air and water:

Effective Z-value for air phase	$Z_A^{Bulk} = v_A^f Z_A + v_Q^f Z_Q$
	$= 1 \times 4.17 \times 10^{-4} + 5.01 \times 10^5 \times 5 \times 10^{-11}$
	$= 4.42 \times 10^{-4}\,\mathrm{mol\,Pa^{-1}\,m^{-3}}$
Effective Z-value for the water phase	$Z_W^{Bulk} = v_W^f Z_W + v_{TSP}^f Z_{TSP}$
	$= 1 \times 2.0 \times 10^{-2} + 615 \times 8.0 \times 10^{-6}$
	$= 2.49 \times 10^{-2}\,\mathrm{mol\,Pa^{-1}\,m^{-3}}$
Fugacity of bulk air phase	$f_A = \dfrac{C_A^{Tot}}{Z_A^{Bulk}} = \dfrac{6\,\mathrm{ng\,m^{-3}} \times 10^{-9}\,\mathrm{g\,ng^{-1}} / 350\,\mathrm{g\,mol^{-1}}}{4.42 \times 10^{-4}\,\mathrm{mol\,Pa^{-1}\,m^{-3}}} = 3.88 \times 10^{-8}\,\mathrm{Pa}$
Fugacity of bulk water phase	$f_W = \dfrac{C_W^{Tot}}{Z_W^{Bulk}} = \dfrac{2.86 \times 10^{-9}\,\mathrm{mol\,m^{-3}}}{2.49 \times 10^{-2}\,\mathrm{mol\,Pa^{-1}\,m^{-3}}} = 1.15 \times 10^{-7}\,\mathrm{Pa}$

The D-values for all transport processes are calculated using the relations from Table 6.2, in conjunction with the application of the Whitman two-resistance approach to determining overall intermedia-diffusion values, where appropriate.

Four processes are considered as shown in Figure 6.1:
1. diffusive exchange by volatilization and the reverse absorption,
2. dry deposition of aerosols,
3. wet dissolution of chemical in rain, and
4. wet deposition of aerosols.

In each case, the D-value (mol Pa^{-1} h^{-1}) is used to characterize the rate, which is Df mol h^{-1}.

For diffusion, the two-resistance approach is used, and D-values are deduced for the air and water boundary layers,

$$D_A^{\text{Diff}} = k_A^M A Z_A$$

$$D_W^{\text{Diff}} = k_W^M A Z_W$$

where k_A^M and k_W^M are mass-transfer coefficients with units of m h^{-1} and A is the area (m^2). Illustrative values of 5 m h^{-1} for k_A^M and 5 cm h^{-1} for k_W^M can be used, but it should be appreciated that environmental values can vary widely, especially with wind speed, and a separate calculation may be needed for the situation being simulated.

The overall resistance is obtained by adding the series resistances as

$$\frac{1}{D_{AW}^{\text{Ov}}} = \left(\frac{1}{D_A^{\text{Diff}}} + \frac{1}{D_W^{\text{Diff}}} \right)$$

The rate of vaporization is then $D_{AW}^{\text{Ov}} f_W$, the rate of absorption is $D_{AW}^{\text{Ov}} f_A$, and the net rate of vaporization is $D_{AW}^{\text{Ov}} (f_W - f_A)$. An overall mass-transfer coefficient may also be calculated.

For dry deposition, a dry deposition velocity U_{Dry} of particles is used, a typical value being 0.3 cm s^{-1} or 10 m h^{-1}. The total dry deposition rate is thus $U_{Dry} v_Q^f A$ m^3 h^{-1}, the corresponding D-value D_Q^{Dry} is $U_{Dry} v_Q^f A Z_Q$, and the rate is $D_Q^{\text{Dry}} f_A$ mol h^{-1}.

For wet dissolution, a rain rate is defined, usually in units of m year^{-1}, a typical value being 0.5 m year^{-1} or 6×10^{-5} m h^{-1}, designated U_{Rain}. The total rain rate is then $U_{Rain} A$ (m^3 h^{-1}), the D-value, D_{Rain}, is $U_{Rain} A Z_W$, and the rate is $D_{Rain} f_A$ mol h^{-1}.

For wet aerosol deposition, a scavenging ratio Q is used, representing the volume of air efficiently scavenged by rain of its aerosol content, per unit volume of rain. A typical value of Q is 200,000. The volume of air scavenged per hour is thus $U_{Rain} A Q$ (m^3 h^{-1}), which will contain $U_{Rain} A Q v_Q^f$ (m^3 h^{-1}) of aerosol (v_Q^f is the volume fraction of aerosol). The D-value D_Q^{Wet} is thus $U_{Rain} A Q v_Q^f Z_Q$, and the rate is $D_Q^{\text{Wet}} f_A$ (mol h^{-1}).

A washout ratio is often employed in such calculations. This is the dimensionless ratio of concentration in rain to total concentration in air, usually on a volumetric (g m^{-3} rain per g m^{-3} air) basis, but occasionally on a gravimetric (mg kg^{-1} per mg kg^{-1}) basis. The total rate of chemical deposition in rain is $(D_{Rain} + D_Q^{\text{Wet}}) f_A$; thus, the concentration in the rain is $(D_{Rain} + D_Q^{\text{Wet}}) f_A / U_{Rain} A$ or $f_A (Z_W + Q v_Q^f Z_Q)$ mol m^{-3}. The total air concentration is $f_A (Z_A + v_Q^f Z_Q)$, and therefore, the volumetric washout ratio is $(Z_W + Q v_Q^f Z_Q)/(Z_A + v_Q^f Z_Q)$. The gravimetric ratio is smaller by the ratio of air to water densities, i.e., approximately 1.2/1000. If the chemical is almost entirely aerosol associated, as is the case with metals such as lead, the volumetric washout ratio approaches Q. These washout ratios are calculated and can be compared with reported values.

The total rates of transfer are thus

Water to air:

$$r_{WA}^{\text{Tot}} = f_W D_{AW}^{\text{Ov}} \text{ mol h}^{-1}$$

Air to water:

$$r_{AW}^{\text{Tot}} = f_W \left(D_{AW}^{\text{Ov}} + D_Q^{\text{Dry}} + D_{Rain} + D_Q^{\text{Wet}} \right) = f_A D_{AW}^{\text{Tot}} \text{ mol h}^{-1}$$

iv. Calculate D-values for all transport processes:

Air–water volatilization /absorption

$$D_{AW}^{Ov} = 1 \Big/ \left(\frac{1}{D_W^{Diff}} + \frac{1}{D_A^{Diff}} \right) 1 \Big/ \left(\frac{1}{k_W^M A_{AW} Z_W} + \frac{1}{k_A^M A_{AW} Z_A} \right)$$

$$= 1 \Big/ \left(\frac{1}{\dfrac{1}{7.2 \times 10^{-2} \, \mathrm{m\,h^{-1}} \times 1 \mathrm{m^2} \times 2.0 \times 10^{-2} \, \mathrm{mol\,Pa^{-1}\,m^{-3}}} + \dfrac{1}{18 \, \mathrm{m\,h^{-1}} \times 1 \mathrm{m^2} \times 4.17 \times 10^{-4} \, \mathrm{mol\,Pa^{-1}\,m^{-3}}}} \right)$$

$$= 1 \Big/ \left(\frac{1}{1.44 \times 10^{-3}} + \frac{1}{7.51 \times 10^{-3}} \right) = 1 \Big/ \left(6.94 \times 10^2 + 1.33 \times 10^2 \right)$$

$$= 1.21 \times 10^{-3} \, \mathrm{mol\,Pa^{-1}\,h^{-1}}$$

Dry aerosol deposition

$$U_Q^{Dry} = 0.3 \, \mathrm{cm\,s^{-1}} = 0.3 \, \mathrm{cm\,s^{-1}} \times 1 \times 10^{-2} \, \mathrm{m\,cm^{-1}} \times 3600 \, \mathrm{s\,h^{-1}} = 10.8 \, \mathrm{m\,h^{-1}}$$

$$D_Q^{Dry} = G_Q^{Dry} Z_Q$$

$$= A_{AW} U_Q^{Dry} v_Q^f Z_Q$$

$$= 1 \mathrm{m^2} \times 10.8 \, \mathrm{m\,h^{-1}} \times 5 \times 10^{-11} \times 5.01 \times 10^5 \, \mathrm{mol\,Pa^{-1}\,m^{-3}}$$

$$= 2.70 \times 10^{-4} \, \mathrm{mol\,Pa^{-1}\,h^{-1}}$$

Rain dissolution

$$U_{Rain} = \left(0.5 \, \mathrm{m\,year^{-1}} / 8760 \, \mathrm{h\,year^{-1}} \right) = 5.71 \times 10^{-5} \, \mathrm{m\,h^{-1}}$$

$$D_{Rain} = U_{Rain} A_{AW} Z_W$$

$$= 5.71 \times 10^{-5} \, \mathrm{m\,h^{-1}} \times 1 \mathrm{m^2} \times 2 \times 10^{-2} \, \mathrm{mol\,m^{-3}\,Pa^{-1}}$$

$$= 1.14 \times 10^{-6} \, \mathrm{mol\,Pa^{-1}\,h^{-1}}$$

Wet deposition

$$D_Q^{Wet} = G_Q^{Wet} Z_Q = A_{AW} U_{Rain} Q v_Q^f Z_Q$$

$$= 1 \mathrm{m^2} \times 5.71 \times 10^{-5} \, \mathrm{m\,h^{-1}} \times 2 \times 10^5 \times 5 \times 10^{-11} \times 5.01 \times 10^5 \, \mathrm{mol\,Pa^{-1}\,m^{-3}}$$

$$= 2.86 \times 10^{-4} \, \mathrm{mol\,Pa^{-1}\,h^{-1}}$$

The individual unidirectional rates for each transfer process are calculated as the product of the appropriate D-value times the fugacity of the source medium. For intermedia diffusive transport (dissolution or volatilization), the overall D-value is used.

v. Calculate (Df) rates of transfer for all processes:

Air–water dissolution

$$r_{AW}^{Dis} = D_{AW}^{Ov} f_A = 1.21 \times 10^{-3} \, \mathrm{mol\,Pa^{-1}\,h^{-1}} \times 3.88 \times 10^{-8} \, \mathrm{Pa} = 4.68 \times 10^{-11} \, \mathrm{mol\,h^{-1}}$$

Water–air volatilization

$$r_{AW}^{Vol} = D_{AW}^{Ov} f_W = 1.21 \times 10^{-3} \, \mathrm{mol\,h^{-1}\,Pa^{-1}} \times 1.15 \times 10^{-7} \, \mathrm{Pa} = 1.39 \times 10^{-10} \, \mathrm{mol\,h^{-1}}$$

Net air–water volatilization

$$r_{AW}^{VolDis} = D_{AW}^{Ov} (f_A - f_W) = 1.21 \times 10^{-3} \, \mathrm{mol\,Pa^{-1}\,h^{-1}} \times \left(3.88 \times 10^{-8} - 1.15 \times 10^{-7} \right) \mathrm{Pa}$$

$$= -9.22 \times 10^{-11} \, \mathrm{mol\,h^{-1}}$$

(transfer is from water to air if negative)

(Continued)

Rain dissolution

$$r_{Rain} = D_{Rain}f_A$$

$$= 1.14 \times 10^{-6} \, mol \, Pa^{-1} \, h^{-1} \times 3.88 \times 10^{-8} \, Pa$$

$$= 4.42 \times 10^{-14} \, mol \, h^{-1}$$

Dry deposition

$$r_Q^{Dry} = D_Q^{Dry} f_A$$

$$= 2.70 \times 10^{-4} \, mol \, h^{-1} \, Pa^{-1} \times 3.88 \times 10^{-8} \, Pa$$

$$= 1.05 \times 10^{-11} \, mol \, h^{-1}$$

Wet deposition

$$r_Q^{Wet} = D_Q^{Wet} f_A$$

$$= 2.86 \times 10^{-4} \, mol \, Pa^{-1} \, h^{-1} \times 3.88 \times 10^{-8} \, Pa$$

$$= 1.11 \times 10^{-11} \, mol \, h^{-1}$$

Air–water transfer (gross)

$$D_{AW}^{Tot} = D_{AW}^{Ov} + D_{Rain} + D_Q^{Dry} + D_Q^{Wet}$$

$$= \left(1.21 \times 10^{-3} + 1.14 \times 10^{-6} + 2.70 \times 10^{-4} + 2.86 \times 10^{-4}\right) mol \, Pa^{-1} \, h^{-1}$$

$$= 1.767 \times 10^{-3} \, mol \, Pa^{-1} \, h^{-1}$$

$$r_{AW}^{Tot} = D_{AW}^{Tot} f_A$$

$$= 1.767 \times 10^{-3} \, mol \, Pa^{-1} \, h^{-1} \times 3.88 \times 10^{-8} \, Pa = 6.86 \times 10^{-11} \, mol \, h^{-1}$$

Water–air transfer (gross)

$$D_{WA}^{Tot} = D_{AW}^{Ov}$$

$$r_{AW}^{Tot} = D_{AW}^{Tot} f_W$$

$$= 1.21 \times 10^{-3} \, mol \, Pa^{-1} \, h^{-1} \times 1.15 \times 10^{-7} \, Pa$$

$$= 1.39 \times 10^{-10} \, mol \, h^{-1}$$

Net air—water transfer

$$r_{AW}^{Net} = r_{AW}^{Tot} - r_{WA}^{Tot}$$

$$= 6.86 \times 10^{-11} \, mol \, h^{-1} - 1.39 \times 10^{-10} \, mol \, h^{-1}$$

$$= -7.04 \times 10^{-11} \, mol \, h^{-1}$$

(transfer is from water to air if negative)

Rate constants and the related half-times for each transfer process are then computed. The rate is given by the D-value, divided by the product of the source medium volume and the Z-value for the chemical in the source medium. Half-times are calculated for first-order processes and are $0.693/k$, as introduced above. These quantities are useful as indicators of the rapidity with which the chemical can be cleared from one phase to the other, thus enabling the significance of these exchange processes to be assessed relative to other processes such as reaction. Inspection of the individual D-values shows which processes are most important.

vi. Calculate rate constants and half-times for all transfers:
The rate constants (h^{-1}) and half-times (h) for transfer from each phase are, respectively,
Air:

$$k_{AW}^{Tot} = \frac{D_{AW}^{Tot}}{V_A Z_A^{Bulk}} = \frac{1.767 \times 10^{-3} \, mol \, Pa^{-1} \, h^{-1}}{1 \times 10^3 \, m^3 \times 4.42 \times 10^{-4} \, mol \, Pa^{-1} \, m^{-3}} = 4.00 \times 10^{-3} \, h^{-1}$$

$$\tau_{1/2}^A = \frac{0.693}{k_{AW}^{Tot}} = \frac{0.693}{4.00 \times 10^{-3}\,h^{-1}} = 250\,h$$

Water:

$$k_{WA}^{Tot} = \frac{D_{WA}^{Tot}}{V_W Z_W^{Bulk}} = \frac{1.21 \times 10^{-3}\,mol\,Pa^{-1}\,h^{-1}}{10\,m^3 \times 2.49 \times 10^{-2}\,mol\,Pa^{-1}\,m^{-3}} = 4.86 \times 10^{-3}\,h^{-1}$$

$$\tau_{1/2}^W = \frac{0.693}{k_{WA}^{Tot}} = \frac{0.693}{4.86 \times 10^{-3}\,h^{-1}} = 206\,h$$

The concentration of the chemical in each phase is then easily determined as the product of the chemical's fugacity in the phase in question times the corresponding Z-value. For aerosols and suspended particles, it is of more interest to know the concentration in each medium, but spread over the entire bulk medium volume. Therefore, for these phases, the true concentration of the chemical in the pure aerosol or suspended particle is multiplied by their respective volume fractions to obtain a concentration within the bulk phase volume due to that constituent.

vii. Calculate concentration of chemical in each phase:

Constituent	Concentration (mol m⁻³)
Air	$C_A = Z_A f_A$
	$= 4.17 \times 10^{-4}\,mol\,Pa^{-1}\,m^{-3} \times 3.87 \times 10^{-8}\,Pa$
	$= 1.61 \times 10^{-11}\,mol\,m^{-3}$
Aerosol	$C_Q = Z_Q f_A$ (particle itself)
	$= 5.01 \times 10^5\,mol\,Pa^{-1}\,m^{-3} \times 3.87 \times 10^{-8}\,Pa$
	$= 1.94 \times 10^{-2}\,mol\,m^{-3}$ (particle itself)
	$C_Q = v_Q^f Z_Q f_A$ (in medium volume)
	$= 1.94 \times 10^{-2}\,mol\,m^{-3} \times 5 \times 10^{-11} = 9.7 \times 10^{-13}\,mol\,m^{-3}$
Water	$C_W = Z_W f_W$
	$= 2.0 \times 10^{-2}\,mol\,Pa^{-1}\,m^{-3} \times 1.15 \times 10^{-7}\,Pa$
	$= 2.3 \times 10^{-9}\,mol\,m^{-3}$
Suspended solids	$C_{TSP} = Z_{TSP} f_W$ (particle itself)
	$= 615\,mol\,Pa^{-1}\,m^{-3} \times 1.15 \times 10^{-7}\,Pa$
	$= 7.07 \times 10^{-5}\,mol\,m^{-3}$
	$C_{TSP} = v_{TSP}^f Z_{TSP} f_W$ (in medium volume)
	$= 8.0 \times 10^{-6}\,mol\,m^{-3} \times 7.07 \times 10^{-5}\,mol\,m^{-3} = 5.65 \times 10^{-10}\,mol\,m^{-3}$

Note that the sum of the concentration of each component in the air or the water is equal to the corresponding bulk concentration, as required:

$$C_A^{\text{Bulk}} = C_A + C_Q = \left(1.61 \times 10^{-11} + 9.7 \times 10^{-13}\right) \text{mol m}^{-3}$$

$$= 1.71 \times 10^{-11} \text{mol m}^{-3}$$

$$C_W^{\text{Bulk}} = C_W + C_{TSP} = \left(2.3 \times 10^{-9} + 5.65 \times 10^{-10}\right) \text{mol m}^{-3}$$

$$= 2.86 \times 10^{-9} \text{mol m}^{-3}$$

A check should be made of the magnitude of f/P^S, where P^S is the solid or liquid vapor pressure. When this ratio equals 1, saturation is achieved. When the ratio exceeds 1, the chemical will *precipitate* as a pure phase, i.e., its solubility in air or water exceeded, and the fugacity will drop to the saturation value indicated by the vapor pressure. Normally, the ratio is much less than unity, because high ratios exceeding 0.01 generally correspond to toxic conditions.

It is noteworthy that a steady-state (i.e., no net transfer) condition may apply in which the air and water fugacities are unequal, i.e., a non-equilibrium, steady-state condition applies. At steady state, the rate of chemical transfer or flux from air to water equals that from water to air:

$$D_{AW}^{\text{Tot}} f_A^{SS} = D_{WA}^{\text{Tot}} f_W^{SS}$$

This fact directly yields an expression for the steady-state ratio of fugacities in air and water as

$$\frac{f_W^{SS}}{f_A^{SS}} = \frac{D_{AW}^{\text{Tot}}}{D_{WA}^{\text{Tot}}} = \frac{\left(D_{AW}^{\text{Ov}} + D_{\text{Rain}} + D_Q^{\text{Dry}} + D_Q^{\text{Wet}}\right)}{D_{WA}^{\text{Ov}}}$$

For our current example, this relation yields a ratio of 1.46, corresponding to higher fugacity in the water with respect to the air:

$$\frac{f_W^{SS}}{f_A^{SS}} = \frac{D_{AW}^{\text{Tot}}}{D_{WA}^{\text{Tot}}} = \frac{\left(1.21 \times 10^{-3} + 1.14 \times 10^{-6} + 2.70 \times 10^{-4} + 2.86 \times 10^{-4}\right) \text{mol Pa}^{-1} \text{h}^{-1}}{1.21 \times 10^{-3} \text{mol Pa}^{-1} \text{h}^{-1}}$$

$$= \frac{1.767 \times 10^{-3}}{1.21 \times 10^{-3}} = 1.46$$

The elevated fugacity in the water with respect to the air is expected, given that dissolution and volatilization processes have the same D-value and could balance at steady state, but the other processes of rain, dry, and wet deposition are all "unidirectional," in that there is no corresponding water-to-air process that can balance them. Therefore, unless there is no rain (and therefore no wet deposition) as well as no dry deposition, there will always be unidirectional transfer processes from air to water that only an excess fugacity in the water can balance.

The steady-state water fugacity and concentration with respect to the air, and those of air with respect to the water, can thus be calculated to give an impression of the extent to which the actual concentration departs from the steady-state values, as distinct from the equilibrium (equifugacity) value. It is noteworthy that, because D_{AW}^{Tot} exceeds D_{WA}^{Tot}, the fugacity in water will tend to exceed the fugacity in air; however, this will be affected by removal processes in water.

The "Inputs—Chemical Properties" page of *AirWater* for the calculation detailed above using the data for the test chemical named "AirWater86" stored in the database is illustrated below:

Chemical Name CAS	AirWater86

Molar Mass (g/mol)	350
Data Temperature (°C)	15
Melting Point (°C)	5
Vapor Pressure (Pa)	5.00E-03
Solubility in Water (g/m³)	3.50E-02
Henry's Law Constant (Pa·m³/mol)	50
$logK_{OW}$	6
K_{OC}	410000

The corresponding "Inputs—Environment" page for the test environment "EnvAirWater86" stored on the "Environment Database" page of *AirWater* is shown in the following figure:

Environment Name	EnvAirWater86

Dimensions

Area (m²)	1
Water Depth (m)	10
Atmosphere Height (m)	1000

Density

Air (kg/m³)	1.23E+00
Aerosol (kg/m³)	2.00E+03
Water (kg/m³)	1.00E+03
Suspended Particles (kg/m³)	1.50E+03

Organic Carbon

Suspended Particles (g/g)	0.05

Particle Concentrations

Suspended Particles (g/m³)	12
Aerosols in Air (µg/m³)	100

Transport Velocities

Water-Side MTC (m/h)	7.20E-02
Air-Side MTC (m/h)	1.80E+01
Dry Deposition Velocity (cm/s)	3.00E-01
Rain Rate (m/year)	5.00E-01
Scavenging Coefficient	2.00E+05

The chemical inputs to both water and air, in the commonly used units of ng L^{-1} and ng m^{-3}, respectively, for the model system, are entered on the "Concentrations" page as follows:

CHEMICAL QUANTITY

Total Chemical Concentration in Water (ng/L)	1
Total Chemical Concentration in Air (ng/m³)	6

The spreadsheet then generates fugacities, steady-state fugacities, transport D-values, and corresponding rates of intermedia transfers for individual processes such as volatilization, on the left-hand side of the "Results" page in the following format:

Chemical Name   AirWater86
Environment Name   EnvAirWater86

Fugacity Ratio	1.00

Subcooled Liq. Vap. Press.	5.000E-03	Pa

Fugacity

	Pa
Air	3.874E-08
Water	1.147E-07

Ratio of Water to Air Fugacities	2.959E+00

Predicition of Fugacities at Steady-State	
Ratio of Water to Air	1.461E+00
Water Fugacity	5.662E-08
Air Fugacity	7.846E-08

Intermedia Transfer

	D Value	Rate of Transfer	
	mol/Pa*h	mol/h	µg/day
Volatilization	1.208E-03	1.385E-10	1.164E+00
Absorption	1.208E-03	4.682E-11	3.933E-01
Rain Dissolution	1.142E-06	4.423E-14	3.715E-04
Dry Deposition	2.705E-04	1.048E-11	8.803E-02
Wet Deposition	2.859E-04	1.108E-11	9.305E-02
Air to Water Transfer (gross)	1.766E-03	6.842E-11	5.747E-01
Water to Air Transfer (gross)	1.208E-03	1.385E-10	1.164E+00
Overall Air to Water Transfer (net)	-	-7.013E-11	-5.891E-01

The right-hand panel of the "Results" page gives the calculated values for total chemical loads in air and water, as well as overall rates of transfer for all combined air-to-water and water-to-air transport processes. Corresponding individual and bulk Z-values and diffusion-related transport parameters are also given, with the chemical concentration in aerosol scaled to a concentration over the entire air volume (written as "TSP × (A/F)"):

System Totals

	Amount of Chemical		Rate of Transfer		Residence Time
	mol	kg	mol/h	kg/h	h
Air	1.714E-08	6.000E-09	6.842E-11	2.395E-11	2.506E+02
Water	2.857E-08	1.000E-08	1.385E-10	4.849E-11	2.062E+02

	Ratio By	
	Volume	Mass
Washout Ratio	1.137E+04	1.398E+01

Phase Properties

	Z Value	Concentration		
	mol/m³*Pa	mol/m³	g/m³	g/m³ of Bulk Phase
Air				
Bulk	4.425E-04	1.714E-11	6.000E-09	-
Air (Gaseous) Vapor	4.174E-04	1.617E-11	5.660E-09	-
Aerosol	5.009E+05	1.941E-02	6.792E+00	3.396E-10
Rain	-	1.948E-07	6.820E-05	-
Water				
Bulk	2.492E-02	2.857E-09	1.000E-06	-
Water (Dissolved)	2.000E-02	2.293E-09	8.026E-07	-
Susp. Particles	6.150E+02	7.051E-05	2.468E-02	1.974E-07

	µg/kg	%
Mass Conc. in Aerosol	3.396E+03	5.66
Mass Conc. in Susp. Particles	1.645E+01	19.74

	ng/m³
TSP x A/F	1.667E+06

Diffusion

	Overall MTC	D Value	Resistance	
	m/h	mol/Pa*h	Pa*h/mol	%
Air Side	2.895E+00	7.514E-03	1.331E+02	16.08
Water Side	6.042E-02	1.440E-03	6.944E+02	83.92
Total	-	1.208E-03	8.275E+02	100

Finally, the "Diagram" page gives a summary of the key concentration and transport properties for air, water, and the transport mechanisms by which they are connected:

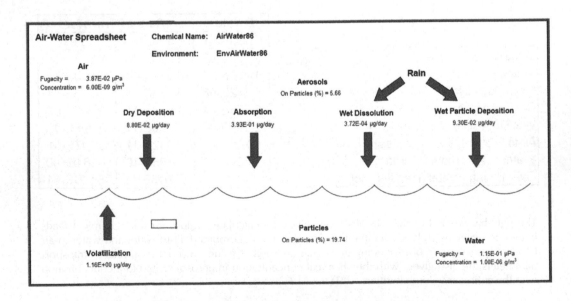

The *AirWater* model is available as an Excel spreadsheet program at the Canadian Environmental Modelling Centre (CEMC) site www.trentu.ca/cemc/resources-and-models. The calculations are carried out as outlined above, and sample chemicals and air and water properties are included in the chemical and environmental database sheets, including AirWater86 and EnvAirWater86.

6.3   SURFACE SOIL RUNOFF AND INFILTRATION: THE *SOIL* MODEL

Chemicals are frequently encountered in surface soils as a result of deliberate application of agrochemicals and sewage sludge, and by inadvertent spillage and leakage. It is often useful to assess the likely fate of the chemical, i.e., how fast the rates of degradation, volatilization, and leaching in water are likely to be, and how long it will take for the soil to "recover" to a specified or acceptable level of contamination. Persistence is an important characteristic for pesticide selection. Expensive remedial measures such as excavation may be needed when recovery times are unacceptably long.

Most modeling efforts in this context have been for agrochemical purposes, the most comprehensive recent effort being described in a series of publications by Jury et al. (1983, 1984a, 1984b, 1984c). Other notable models are reviewed in these papers. The *Soil* model is essentially a very simplified version of the Jury model (1983) and is a modification of a published herbicide fate model (Mackay and Stiver, 1989). The reader is referred to the texts by Sposito (1989) and Sawhney and Brown (1989), the chapter by Green (1988) and the work of Thibodeaux and Mackay (2011) for fuller accounts of chemical fate in soils. Cousins et al. (1999a,b) have reviewed and modeled these processes.

In the *Soil* model, only soil-to-air processes are treated; no air-to-soil transport is considered. A second, more complex fugacity model *SoilFug* was developed by Di Guardo et al. (1994a), which allows the user to calculate the fate of the pesticide in a defined agricultural area over time with changing rainfall. The model gave satisfactory predictions of pesticide runoff in agricultural regions in Italy and the U.K. (Di Guardo et al., 1994a, 1994b).

In the *Soil* model, the soil matrix illustrated in Figure 6.2 is considered to consist of four phases: pore air, pore water, OM, and MM. As noted above, the OM is considered to be 56% OC. The volume fractions of pore air and pore water are defined, either by the user or by default values, as is the ratio by mass of OC content to total soil mass. Density assumptions are made for pore air

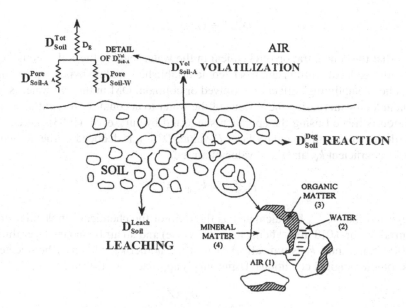

FIGURE 6.2   Chemical transport and transformation processes in a surface soil.

(1.19 kg m^{-3}), pore water (1000 kg m^{-3}), OM including roots (1000 kg m^{-3}), and MM (2500 kg m^{-3}), enabling the mass and volume fractions of each phase, and the overall soil density, to be calculated.

The soil area and depth are specified, thus enabling the total volumes and mass of soil and its component phases to be deduced. The amount of chemical present in the soil is specified as a concentration or as an amount in units of kg ha^{-1}, which is a convenient unit for agrochemicals. The chemical is assumed to be homogenously distributed throughout the entire soil volume.

The individual phase Z-values are calculated, and then the bulk Z-value of the soil, $Z_{\text{Soil}}^{\text{Bulk}}$, is deduced. From the concentration, the fugacity is deduced, and then the individual phase quantities and concentration are calculated.

It is prudent to examine the fugacity to check that it is less than the vapor pressure. If it exceeds the vapor pressure, phase separation of pure chemical will occur; i.e., the capacity of all phases to "dissolve" chemical exceeded. This can occur in heavily contaminated soils that have been subject to spills, or when there is heavy application of a pesticide. Essentially, the "solubility" of the chemical in the soil is exceeded. This calculation of partitioning behavior provides an insight into the amounts present in the air and water phases. It also shows the extent to which OM dominates the sorptive capacity of the soil.

Three loss processes are considered: degrading reactions, volatilization, and leaching, each rate being characterized by a D-value. An overall degradation half-life $\tau_{\text{Deg}}^{1/2}$ is specified in hours, from which an overall rate constant k_{Deg} in h^{-1} units is deduced as $0.693/\tau_{\text{Deg}}^{1/2}$. The reaction D-value $D_{\text{Soil}}^{\text{Deg}}$ is then calculated from the total soil volume and the bulk Z-value as

$$D_{\text{Soil}}^{\text{Deg}} = k_{\text{Soil}}^{\text{Deg}} V_{\text{Soil}} Z_{\text{Soil}}^{\text{Bulk}}$$

In principle, if a rate constant k_i is known for a specific phase in the soil, the phase-specific D-value can be deduced as $k_i V_i Z_i$, but the normal practice is to report an overall rate constant applicable to the total amount of chemical in the entire soil matrix. If no reaction occurs, an arbitrarily large value for the half-life, such as 10^{10} h, should be input to avoid model instability due to arithmetical operations with the default value of 0.

A water leaching rate is specified in units of mm day^{-1}. This may represent rainfall (which is typically 1–2 mm day^{-1}) or irrigation. This rate is converted into a total water flow rate $G_{\text{Soil}}^{\text{Leach}}$ (m^3 h^{-1}), which is combined with the water Z-value to give the advection leaching D-value:

$$D_{\text{Soil}}^{\text{Leach}} = G_{\text{Soil}}^{\text{Leach}} Z_W$$

This assumes that the concentration of chemical in the water leaving the soil is equal to that in the water in the soil; i.e., local equilibrium has become established, and no bypassing or "short circuiting" occurs. The "solubilizing" effect of dissolved or colloidal OM in the soil water is ignored, but it could be included by increasing the Z-value of the water to account for this extra capacity.

Volatilization is treated using the approach suggested by Jury et al. (1983). Three contributing D-values are deduced. An air boundary layer D-value, $D_{\text{Soil}-A}^{\text{Vol}}$, is deduced as the product of area A, a mass-transfer coefficient k_V, and the Z-value of air, i.e.:

$$D_{\text{Soil}-A}^{\text{Vol}} = k_{\text{Soil}-A}^{M} A Z_A$$

Jury has suggested that $k_{\text{Soil}-A}^{\text{Vol}}$ be calculated as the ratio of the chemical's molecular diffusivity in air B_A (0.43 m^2 day^{-1} or 0.018 m^2 h^{-1} being a typical value) and an air boundary layer thickness Y_A of 4.75 mm (0.00475 m); thus, k_V is typically 3.77 m h^{-1}. Another k_V value may be selected to reflect different micrometerological conditions. Using Jury's approach, we therefore have

$$D_{\text{Soil}-A}^{\text{Vol}} = \frac{B_A A Z_A}{Y_A}$$

A pore air diffusion D-value characterizes the rate of transfer of chemical vapor through the soil in the interstitial air phase. The Millington–Quirk equation is used to deduce an effective diffusivity B_{EA} from the pore air phase molecular diffusivity B_A, as outlined in Chapter 4, namely,

$$B_{EA} = \frac{B_A v_A^{f(10/3)}}{\left(v_A^f + v_W^f\right)^2}$$

where v_A^f is the volume fraction of pore air and v_W^f is the volume fraction of pore water. If v_W^f is small, this reduces to a dependence on v_A^f to the power 1.33. A diffusion path length Y must be specified, which is the vertical distance from the position of the chemical of interest to the soil surface; i.e., it is not the "tortuous" distance. The pore air diffusion D-value is then

$$D_{Soil-A}^{Pore} = \frac{B_{EA} A Z_A}{Y}$$

A similar approach is used to calculate the D-value for chemical diffusion in the pore water phase in the soil, except that the molecular diffusivity in water B_W is used (a value of 4.3×10^{-5} m^2 day^{-1} being assumed), and the water volume fraction and Z-value being used, namely

$$D_{Soil-W}^{Pore} = \frac{B_{EW} A Z_W}{Y} \quad \text{where} \quad B_{EW} = \frac{B_W v_W^{f(10/3)}}{\left(v_A^f + v_W^f\right)^2}$$

Since the diffusion D-values D_{Soil-A}^{Pore} and D_{Soil-W}^{Pore} apply in parallel, the total D-value for chemical transfer from bulk soil to the soil surface is $D_{Soil-A}^{Pore} + D_{Soil-W}^{Pore}$. The boundary layer D-value then applies in series so that the overall volatilization D-value, D_{Soil}^{Tot}, is given as illustrated in Figure 6.2 as

$$\frac{1}{D_{Soil}^{TotVol}} = \frac{1}{D_{Soil-A}^{Vol}} + \frac{1}{\left(D_{Soil-A}^{Pore} + D_{Soil-W}^{Pore}\right)}$$

This equation is a good example of the way in which series and parallel diffusion processes are combined to obtain an overall D-value for transport. The parallel process D-values are summed, and the series D-values are combined in a sum of reciprocals.

Selection of the diffusion path length Y involves an element of judgment. If, for example, the chemical is equally distributed in the top 20 cm of soil, an average value of 10 cm for Y may be appropriate as a first estimate. This will greatly underestimate the volatilization rate of chemical at the surface. Since the rate is inversely proportional to Y, a more appropriate single value of Y as the average between two depths Y_1 and Y_2 is the log mean of Y_1 and Y_2, i.e., $(Y_1 - Y_2)/\ln(Y_1/Y_2)$. Unfortunately, a zero (surface) value of Y cannot be used when calculating the log mean. For chemical between depths of 1 and 10 cm, a log mean depth of 3.9 cm is more appropriate than the arithmetic mean of 5.5 cm. It may be useful to consider layers of soil separately, e.g., 2–4 cm and 4–6 cm, and calculate separate volatilization rates for each. Chemical present at greater depths will thus volatilize more slowly, leaving the remaining chemical more susceptible to other removal processes. It is acceptable to specify a mean Y of, say, 10 cm to examine the fate of chemical in the 2 cm depth region from 9 to 11 cm. This depth issue is irrelevant to reaction or leaching, but it must be appreciated that, if the soil is treated as separate layers, the leaching rate is applicable to the total soil, not to each layer independently. The total rate of chemical removal is then $D_{Soil}^{Ov} \times f_{Sys}$, where the total D-value is

$$D_{Soil}^{Ov} = D_{Soil}^{TotVol} + D_{Soil}^{Deg} + D_{Soil}^{Leach}$$

the individual rates being $D_{\text{Soil}}^{\text{TotVol}} f_{\text{Sys}}$, $D_{\text{Soil}}^{\text{Deg}} f_{\text{Sys}}$, and $D_{\text{Soil}}^{\text{Leching}} f_{\text{Sys}}$. The overall rate constant is given by

$$k_{\text{Soil}}^{\text{Ov}} = \frac{D_{\text{Soil}}^{\text{Ov}}}{V_{\text{Soil}} Z_{\text{Soil}}^{\text{Bulk}}}$$

Here the denominator is the sum of individual soil component $V_i Z_i$ products, and the overall half-life is

$$\tau_{\text{Soil}}^{\text{Ov}} = \frac{0.693}{k_{\text{Soil}}^{\text{Ov}}} = \frac{0.693 \times V_{\text{Soil}} Z_{\text{Soil}}^{\text{Bulk}}}{D_{\text{Soil}}^{\text{Ov}}}$$

The half-life τ_i attributable to each process individually is

$$\tau_{\text{Soil}}^{i} = \frac{0.693}{k_{\text{Soil}}^{i}} = \frac{0.693 \times V_{\text{Soil}} Z_{\text{Soil}}^{\text{Bulk}}}{D_{\text{Soil}}^{i}}$$

Thus,

$$\frac{1}{\tau_{\text{Soil}}^{\text{Ov}}} = \frac{1}{\tau_{\text{Soil}}^{\text{Deg}}} + \frac{1}{\tau_{\text{Soil}}^{\text{Leach}}} + \frac{1}{\tau_{\text{Soil}}^{\text{TotVol}}}$$

It is illuminating to calculate the rates of each process, the percentages, and the individual half-lives. Obviously, the shorter half-lives dominate. The situation being simulated is essentially the first-order decay of chemical in the soil by three simultaneous processes, thus the amount remaining from an initial amount M (mol) at any time t (h) will be

$$M_t = M_0 \, \exp\left(-\frac{D_{\text{Soil}}^{\text{Ov}} t}{V_{\text{Soil}} Z_{\text{Soil}}^{\text{Bulk}}}\right) = M_0 \, \exp\left(-k_{\text{Soil}}^{\text{Ov}} t\right)$$

This relatively simple calculation can be used to assess the potential for volatilization or for groundwater contamination. Implicit in this calculation is the assumption that the chemical concentration in the air, and in the entering leaching water, is zero. If this is not the case, an appropriate correction must be included. In principle, it is possible to estimate atmospheric deposition rates as was done in the air–water example and couple these processes to the soil fate processes in a more comprehensive air–soil exchange model. It may prove desirable to segment the soil into multiple layers, especially if evaporation or input from the atmosphere is important. Models of this type have been reported by Cousins et al. (1999a) for PCBs in soils.

Illustrated below are the details of a typical *Soil* model calculation for a persistent chemical applied once to a model field.

Worked Example 6.2: Model soil calculation.

Consider the fate and distribution of the fictitious molecule pesticene with the properties given below, after a single application of kg ha⁻¹ to the model environment called "Carrot field" within the *Soil* model. What are the estimated concentrations and amounts in each of the following compartments: Pore air, pore water, OM, MM, and roots, and what are the half-times for volatilization, leaching and degradation in hours, and the corresponding residence times?

For this example, pesticene is considered to have properties similar to a volatile organochlorine pesticide. These properties are summarized in the following reproduction of the "Chemical Properties" input page of the *Soil* spreadsheet model.

Chemical Name CAS	Pesticine

Molar Mass (g/mol)	3.55E+02
Data Temperature (°C)	2.50E+01
Melting Point (°C)	1.09E+02
Vapor Pressure (Pa)	2.00E-02
Solubility in Water (g/m³)	3.10E-03
Henry's Law Constant (Pa·m³/mol)	2.29E+03
$logK_{OW}$	6.19E+00
K_{OC}	2.40E+05
K_{MW} (L/kg)	1.00E+00
Degradation Half-Life (h)	9.21E+04

Note that, like a volatile organochlorine pesticide, "pesticene" has a moderate vapor pressure and a high log K_{OW} value, which means that it will have a high propensity for accumulation in OM compared with air and water.

The environmental properties for the model field are the same as the "Carrot field" parameters which are already stored on the *Soil* model "Environmental Database" page. Below is an image of the "Inputs—Environment" page.

ENVIRONMENTAL PROPERTIES

Please complete all fields or select an environment from the database:

Environment Name	Carrot Field

System Criteria

Area (m²)	1.00E+04
Depth (m)	1.00E-01
Diffusion Path Length (m)	5.00E-02
Volume Fraction of Air	2.00E-01
Volume Fraction of Water	3.00E-01
Leaching Rate (mm/day)	5.00E+00
Air Boundary Layer Thickness (m)	4.75E-03
Molecular Diffusivity in Air (m²/day)	4.30E-01
Molecular Diffusivity in Water (m²/day)	4.30E-05
Volume Fraction of Roots	1.00E-02
Fraction of Lipid in Roots	2.50E-02

Density

Air (kg/m³)	1.19E+00
Water (kg/m³)	1.00E+03
Organic Matter (kg/m³)	1.00E+03
Mineral Matter (kg/m³)	2.50E+03
Roots (kg/m³)	1.00E+03

Mass Fractions of Organic Carbon

In Dry Soil	2.00E-02
In Organic Matter	5.60E-01

The "Chemical Dosage" page reflects the one-time application of 1 kg ha⁻¹:

Chemical Dosage (kg/ha)	1.0

With these input data in hand, we may begin the detailed calculation.

i. Determine Z-values for all relevant media:

$$Z_{\text{Soil}}^{\text{PoreA}} = \frac{1}{RT} = \frac{1}{8.3145\,\text{Pa}\,\text{m}^3\,\text{mol}^{-1}\,\text{K}^{-1} \times (25 + 273.15)\,\text{K}} = 4.034 \times 10^{-4}\,\text{mol}\,\text{Pa}^{-1}\,\text{m}^{-3}$$

$$Z_{\text{Soil}}^{\text{PoreW}} = \frac{1}{H} = \frac{C_W^{\text{Sat}}}{P_A^{\text{Sat}}} = \frac{\left(0.0031\,\text{g}\,\text{m}^{-3} / 354.5\,\text{g}\,\text{mol}^{-1}\right)}{2.00 \times 10^{-2}\,\text{Pa}} = 4.372 \times 10^{-4}\,\text{mol}\,\text{Pa}^{-1}\,\text{m}^{-3}$$

$$Z_{\text{Soil}}^{OM} = m_{OC}^f \times Z_{OC} = m_{OC}^f \left(\frac{K_{OC} Z_W \rho_{OC}\,(\text{kg}\,\text{m}^{-3})}{1000} \right)$$

$$= 0.56 \times \left(\frac{2.4 \times 10^5 \times 4.372 \times 10^{-4}\,\text{mol}\,\text{Pa}^{-1}\,\text{m}^{-3} \times 1000}{1000} \right)$$

$$= 58.76\,\text{mol}\,\text{Pa}^{-1}\,\text{m}^{-3}$$

$$Z_{\text{Soil}}^{MM} = \left(\frac{K_{\text{Min}-W}\,(\text{L}\,\text{kg}^{-1}) Z_W \rho_{MM}\,(\text{kg}\,\text{m}^{-3})}{1000} \right)$$

$$= \left(\frac{1 \times 4.372 \times 10^{-4}\,\text{mol}\,\text{Pa}^{-1}\,\text{m}^{-3} \times 2500\,\text{kg}\,\text{m}^{-3}}{1000} \right)$$

$$= 1.093 \times 10^{-3}\,\text{mol}\,\text{Pa}^{-1}\,\text{m}^{-3}$$

$$Z_{\text{Soil}}^{\text{Roots}} = v_L^{f-\text{Roots}} Z_W K_{OW}$$

$$= 0.025 \times 4.372 \times 10^{-4}\,\text{mol}\,\text{Pa}^{-1}\,\text{m}^{-3} \times 10^{6.19}$$

$$= 16.93\,\text{mol}\,\text{Pa}^{-1}\,\text{m}^{-3}$$

ii. Deduce the bulk Z-value for the soil:

As a one compartment model, we need only determine the bulk Z-value for the soil, and use the bulk soil concentration to determine the prevailing system fugacity. Recall that the bulk Z-value for any medium composed of more than one component, all of which are assumed to be at the same fugacity, is given by

$$Z_{\text{Soil}}^{\text{Bulk}} = \sum_{i=1}^{n} v_i^f Z_i$$

We first need to determine the various volume fractions, if not already known, before we can know the bulk Z-value. In our case, most of these values form part of the input data, but we need to calculate the volume fractions of the OM and MM in the soil.

The volume fraction for MM is given by the total volume of MM, divided by the total of the MM and OM volumes in the dry soil. The volume fraction associated with total dry matter, i.e., the sum of the volumes of the MM and OM, is determined by "removing" the volume fractions of the other soil components, due to air, water, and roots:

$$v_{MM}^f + v_{OM}^f = 1 - \left(v_A^f - v_W^f - v_{\text{Roots}}^f \right) = 1 - (0.2 + 0.3 + 0.01) = 0.49$$

To determine the individual volume fractions of MM and OM alone with the data on hand, we need to relate volume fractions to mass fractions. The following conversion demonstrates this relationship for the MM:

$$v_{MM}^f = \frac{V_{MM}}{V_{MM} + V_{OM}} = \left[\frac{1}{1 + \left(\dfrac{V_{OM}}{V_{MM}} \right)} \right] = \left[\frac{1}{1 + \left(\dfrac{m_{OM}/\rho_{OM}}{m_{MM}/\rho_{MM}} \right)} \right] = \left[\frac{1}{1 + \left(\dfrac{m_{DM} m_{OM}^f/\rho_{OM}}{m_{DM} m_{MM}^f/\rho_{MM}} \right)} \right] = \left[\frac{1}{1 + \left(\dfrac{m_{OM}^f/\rho_{OM}}{m_{MM}^f/\rho_{MM}} \right)} \right]$$

Here m_{DM} is the actual mass of dry soil, which does not need to be known, since it cancels out.

To use this relationship, we need to know the mass fractions of OM and MM. We know the mass fraction of OC *in dry soil* and the % OC in the organic matter. In our case, the mass of OC and OM in dry soil are related as

$$m_{OC} = 0.56 \times m_{OM} \text{ or } m_{OM} = \frac{m_{OC}}{0.56}$$

The same relationship holds for the mass fractions, so for the OM in dry soil:

$$m_{OM}^f = \frac{m_{OC}^f}{0.56} = \frac{0.02}{0.56} = 0.03571$$

Based on the fact that the mass fractions of all matter in the dry soil must sum to 1.0, we have directly that the mass fraction of MM in the soil must be

$$m_{MM}^f = 1 - m_{OM}^f = 1 - 0.03571 = 0.96429$$

Using these mass fractions, we may now calculate the volume fractions of MM and OM in the dry soil:

$$v_{MM}^{f-dry} = \left[\frac{1}{1 + \left(\dfrac{m_{OM}^{f-dry}/\rho_{OM}}{m_{MM}^{f-dry}/\rho_{MM}} \right)} \right] = \left[\frac{1}{1 + \left(\dfrac{0.03571/1000\,\text{kg m}^{-3}}{0.96429/2500\,\text{kg m}^{-3}} \right)} \right] = 0.91526$$

$$v_{OM}^{f-dry} = 1 - 0.91526 = 0.084736$$

It is important to recognize at this point that these volume fractions apply to the dry soil, but that we have determined already that the volume fraction of dry soil is only 49% of the sample. So, the volume fractions for MM and OM in the actual soil itself is

$$v_{OM}^f = v_{OM}^{f-dry} \times v_{DrySoil}^f = 0.084736 \times 0.49 = 0.04152$$

$$v_{MM}^f = v_{MM}^{f-dry} \times v_{DrySoil}^f = 0.91526 \times 0.49 = 0.44848$$

Finally, having all Z-values and volume fractions for each soil component in hand, we can now determine the Z-value for the bulk soil as

$$Z_{Soil}^{Bulk} = v_A^f Z_{Soil}^{PoreAir} + v_W^f Z_{Soil}^{PoreW} + v_{OM}^f Z_{Soil}^{OM} + v_{MM}^f Z_{Soil}^{MM} + v_{Roots}^f Z_{Soil}^{Roots}$$

$$Z_{Soil}^{Bulk} = 0.2 \times 4.034 \times 10^{-4} + 0.3 \times 4.372 \times 10^{-4} + 0.04152 \times 58.76 + 0.44848 \times 1.093 \times 10^{-3} + 0.01 \times 16.93$$

$$Z_{Soil}^{Bulk} = \left(8.068 \times 10^{-5} + 1.312 \times 10^{-4} + 2.4397 + 4.9019 \times 10^{-4} + 1.693 \times 10^{-1} \right) \text{mol Pa}^{-1}\,\text{m}^{-3}$$

$$Z_{Soil}^{Bulk} = 2.6097 \, \text{mol Pa}^{-1}\,\text{m}^{-3}$$

With the bulk soil fugacity capacity in hand, we can now easily calculate the system fugacity:

$$f_{\text{Sys}} = \frac{C_{\text{Soil}}^{\text{Bulk}}}{Z_{\text{Soil}}^{\text{Bulk}}}$$

To determine the bulk concentration of pesticene added, we need to determine the total number of moles added to the system:

$$m = \frac{1000\,\text{g}\,\text{ha}^{-1} \times 1 \times 10^4\,\text{m}^2}{1 \times 10^4\,\text{m}^2\,\text{ha}^{-1} \times 354.5\,\text{g}\,\text{mol}^{-1}} = 2.821\,\text{mol}$$

This amount divided by the volume of the soil gives the bulk concentration:

$$C_{\text{Soil}}^{\text{Bulk}} = \frac{2.821\,\text{mol}}{1 \times 10^4\,\text{m}^2 \times 0.1\,\text{m}} = 2.821 \times 10^{-3}\,\text{mol}\,\text{m}^{-3}$$

The corresponding system fugacity is therefore

$$f_{\text{Sys}} = \frac{C_{\text{Soil}}^{\text{Bulk}}}{Z_{\text{Soil}}^{\text{Bulk}}} = \frac{2.821 \times 10^{-3}\,\text{mol}\,\text{m}^{-3}}{2.6097\,\text{mol}\,\text{Pa}^{-1}\,\text{m}^{-3}} = 1.0807 \times 10^{-3}\,\text{Pa}$$

With the bulk soil fugacity in hand, we can now directly calculate the concentrations of pesticene in each of the soil constituent phases:

$$C_{\text{Soil}}^{\text{PoreA}} = Z_{\text{Soil}}^{\text{PoreA}} f_{\text{Sys}} = 4.034 \times 10^{-4}\,\text{mol}\,\text{Pa}^{-1}\,\text{m}^{-3} \times 1.0807 \times 10^{-3}\,\text{Pa} = 4.360 \times 10^{-7}\,\text{mol}\,\text{m}^{-3}$$

$$C_{\text{Soil}}^{\text{PoreW}} = Z_{\text{Soil}}^{\text{PoreW}} f_{\text{Sys}} = 4.372 \times 10^{-4}\,\text{mol}\,\text{Pa}^{-1}\,\text{m}^{-3} \times 1.0807 \times 10^{-3}\,\text{Pa} = 4.725 \times 10^{-7}\,\text{mol}\,\text{m}^{-3}$$

$$C_{\text{Soil}}^{MM} = Z_{\text{Soil}}^{MM} f_{\text{Sys}} = 1.093 \times 10^{-3}\,\text{mol}\,\text{Pa}^{-1}\,\text{m}^{-3} \times 1.0807 \times 10^{-3}\,\text{Pa} = 1.181 \times 10^{-6}\,\text{mol}\,\text{m}^{-3}$$

$$C_{\text{Soil}}^{OM} = Z_{\text{Soil}}^{OM} f_{\text{Sys}} = 5.876 \times 10^{1}\,\text{mol}\,\text{Pa}^{-1}\,\text{m}^{-3} \times 1.0807 \times 10^{-3}\,\text{Pa} = 6.351 10^{-2}\,\text{mol}\,\text{m}^{-3}$$

$$C_{\text{Soil}}^{\text{Roots}} = Z_{\text{Soil}}^{\text{Roots}} f_{\text{Sys}} = 16.93\,\text{mol}\,\text{Pa}^{-1}\,\text{m}^{-3} \times 1.0807 \times 10^{-2}\,\text{Pa} = 1.830 \times 10^{-2}\,\text{mol}\,\text{m}^{-3}$$

These results nicely demonstrate that it is the various fugacity capacities that "distribute" the chemical and its associated fugacity amongst the various subcompartments. Thus, the relatively high fugacity capacities of soil OM and roots result in much higher concentrations in these subcompartments compared with pore air, pore water, and MM. These latter phases have low OC content and are not where the hydrophobic or "water-hating" chemical pesticene "wants" to go.

The actual amounts of chemical in each subcompartment are determined by the product of the component volume and component concentration. For convenience here, the component volume is determined from the product of the total phase volume and the volume fraction of the component in that phase:

$$v_{\text{Soil}}^{f\text{-Phase}} = v_{\text{Phase}}^{f} V_{\text{Soil}}^{\text{Bulk}}$$

From the data on hand, we have that the total soil volume is given by

$$V_{\text{Soil}}^{\text{Bulk}} = A_{\text{Soil}} Y_{\text{Soil}} = 1.0 \times 10^4\,\text{m}^2 \times 0.1\,\text{m} = 1.0 \times 10^3\,\text{m}^3$$

Therefore, for each subcompartment, the total amount of pesticene in moles is given by

$$m_{\text{Soil}}^{\text{PoreA}} = v_{\text{PoreAir}}^{f} V_{\text{Soil}}^{\text{Bulk}} C_{\text{Soil}}^{\text{PoreA}} = 0.2 \times 1 \times 10^3\,\text{m}^3 \times 4.360 \times 10^{-7}\,\text{mol}\,\text{m}^{-3} = 8.720 \times 10^{-5}\,\text{mol}$$

$$m_{Soil}^{PoreW} = v_{PoreWater}^f V_{Soil}^{Bulk} C_{Soil}^{PoreW} = 0.3 \times 1 \times 10^3 \, m^3 \times 4.725 \times 10^{-7} mol \, m^{-3} = 1.418 \times 10^{-4} mol$$

$$m_{Soil}^{MM} = v_{MM}^f V_{Soil}^{Bulk} C_{Soil}^{MM} = 0.44848 \times 1 \times 10^3 \, m^3 \times 1.181 \times 10^{-6} \, mol \, m^{-3} = 5.298 \times 10^{-4} mol$$

$$m_{Soil}^{OM} = v_{OM}^f V_{Soil}^{Bulk} C_{Soil}^{OM} = 0.04152 \times 1 \times 10^3 \, m^3 \times 6.351 \times 10^{-2} \, mol \, m^{-3} = 2.638 \, mol$$

$$m_{Soil}^{Roots} = v_{Roots}^f V_{Soil}^{Bulk} C_{Soil}^{Roots} = 0.01 \times 1 \times 10^3 \, m^3 \times 1.830 \times 10^{-2} \, mol \, m^{-3} = 0.1830 \, mol$$

The percent mass of chemical in each subcompartment is given by

$$\%Mass = \left(\frac{moles}{molar \; mass} \right) \times 100\%$$

Such calculations lead to the following percent mass contents for each subcompartment:
Pore air: 3.1×10^{-3}
Pore water: 5.0×10^{-3}
MM: 0.019%
OM: 93.5%
Roots: 6.5%

It is clear that the high fugacity capacity of the soil OM and the roots results in most of the hydrophobic chemical pesticene partitioning to these subcompartment. Since the volume fraction of the OM is about four times that of the roots and the organic content of the roots is only about half that of the soil OM, the partitioning is mainly to the soil OM, which acts as a "fugacity reservoir" for the chemical, allowing it to transfer primarily to the roots with each subsequent planting of the field.

A key question is that of the length of time the chemical will persist in the field. Given the degradation half-lives we can determine the characteristic or residence time for the chemical in the soil as an overall compartment by appropriately combining the various D-values for loss processes. Therefore, we need to determine the various D-values for the processes considered in the model:

For diffusion in the pore air, we need first to determine the pore air phase molecular diffusivity:

$$B_{EA} = \frac{B_A v_A^{f(10/3)}}{\left(v_A^f + v_W^f\right)^2} = \frac{0.43 \, m^2 \, day^{-1} \times 0.2^{(10/3)}}{24 \, h \, day^{-1} \times (0.2+0.3)^2} = 3.35 \times 10^{-4} \, m^2 \, h^{-1}$$

$$D_{PoreA}^{Diff} = \frac{B_{EA} A Z_A}{Y}$$

$$= \frac{3.353 \times 10^{-4} \, m^2 \, h^{-1} \times 1 \times 10^4 \, m^2 \times 4.034 \times 10^{-4} \, mol \, Pa^{-1} \, m^{-3}}{5 \times 10^{-2} \, m}$$

$$= 2.705 \times 10^{-2} \, mol \, Pa^{-1} \, h^{-1}$$

Similarly for diffusion within the pore water, we have

$$B_{EW} = \frac{B_W v_W^{f(10/3)}}{\left(v_A^f + v_W^f\right)^2} = \frac{4.3 \times 10^{-5} \, m^2 \, day^{-1} \times 0.3^{(10/3)}}{24 \, h \, day^{-1} \times (0.2+0.3)^2} = 1.295 \times 10^{-7} \, m^2 \, h^{-1}$$

$$D_{PoreW}^{Diff} = \frac{B_{EW} A Z_W}{Y} = \frac{1.295 \times 10^{-7} \, m^2 \, h^{-1} \times 1 \times 10^4 \, m^2 \times 4.372 \times 10^{-4} \, mol \, Pa^{-1} \, m^{-3}}{5 \times 10^{-2} \, m}$$

$$= 1.133 \times 10^{-5} \, mol \, Pa^{-1} \, h^{-1}$$

As discussed above, the pore–air and pore–water diffusion processes are parallel, and their combination is in series with the soil–air volatilization process. For the soil–air boundary layer, we have

$$
\begin{aligned}
D_{\text{Soil}-A}^{\text{Vol}} &= \frac{B_{\text{Soil}-A}^{\text{Vol}} A_{\text{Soil}-A} Z_A}{Y} \\
&= \frac{0.43\,\text{m}^2\,\text{day}^{-1} \times 1 \times 10^4\,\text{m}^2 \times 4.034 \times 10^{-4}\,\text{mol}\,\text{Pa}^{-1}\,\text{m}^{-3}}{24\,\text{h}\,\text{day}^{-1} \times 4.75 \times 10^{-3}} \\
&= 15.22\,\text{mol}\,\text{Pa}^{-1}\,\text{h}^{-1}
\end{aligned}
$$

The appropriate combination of D-value for overall volatilization is then

$$
\begin{aligned}
\frac{1}{D_{\text{Soil}}^{\text{TotVol}}} &= \frac{1}{D_{\text{Soil}-A}^{\text{Vol}}} + \frac{1}{\left(D_{\text{PoreA}}^{\text{Diff}} + D_{\text{PoreW}}^{\text{Diff}}\right)} \\
&= \frac{1}{15.22\,\text{mol}\,\text{Pa}^{-1}\,\text{h}^{-1}} + \frac{1}{\left(2.705 \times 10^{-2}\,\text{mol}\,\text{Pa}^{-1}\,\text{h}^{-1} + 1.133 \times 10^{-5}\,\text{mol}\,\text{Pa}^{-1}\,\text{h}^{-1}\right)} \\
&= \frac{1}{15.22\,\text{mol}\,\text{Pa}^{-1}\,\text{h}^{-1}} + \frac{1}{\left(2.70613 \times 10^{-2}\,\text{mol}\,\text{Pa}^{-1}\,\text{h}^{-1}\right)} \\
&= \left(6.57 \times 10^{-2} + 36.95\right) = 37.02\,\text{Pa}\,\text{mol}^{-1}\,\text{h}^{-1}
\end{aligned}
$$

Therefore we have

$$
D_{\text{Soil}}^{\text{TotVol}} = \frac{1}{37.02\,\text{Pa}\,\text{mol}^{-1}\,\text{h}^{-1}} = 2.702 \times 10^{-2}\,\text{mol}\,\text{Pa}^{-1}\,\text{h}^{-1}
$$

Now that we have the D-value for diffusion and volatilization, we can combine these three "in-soil" activities, i.e., diffusion, leaching, and degradation. The total D-value is simply the sum of the D-values for each of these processes.

The leaching rate is given by

$$
\begin{aligned}
D_{\text{Soil}}^{\text{Leach}} &= G_{\text{Soil}}^{\text{Leach}} Z_{\text{Soil}}^{\text{PoreW}} \\
&= \left(\frac{5.0\,\text{mm}\,\text{day}^{-1}}{1 \times 10^3\,\text{mm}\,\text{m}^{-1} \times 24\,\text{h}\,\text{day}^{-1}}\right) \times 1 \times 10^4\,\text{m}^2 \times 4.372 \times 10^{-4}\,\text{mol}\,\text{Pa}^{-1}\,\text{m}^{-3} \\
&= 9.108 \times 10^{-4}\,\text{mol}\,\text{Pa}^{-1}\,\text{h}^{-1}
\end{aligned}
$$

The degradation rate constant is determined in part from the degradation half-time, which is

$$
k_{\text{Soil}}^{\text{Deg}} = \frac{-\ln(0.5)}{\tau_{\text{Soil}}^{0.5}} = \frac{0.693}{9.21 \times 10^4\,\text{h}} = 7.526 \times 10^{-6}\,\text{h}^{-1}
$$

So that:

$$
\begin{aligned}
D_{\text{Soil}}^{\text{Deg}} &= k_{\text{Soil}}^{\text{Deg}} V_{\text{Soil}} Z_{\text{Soil}}^{\text{Bulk}} \\
&= 7.526 \times 10^{-6}\,\text{h}^{-1} \times 1 \times 10^4\,\text{m}^2 \times 0.1\,\text{m} \times 2.6097\,\text{mol}\,\text{Pa}^{-1}\,\text{m}^{-3} \\
&= 1.965 \times 10^{-2}\,\text{mol}\,\text{Pa}^{-1}\,\text{h}^{-1}
\end{aligned}
$$

We then arrive at the total D-value for all losses from the soil as

$$D_{\text{Soil}}^{\text{Overall}} = D_{\text{Soil}-A}^{\text{TotVol}} + D_{\text{Soil}}^{\text{Deg}} + D_{\text{Soil}}^{\text{Leach}}$$

$$= \left(2.7016 \times 10^{-2} + 1.965 \times 10^{-2} + 9.108 \times 10^{-4}\right) \text{mol Pa}^{-1}\,\text{h}^{-1}$$

$$= 4.7576 \times 10^{-2}\,\text{mol Pa}^{-1}\,\text{h}^{-1}$$

With the overall D-value in hand, we can calculate the half-time for the total of all losses as

$$\tau_{\text{Soil}}^{1/2} = \frac{0.693}{k_{\text{Soil}}^{\text{Ov}}} = \frac{0.693 \times V_{\text{Soil}} Z_{\text{Soil}}^{\text{Bulk}}}{D_{\text{Soil}}^{\text{Overall}}} = \frac{0.693 \times 1.0 \times 10^3\,\text{m}^3 \times 2.6097\,\text{mol Pa}^{-1}\,\text{m}^{-3}}{4.7576 \times 10^{-2}\,\text{mol Pa}^{-1}\,\text{h}^{-1}} = 3.80 \times 10^4\,\text{h}$$

This total loss half-time corresponds to slightly over 4 years. The individual half-time for each process, if it were to occur in the absence of all other loss rates, is given by

$$\tau_{1/2}^{\text{TotVol}} = \frac{0.693}{k_{\text{Soil}}^{\text{TotVol}}} = \frac{0.693 \times V_{\text{Soil}} Z_{\text{Soil}}^{\text{Bulk}}}{D_{\text{Soil}}^{\text{TotVol}}} = \frac{0.693 \times 1.0 \times 10^3\,\text{m}^3 \times 2.6097\,\text{mol Pa}^{-1}\,\text{m}^{-3}}{2.7016 \times 10^{-2}\,\text{mol Pa}^{-1}\,\text{h}^{-1}} = 6.69 \times 10^4\,\text{h}$$

$$\tau_{1/2}^{\text{Deg}} = \frac{0.693}{k_{\text{Soil}}^{\text{Deg}}} = \frac{0.693 \times V_{\text{Soil}} Z_{\text{Soil}}^{\text{Bulk}}}{D_{\text{Soil}}^{\text{Deg}}} = \frac{0.693 \times 1.0 \times 10^3\,\text{m}^3 \times 2.6097\,\text{mol Pa}^{-1}\,\text{m}^{-3}}{1.965 \times 10^{-2}\,\text{mol Pa}^{-1}\,\text{h}^{-1}} = 9.21 \times 10^4\,\text{h}$$

$$\tau_{1/2}^{\text{Leach}} = \frac{0.693}{k_{\text{Soil}}^{\text{Leach}}} = \frac{0.693 \times V_{\text{Soil}} Z_{\text{Soil}}^{\text{Bulk}}}{D_{\text{Soil}}^{\text{Leach}}} = \frac{0.693 \times 1.0 \times 10^3\,\text{m}^3 \times 2.6097\,\text{mol Pa}^{-1}\,\text{m}^{-3}}{9.108 \times 10^{-4}\,\text{mol Pa}^{-1}\,\text{h}^{-1}} = 1.98 \times 10^6\,\text{h}$$

Residence times are calculated as the total amount of chemical $V_{\text{Soil}} Z_{\text{Soil}}^{\text{Bulk}} f_{\text{Sys}}$ divided by the rate of the loss process for individual loss mechanisms $D_{\text{Soil}}^i f_{\text{Sys}}$, or by the total loss rate for the overall residence time of the chemical (f_{Sys} cancels here):

$$\tau_{\text{Soil}} = \frac{V_{\text{Soil}} Z_{\text{Soil}}^{\text{Bulk}}}{D_{\text{Soil}}^{\text{Overall}}} = \frac{1.0 \times 10^3\,\text{m}^3 \times 2.6097\,\text{mol Pa}^{-1}\,\text{m}^{-3}}{4.7576 \times 10^{-2}\,\text{mol Pa}^{-1}\,\text{h}^{-1}} = 5.49 \times 10^4\,\text{h}$$

$$\tau_{\text{TotVol}} = \frac{V_{\text{Soil}} Z_{\text{Soil}}^{\text{Bulk}}}{D_{\text{Soil}}^{\text{TotVol}}} = \frac{1.0 \times 10^3\,\text{m}^3 \times 2.6097\,\text{mol Pa}^{-1}\,\text{m}^{-3}}{2.7016 \times 10^{-2}\,\text{mol Pa}^{-1}\,\text{h}^{-1}} = 9.66 \times 10^4\,\text{h}$$

$$\tau_{\text{Deg}} = \frac{V_{\text{Soil}} Z_{\text{Soil}}^{\text{Bulk}}}{D_{\text{Soil}}^{\text{Deg}}} = \frac{1.0 \times 10^3\,\text{m}^3 \times 2.6097\,\text{mol Pa}^{-1}\,\text{m}^{-3}}{1.965 \times 10^{-2}\,\text{mol Pa}^{-1}\,\text{h}^{-1}} = 1.33 \times 10^5\,\text{h}$$

$$\tau_{\text{Leach}} = \frac{V_{\text{Soil}} Z_{\text{Soil}}^{\text{Bulk}}}{D_{\text{Soil}}^{\text{Leach}}} = \frac{1.0 \times 10^3\,\text{m}^3 \times 2.6097\,\text{mol Pa}^{-1}\,\text{m}^{-3}}{9.108 \times 10^{-4}\,\text{mol Pa}^{-1}\,\text{h}^{-1}} = 2.87 \times 10^6\,\text{h}$$

The transfer rates for pesticene associated with the various processes is simply given by the product of the corresponding D-value and the overall soil fugacity:

$$r_{\text{Soil}}^{\text{TotVol}} = D_{\text{Soil}}^{\text{TotVol}} f_{\text{Sys}} = 2.7016 \times 10^{-2}\,\text{mol Pa}^{-1}\,\text{h}^{-1} \times 1.0807 \times 10^{-3}\,\text{Pa} = 2.92 \times 10^{-5}\,\text{mol h}^{-1}$$

$$r_{\text{Soil}}^{\text{Deg}} = D_{\text{Soil}}^{\text{Deg}} f_{\text{Sys}} = 1.965 \times 10^{-2}\,\text{mol Pa}^{-1}\,\text{h}^{-1} \times 1.0807 \times 10^{-3}\,\text{Pa} = 2.12 \times 10^{-5}\,\text{mol h}^{-1}$$

$$r_{\text{Soil}}^{\text{Leach}} = D_{\text{Soil}}^{\text{Leach}} f_{\text{Sys}} = 9.108 \times 10^{-4}\,\text{mol Pa}^{-1}\,\text{h}^{-1} \times 1.0807 \times 10^{-3}\,\text{Pa} = 9.84 \times 10^{-7}\,\text{mol h}^{-1}$$

$$r_{\text{Soil}}^{\text{Overall}} = D_{\text{Soil}}^{\text{Overall}} f_{\text{Sys}} = 4.7576 \times 10^{-2}\,\text{mol Pa}^{-1}\,\text{h}^{-1} \times 1.0807 \times 10^{-3}\,\text{Pa} = 5.14 \times 10^{-5}\,\text{mol h}^{-1}$$

The rates for the various processes comprise 0.2% (Volatilization), 4.4% (Leaching), and 95.4% (Degradation) of the sum of losses. These relative magnitudes reflect pesticene's modest vapor pressure and relatively high K_{OW}, both of which resulting in losses due to volatilization and leaching being relatively small compared with decomposition.

The output pages from the *Soil* model spreadsheet for this Worked Example are shown below:

Soil Results

Chemical Name   Pesticine
Environment Name   Carrot Field

Summary

	kg/ha
Chemical Dosage	1.00E+00

	g
Mass of Chemical	1.00E+03

	g/m³	µg/g
Conc. in Bulk Soil	9.446E-01	6.392E-07
Conc. (amount/mass of dry soil)	-	8.042E-01

	Pa
Fugacity in Soil	1.08070E-03

Residence Time

	Half-time	Half-time	Residence Time	Residence Time
	hours	days	hours	days
Volatilization	6.697E+04	2.790E+03	9.662E+04	4.026E+03
Leaching	1.986E+06	8.276E+04	2.866E+06	1.194E+05
Reaction	9.209E+04	3.837E+03	1.329E+05	5.536E+03
Total	3.803E+04	1.585E+03	5.487E+04	2.286E+03

Phase Properties

	Z Value	Concentration			Amount		
	mol/m³*Pa	mol/m³	g/m³	µg/g	mol	g	%
Pore Air	4.034E-04	4.360E-07	1.546E-04	1.299E-01	8.720E-05	3.091E-02	3.091E-03
Pore Water	4.372E-04	4.725E-07	1.675E-04	1.675E-04	1.418E-04	5.025E-02	5.025E-03
Organic Matter	5.876E+01	6.351E-02	2.251E+01	2.251E+01	2.637E+00	9.349E+02	9.349E+01
Mineral Matter	1.093E-03	1.181E-06	4.188E-04	1.675E-04	5.298E-04	1.878E-01	1.878E-02
Bulk Soil	-	2.665E-03	9.446E-01	6.392E-07	2.638E+00	9.351E+02	9.351E+01
Roots	1.693E+01	1.830E-02	6.486E+00	6.486E+00	1.830E-01	6.486E+01	6.486E+00

| System Totals | | | | | 2.82E+00 | 1.00E+03 | 100 |

Transfer and Loss Processes

	D Value
	mol/Pa*h
Air Diffusion	2.705E-02
Water Diffusion	1.133E-05
Air Boundary Layer	1.522E+01

	D Value	Rate			
	mol/Pa*h	mol/h	g/h	µg/day	%
Volatilization	2.702E-02	2.920E-05	1.035E-02	2.484E+05	56.787
Leaching	9.109E-04	9.844E-07	3.490E-04	8.375E+03	1.915
Reaction	1.965E-02	2.123E-05	7.527E-03	1.806E+05	41.299
Total	4.757E-02	5.141E-05	1.823E-02	4.374E+05	100.000

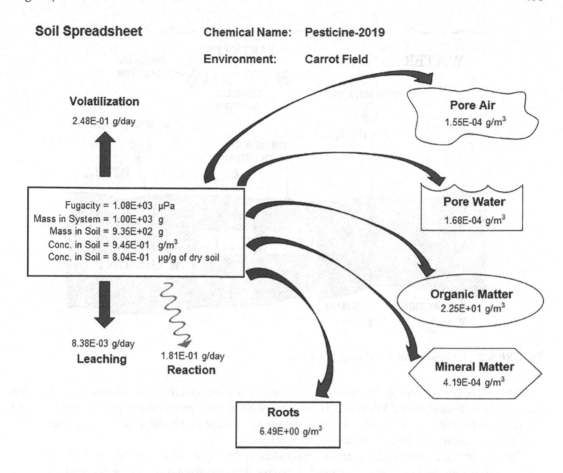

Soil Spreadsheet

Chemical Name: Pesticine-2019

Environment: Carrot Field

Volatilization

2.48E-01 g/day

Pore Air

1.55E-04 g/m³

Fugacity = 1.08E+03 μPa
Mass in System = 1.00E+03 g
Mass in Soil = 9.35E+02 g
Conc. in Soil = 9.45E-01 g/m³
Conc. in Soil = 8.04E-01 μg/g of dry soil

Pore Water

1.68E-04 g/m³

Organic Matter
2.25E+01 g/m³

8.38E-03 g/day
Leaching

1.81E-01 g/day
Reaction

Mineral Matter
4.19E-04 g/m³

Roots
6.49E+00 g/m³

The *Soil* model is available from the CEMC website as a spreadsheet model, similar to the *AirWater* model. It should be noted again that varying the input temperature will not vary physical chemical properties such as vapor pressure. Temperature dependence must be entered "by hand."

6.4 MODELING SEDIMENTS: THE *SEDIMENT* MODEL

Exchange of chemical at the sediment–water interface can be important for the estimation of (1) the rates of accumulation or release from sediments, (2) the concentration of chemicals in organisms living in, or feeding from, the benthic region, (3) which transfer processes are most important in a given situation, and (4) the likely recovery times in the case of "in-place" sediment contamination. The complexity of the system and the varying properties of chemicals of possible concern lead to a situation in which a specific chemical's behavior is not necessarily obvious. This situation treated here and the resulting model are largely based on a discussion of sediment–water exchange by Reuber et al. (1987), Eisenreich (1987), Diamond et al. (1990), and in part on a report by Formica et al. (1988) and the work of DiToro (2001). It is depicted in Figure 6.3.

The water phase area and depth (and hence volume) are defined, it being assumed that the water is well mixed. The water contains suspended particulate matter, which may contain mineral and organic material. The concentration (mg L⁻¹ or g m⁻³) of suspended matter is defined, as is its OC content as a mass fraction (g OC per g dry particulates). The volume fractions are calculated similar to those for air–water exchange.

The sediment phase is treated similarly, having the same area, a defined well-mixed depth, and a specified concentration of solids and interstitial or pore water.

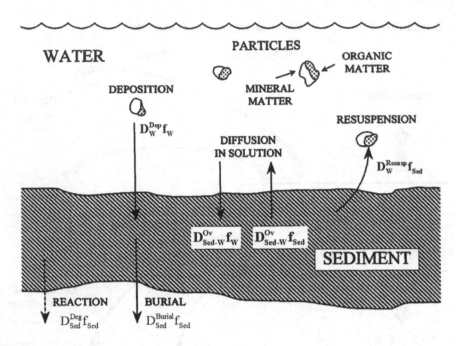

FIGURE 6.3 Sediment–water exchange processes.

Rates of sediment deposition, resuspension, and burial are specified, as are first-order reaction rates in the sediment phase. Allowance is made for infiltration of groundwater through the sediment in either vertical direction. Lipid contents of organisms present in the water and sediment are specified for later illustrative bioconcentration calculations.

The equilibrium partitioning distribution is calculated using Z-values for the water and sediment phases using specified total g m^{-3} or mg L^{-1} chemical concentrations in the water, and mg g^{-1} of dry sediment solids in the sediment. Since no air phase appears in the calculation, the vapor pressure is not strictly necessary. Identical concentration, but not fugacity, results are obtained when an arbitrary vapor pressure is used. This convenience arises from the fact that all Z-values eventually appear as ratios in the subsequent calculations, and an arbitrary choice of fugacity does not change such ratios. The only point at which the absolute fugacity becomes necessary in multimedia calculations is in the determination of vapor pressures and other gas-phase properties. It is on the strength of this fact that the *aquivalence* concept has been proposed by Mackay and Diamond (1989), as discussed earlier.

Illustrative biotic Z-values can be deduced for both water and sediment as

$$Z_{\text{Biota}} = K_{\text{Biota}-W} Z_W$$

where the bioconcentration factor (BCF) $K_{\text{Biota}-W}$ is estimated from the product of lipid fraction L_{Biota} (e.g., 0.05) and K_{OW}:

$$K_{\text{Biota}-W} \simeq v_L^{f-\text{Biota}} K_{OW}$$

Brought together, these two equations demonstrate that Z_{Biota} is really just the fugacity capacity of octanol, "diluted" by the lipid fraction:

$$Z_{\text{Biota}} = K_{\text{Biota}-W} Z_W \simeq \left(v_L^{f-\text{Biota}} K_{OW} \right) Z_W = v_L^{f-\text{Biota}} \left(\frac{Z_O}{Z_W} \right) Z_W = v_L^{f-\text{Biota}} Z_O$$

This approach is the most common manner by which Z-values for complex media are estimated: They are assumed to be comprised of inert matter and OM that is active in partitioning. Generally, such OM may be effectively represented by octanol. Note that in the *Sediment* model, biota are included only for illustrative purposes and are not included in the mass balance.

The total and contributing concentrations in all phases and the fugacities can thus be deduced without the use of a mass balance. From the biotic Z-values, the corresponding concentrations can also be deduced for biota resident in water and sediment.

Several transport and transformation processes are treated:

1. sediment deposition,
2. sediment resuspension,
3. sediment burial,
4. diffusive exchange of water between the water column and the pore water, and
5. sediment reaction.

Irrigation, i.e., net flow of groundwater into or out of the sediment, can be added as a sixth process if present.

The necessary D-values for the *Sediment* Model are defined as follows:

Process	D-Value
Deposition	$D_W^{Dep} = G_W^{Dep} Z_{TSP}$
Resuspension	$D_{Sed}^{Resusp} = G_{Sed}^{Resusp} Z_{Sed}$
Burial	$D_{Sed}^{Burial} = G_{Sed}^{Burial} Z_{Sed}$
Diffusion	$D_{Sed-W}^{Ov} = \dfrac{1}{\left[\dfrac{1}{D_{Sed-W}^{Diff}} + \dfrac{1}{D_{W-Sed}^{Diff}} \right]}$
Reaction	$D_{Sed}^{Deg} = k_{Sed}^{Deg} V_{Sed} Z_{Sed}$
Water inflow	$D_{Sed}^{Inf} = U_{Sed}^{Inf} A_{Sed} Z_W$

The individual and total rates of transfer can be calculated as Df products.

The *Sediment* model is available in spreadsheet format from the CEMC website. Input data are requested on the properties of the chemical, the dimensions and properties of the media, and the prevailing concentrations. The Z- and D-values are calculated, followed by fugacities and fluxes.

It is also of interest to calculate the overall steady-state mass balance, which is given by

$$f_W \left(D_W^{Dep} + D_{Sed-W}^{Ov} \right) = f_S \left(D_{Sed}^{Resusp} + D_{Sed-W}^{Ov} + D_{Sed}^{Burial} + D_{Sed}^{Deg} \right)$$

The steady-state water and sediment fugacities corresponding to the defined sediment and water fugacities may be deduced. Response times can be calculated for each medium if the volumes are known.

The steady-state expression above leads directly to a statement of the ratio of the two fugacities of interest:

$$\frac{f_S}{f_W} = \frac{\left(D_W^{Dep} + D_{Sed-W}^{Ov} \right)}{\left(D_{Sed}^{Resusp} + D_{Sed-W}^{Ov} + D_{Sed}^{Burial} + D_{Sed}^{Deg} \right)}$$

It is noteworthy that, for a persistent, hydrophobic substance, it is likely that the steady-state sediment fugacity will exceed that of the water. The principal loss process of a persistent chemical from the sediment is likely to be D_{Sed}^{Resusp}, which must be less than D_W^{Dep}, because some sediment is buried, and the OC content of the resuspended material will generally be less than the deposited material because of OC mineralization. Such a condition ensures that the denominator in the above expression will be greater than 1.0 for a persistent chemical under typical burial rates. As a result, a benthic organism that respires sediment pore water may reach a higher fugacity and concentration than a corresponding organism in the water column above. A compelling case can be made for monitoring benthic organisms, because they are less mobile than fish and they are likely to build up higher tissue concentrations of contaminants.

These sediment–water calculations can be invaluable for estimating the rate at which "in-place" sediment concentrations, resulting from past discharges of persistent substances, are falling. Often, the memory of past transgressions lingers longer in sediments than in the water column.

Worked Example 6.3: Application of the sediment model to hypothene, with properties similar to DDT.

Hypothene is present in a water body at a concentration of 1×10^{-4} g m^{-3} and in the associated sediment at 10 µg g^{-1} of dry sediment solids weight. Given the chemical and environmental properties below, determine the initial and steady-state fugacities of hypothene in the water and sediment at steady state, the concentration of hypothene in each of the components of the water and sediment phases, and the transport rate and associated half-times for all relevant transport processes.

Here are the chemical properties of hypothene and the environmental properties, as they appear on the input pages of the *Sediment* spreadsheet model:

Chemical Name	Hypothene
CAS	

Molar Mass (g/mol)	300.00
Data Temperature (°C)	25.00
Melting Point (°C)	0.00
Vapor Pressure (Pa)	1.00
Solubility in Water (g/m³)	1.00
Henry's Law Constant (Pa·m³/mol)	300.00
$\log K_{OW}$	6.00
K_{OC}	410000.00
K_{MW} (L/kg)	1.00
Degradation Half-Life (h)	120000.00

Environment Name	Hypothene

System Criteria

Area (m²)	1000.00
Water Depth (m)	20.00
Sediment Depth (m)	0.0300
Volume Fraction of Pore Water in Sed.	0.80
Conc. of Susp. Particles in Water (g/m³)	5.00

Transport

Diffusion Path Length in Sed. (m)	0.0150
Molecular Diffusivity of Chemical in Water (m²/h)	2.00E-06

Transfer Fluxes

Sediment Deposition (g/m²/day)	3.00
Sediment Resuspension (g/m²/day)	1.00
Burial (g/m²/day)	1.50

Density	
Air (kg/m³)	1.185
Water (kg/m³)	1000.00
Organic Matter (kg/m³)	1000.00
Mineral Matter (kg/m³)	2500.00

Mass Fractions of Organic Carbon	
Susp. Particles	0.25
Sediment Soils	0.05
Organic Matter	0.56

Organic Lipid Fraction	
Water Organsims	0.05
Benthic Organsims	0.03

The assumed concentrations to be used are shown as entered on the "Concentrations" input sheet:

Total Chemical Conc. in Water (g/m³)	1.00E-04
Chemical Conc. in Sediment (µg/g)	1.00E+01

i. Calculate Z-values for hypothene in water:

The Z-value for hypothene in water cannot be determined here from the fugacity capacity of air, as we do not have a partition ratio involving air. We can however determine Z_W from the Henry's law constant, which is the ratio of the saturation vapor pressure with respect to the saturation concentration in water (sub-cooled if the chemical is solid at room temperature):

$$Z_W = \frac{1}{H} = \frac{C_W^{Sat}}{P_A^{Sat}} = \frac{\left[m_W^{Sat} \middle/ M^m \right]}{P_A^{Sat}} = \frac{\left[1\,\mathrm{g\,m^{-3}} \middle/ 300\,\mathrm{g\,mol^{-1}} \right]}{1\,\mathrm{Pa}} = 3.33 \times 10^{-3}\,\mathrm{mol\,Pa^{-1}\,m^{-3}}$$

ii. Calculate Z-values for all components in the sediment phase:

The Z-values for OM and MM are required to determine the overall Z-value for the sediment particles, and later below the suspended particles in water. Note that the mass fractions and therefore volume fractions of organic and mineral components are different in suspended particles and settled sediment, so we will have two different overall Z-values for particles suspended in the water phase and those that comprise the solid material in the sediment.

The Z-value for OM is obtained from the mass fraction of OC in OM, and the Z-value for OC. The latter is obtained from the product of $K_{OC}Z_W$, where we are once again taking advantage of the fact that partition ratios are ratios of Z-values, and in this instance $K_{OC} = Z_{OC}/Z_W$:

$$K_{OC}Z_W = \frac{Z_{OC}Z_W}{Z_W} = Z_{OC}$$

We include a redundant adjustment for relative density of OM and water, both of which happen to have the same value in this particular case, but in general this need not apply.

$$Z_{OM} = \frac{m_{OM}^f K_{OC} Z_W \rho_{OM}}{1000} = \frac{0.56 \times 4.1 \times 10^5 \times 3.33 \times 10^{-3}\,\mathrm{mol\,Pa^{-1}\,m^{-3}} \times 1000}{1000}$$

$$= 765.3\,\mathrm{mol\,Pa^{-1}m^{-3}}$$

The Z-value for MM in both phases is calculated in a similar manner, but now the density difference does impact the calculation, and there is no correction needed that is analogous to the adjustment for the OC content in OM, since the MM is assumed to be 100% "active" matter:

$$Z_{MM} = \frac{K_{MM-W} Z_W \rho_{MM}}{1000}$$

$$= \frac{1.00 \times 3.33 \times 10^{-3} \, \text{mol Pa}^{-1} \, \text{m}^{-3} \times 2500}{1000}$$

$$= 8.33 \times 10^{-3} \, \text{mol Pa}^{-1} \, \text{m}^{-3}$$

As with the Soil model calculations above, the overall or total Z-values for both bulk water and sediment require volume fractions for each component, which must be determined from mass fractions as in Worked Example 6.2.

The volume fraction of solids in the sediment is obtained from the volume fraction of sediment water, which is given as input data:

$$v_W^{f-\text{Sed}} = 0.8$$

$$v_{\text{Solid}}^{f-\text{Sed}} = 1 - v_W^{f-\text{Sed}} = 1 - 0.8 = 0.2$$

The mass fraction of OC in the sediment solids is selected as 0.05, and the mass fraction of OC in OM is 0.56. Therefore, the mass fraction of OM in the sediment solids is

$$m_{OM}^{f-\text{SedDry}} = \frac{m_{OC}^{f-\text{SedDry}}}{m_{OM}^{f-OC}} = \frac{0.05}{0.56} = 8.928 \times 10^{-2}$$

As before, we can determine the mass fraction of MM in the sediment solids by mass balance, such that

$$m_{MM}^{f-\text{SedDry}} = 1 - m_{OM}^{f-\text{SedDry}} = 1 - 8.928 \times 10^{-2} = 9.1072 \times 10^{-1}$$

We now need to convert these to volume fractions:

$$v_{MM}^{f-\text{SedDry}} = \left[\frac{1}{1 + \left(\dfrac{m_{OM}^{f-\text{SedDry}} / \rho_{OM}}{m_{MM}^{f-\text{SedDry}} / \rho_{MM}} \right)} \right] = \left[\frac{1}{1 + \left(\dfrac{0.08928/1000 \, \text{kg m}^{-3}}{0.91072/2500 \, \text{kg m}^{-3}} \right)} \right] = 0.80316$$

It follows by a volume balance that

$$v_{OM}^{f-\text{SedDry}} = 1 - v_{MM}^{f-\text{SedDry}} = 1 - 0.80316 = 0.1968$$

Finally, we must recognize that the solids comprise only 20% of the sediment by volume (by volume fraction 1.0 − 0.8), such that the actual volume fractions of OM and MM in the sediment are

$$v_{MM}^{f-\text{Sed}} = \left(1 - v_W^{f-\text{Sed}} \right) v_{MM}^{\text{SedDry}} = (1 - 0.8) \times 0.80316 = 0.1606$$

$$v_{OM}^{f-\text{Sed}} = \left(1 - v_W^{f-\text{Sed}} \right) v_{OM}^{\text{SedDry}} = (1 - 0.8) \times 0.19684 = 0.03937$$

Therefore, for the sediment, we can determine the bulk Z-value from the volume fraction-weighted Z-values of the three components, these being water, OM, and MM:

$$Z_{Sed}^{Bulk} = v_W^{f-Sed} Z_W + v_{MM}^{f-Sed} Z_{MM} + v_{OM}^{f-Sed} Z_{OM}$$

$$= 0.8 \times 3.33 \times 10^{-3} \, mol \, Pa^{-1} \, m^{-3} + 0.1606 \times 8.33 \times 10^{-3} \, mol \, Pa^{-1} \, m^{-3} + 0.03937 \times 765.3 \, mol \, Pa^{-1} \, m^{-3}$$

$$= 30.14 \, mol \, Pa^{-1} \, m^{-3}$$

Note that we could also do this calculation by first calculating the bulk Z-value for the sediment solids as

$$Z_{SedDry}^{Bulk} = v_{MM}^{f-SedDry} Z_{MM} + v_{OM}^{f-SedDry} Z_{OM}$$

$$= 0.80316 \times 8.33 \times 10^{-3} \, mol \, Pa^{-1} \, m^{-3} + 0.1968 \times 765.3 \, mol \, Pa^{-1} \, m^{-3}$$

$$= 150.65 \, mol \, Pa^{-1} \, m^{-3}$$

Now, the same bulk sediment value would be obtained from the simpler volume fractions of just water and all solids:

$$Z_{Sed}^{Bulk} = v_W^{f-Sed} Z_W + v_{SedDry}^{f-Sed} Z_{SedDry}^{Bulk}$$

$$= 0.8 \times 3.33 \times 10^{-3} \, mol \, Pa^{-1} \, m^{-3} + (1 - 0.8) \times 150.65 \, mol \, Pa^{-1} \, m^{-3}$$

$$= 30.14 \, mol \, Pa^{-1} \, m^{-3}$$

The latter approach would leave us with the dry sediment bulk Z-value on hand, should we want to use it to determine dry sediment concentrations and other properties.

iii. Calculate Z-values for all components in the water phase:
    The analogous calculations for the suspended solids in water are as follows:
    Using the same Z-values for OM and MM, we may now determine the bulk Z-value for the total (dry) suspended solids (TSP). Again, we must use the known mass fraction of OC in the suspended solids and the mass fraction of OC in OM:

$$m_{OM}^{f-TSP} = \frac{m_{OC}^{f-TSP}}{m_{OM}^{f-OC}} = \frac{0.25}{0.56} = 0.4464$$

As before, we can determine the mass fraction of MM in the sediment solids by mass balance, such that

$$m_{MM}^{f-TSP} = 1 - m_{OM}^{f-TSP} = 1 - 0.4464 = 0.5536$$

Convert these to volume fractions:

$$v_{MM}^{f-TSP} = \left[\frac{1}{1 + \left(\dfrac{m_{OM}^{f-TSP} / \rho_{OM}}{m_{MM}^{f-TSP} / \rho_{MM}} \right)} \right] = \left[\frac{1}{1 + \left(\dfrac{0.4464 / 1000 \, kg \, m^{-3}}{0.5536 / 2500 \, kg \, m^{-3}} \right)} \right] = 0.3316$$

It follows by volume balance that

$$v_{OM}^{f-TSP} = 1 - v_{MM}^{f-TSP} = 1 - 0.3316 = 0.6684$$

The solids comprise only a fraction of the total suspended particles by volume. To determine the actual volume fractions of OM and MM in the sediment, we must first determine the volume fractions of water and solid matter in the bulk water. We begin by determining the density of solid particles from a volume fraction-weighted combination of the densities of OM and MM:

$$\rho_{TSP} = v_{OM}^{f-TSP} \rho_{OM} + v_{MM}^{f-TSP} \rho_{MM}$$

$$= 0.6684 \times 1000 \, \text{g m}^{-3} + 0.3316 \times 2500 \, \text{g m}^{-3}$$

$$= 1497.4 \, \text{g m}^{-3}$$

With the solid matter density in hand, we may now convert the mass concentration of solid matter in the suspended particles into a volume fraction:

$$v_{TSP}^{f} = \frac{C_{TSP}}{\rho_{TSP}} = \frac{5 \, \text{g m}^{-3} / 1000 \, \text{g kg}^{-1}}{1497.4 \, \text{kg m}^{-3}} = 3.339 \times 10^{-6}$$

Now using this volume fraction, we can determine the volume fractions of both OM and MM in the bulk water:

$$v_{OM}^{f-W} = v_{TSP}^{f} \times v_{OM}^{f-TSP} = 3.339 \times 10^{-6} \times 0.6684 = 2.232 \times 10^{-6}$$

$$v_{MM}^{f-W} = v_{TSP}^{f} \times v_{MM}^{f-TSP} = 3.339 \times 10^{-6} \times 0.3316 = 1.107 \times 10^{-6}$$

The volume fraction of water in the bulk water phase is obtained by mass balance:

$$v_{W}^{f-W} = 1 - v_{TSP}^{f} = 1 - 3.339 \times 10^{-6} \approx 1$$

With this information in hand, we may determine the Z-value for the dry suspended particles (TSP):

$$Z_{TSP} = v_{OM}^{f-TSP} Z_{OM} + v_{MM}^{f-TSP} Z_{MM}$$

$$= 0.6684 \times 765.3 \, \text{mol Pa}^{-1} \, \text{m}^{-3} + 0.3316 \times 8.33 \times 10^{-3} \, \text{mol Pa}^{-1} \, \text{m}^{-3}$$

$$= \left(511.53 + 2.76 \times 10^{-3} \right) \text{mol Pa}^{-1} \, \text{m}^{-3}$$

$$= 511.53 \, \text{mol Pa}^{-1} \, \text{m}^{-3}$$

Finally, for the bulk water phase, we can again determine the bulk Z-value from the volume fraction-weighted Z-values of the three components, these being water, OM, and MM:

$$Z_{W}^{Bulk} = v_{W}^{f-W} Z_{W} + v_{OM}^{f-W} Z_{OM} + v_{MM}^{f-W} Z_{MM}$$

$$= 1 \times 3.33 \times 10^{-3} \, \text{mol Pa}^{-1} \, \text{m}^{-3} + 2.232 \times 10^{-6} \times 765.3 \, \text{mol Pa}^{-1} \, \text{m}^{-3}$$

$$+ 1.107 \times 10^{-6} \times 8.33 \times 10^{-3} \, \text{mol Pa}^{-1} \, \text{m}^{-3}$$

$$= \left(3.333 \times 10^{-3} + 1.708 \times 10^{-3} + 9.22 \times 10^{-9} \right) \text{mol Pa}^{-1} \, \text{m}^{-3}$$

$$= 5.042 \times 10^{-3} \, \text{mol Pa}^{-1} \, \text{m}^{-3}$$

iv. Calculate fugacities in water and sediment phases:

Now having all the needed Z-values, the fugacity in the water and the sediment phases may be calculated from the bulk concentrations of each phase with their corresponding bulk Z-values. However, the concentration of hypothene in the sediment is only available as a ratio of mass of chemical to mass of bulk dry sediment, and therefore first needs to be converted to molar concentration units of moles of hypothene per m³ of bulk dry sediment. First we determine the number of moles per gram of dry sediment:

$$m_{Sed}^{1g} = \frac{mass_{Sed}^{1g}}{M^m} = \frac{10\,\mu g \times 10^{-6}\,g\mu g^{-1}}{300\,g\,mol^{-1}} = 3.33 \times 10^{-8}\,mol$$

The volume of 1 g of dry sediment must be determined from the bulk dry sediment density, which may be determined in the same manner as for the dry suspended solids above.

The mass fraction of OM in the dry sediment is obtained from the mass fraction of OC in the dry sediment and the mass fraction of OC in OM:

$$m_{OM}^{f-SedDry} = \frac{m_{OC}^{f-SedDry}}{m_{OM}^{f-OC}} = \frac{0.05}{0.56} = 8.928 \times 10^{-2}$$

Again, we can determine the mass fraction of MM in the dry sediment solids by mass balance, such that

$$m_{MM}^{f-SedDry} = 1 - m_{OM}^{f-SedDry} = 1 - 8.928 \times 10^{-2} = 9.1072 \times 10^{-1}$$

Convert these to volume fractions:

$$v_{MM}^{f-SedDry} = \left[\frac{1}{1 + \left(\dfrac{m_{OM}^{f-SedDry}/\rho_{OM}}{m_{MM}^{f-SedDry}/\rho_{MM}} \right)} \right] = \left[\frac{1}{1 + \left(\dfrac{8.928 \times 10^{-2}/1000\,kg\,m^{-3}}{9.1072 \times 10^{-1}/2500\,kg\,m^{-3}} \right)} \right] = 0.80316$$

It follows by volume balance that

$$v_{OM}^{f-SedDry} = 1 - v_{MM}^{f-SedDry} = 1 - 0.80316 = 0.19684$$

The solids comprise only a fraction of the total suspended particles by volume. To determine the actual volume fractions of OM and MM in the sediment, we must first determine the volume fractions of water and solid matter in the bulk water. We begin by determining the density of solid particles from a volume fraction-weighted combination of the densities of OM and MM:

$$\rho_{TSP}^{Dry} = v_{OM}^{f-SedDry}\rho_{OM} + v_{MM}^{f-SedDry}\rho_{MM}$$

$$= 0.19684 \times 1000\,g\,m^{-3} + 0.80316 \times 2500\,g\,m^{-3}$$

$$= 2204.72\,g\,m^{-3}$$

$$V_{Sed}^{1g} = \frac{m_{Sed}}{\rho_{Sed}^{Bulk}} = \frac{\left(1\,g/1\times10^3\,g\,kg^{-1}\right)}{2204.72\,kg\,m^{-3}} = 4.5357 \times 10^{-7}\,m^3$$

Finally, we arrive at the concentration of hypothene in the sediment in mol m⁻³ units as

$$C_{Sed} = \frac{m_{Sed}^{1g}}{V_{Sed}^{1g}} = \frac{3.33\overline{3} \times 10^{-8}\,mol}{4.5357 \times 10^{-7}\,m^3} = 7.349 \times 10^{-2}\,mol\,m^{-3}$$

The fugacities in the two phases may now be determined under the initial concentration conditions:

$$f_W = \frac{C_W^{Bulk}}{Z_W^{Bulk}} = \frac{\left(1 \times 10^{-4}\,\text{g m}^{-3}/300\,\text{g mol}^{-1}\right)}{5.042 \times 10^{-3}\,\text{mol Pa}^{-1}\,\text{m}^{-3}} = 6.612 \times 10^{-5}\,\text{Pa}$$

$$f_{Sed} = \frac{C_{Sed}^{Bulk}}{Z_{SedDry}^{Bulk}} = \frac{7.349 \times 10^{-2}\,\text{mol m}^{-3}}{150.65\,\text{mol Pa}^{-1}\,\text{m}^{-3}} = 4.878 \times 10^{-4}\,\text{Pa}$$

v. Calculate concentrations of hypothene in all components of the water and sediment phases:

With the bulk fugacities in hand, we can now directly calculate the concentrations of hypothene in each of the constituents of both water and sediment phases.

Beginning with the water phase:

$$C_W^{Bulk} = Z_W^{Bulk} f_W = 5.024 \times 10^{-3}\,\text{mol Pa}^{-1}\,\text{m}^{-3} \times 6.612 \times 10^{-5}\,\text{Pa} = 3.32 \times 10^{-7}\,\text{mol m}^{-3}$$

$$C_W^W = Z_W f_W = 3.333 \times 10^{-3}\,\text{mol Pa}^{-1}\,\text{m}^{-3} \times 6.612 \times 10^{-5}\,\text{Pa} = 2.2038 \times 10^{-7}\,\text{mol m}^{-3}$$

$$C_W^{OM} = Z_{OM} f_W = 765.3\,\text{mol Pa}^{-1}\,\text{m}^{-3} \times 6.612 \times 10^{-5}\,\text{Pa} = 5.060 \times 10^{-2}\,\text{mol m}^{-3}$$

$$C_W^{MM} = Z_{MM} f_W = 8.33 \times 10^{-3}\,\text{mol Pa}^{-1}\,\text{m}^{-3} \times 6.612 \times 10^{-5}\,\text{Pa} = 5.508 \times 10^{-7}\,\text{mol m}^{-3}$$

$$C_W^{TSP} = Z_{TSP} f_W = 511.53\,\text{mol Pa}^{-1}\,\text{m}^{-3} \times 6.612 \times 10^{-5}\,\text{Pa} = 3.382 \times 10^{-2}\,\text{mol m}^{-3}$$

Similarly for the sediment phase:

$$C_{Sed}^{Bulk} = Z_{Sed}^{Bulk} f_{Sed} = 30.13\,\text{mol Pa}^{-1}\,\text{m}^{-3} \times 4.878 \times 10^{-4}\,\text{Pa} = 1.470 \times 10^{-2}\,\text{mol m}^{-3}$$

$$C_{Sed}^W = Z_W f_{Sed} = 3.333 \times 10^{-3}\,\text{mol Pa}^{-1}\,\text{m}^{-3} \times 4.878 \times 10^{-4}\,\text{Pa} = 1.626 \times 10^{-6}\,\text{mol m}^{-3}$$

$$C_{SedDry}^{Bulk} = Z_{SedDry}^{Bulk} f_{Sed} = 150.65\,\text{mol Pa}^{-1}\,\text{m}^{-3} \times 4.878 \times 10^{-4}\,\text{Pa} = 7.341 \times 10^{-2}\,\text{mol m}^{-3}$$

$$C_{Sed}^{OM} = Z_{OM} f_{Sed} = 765.3\,\text{mol Pa}^{-1}\,\text{m}^{-3} \times 4.878 \times 10^{-4}\,\text{Pa} = 3.733 \times 10^{-1}\,\text{mol m}^{-3}$$

$$C_{Sed}^{MM} = Z_{MM} f_{Sed} = 8.33 \times 10^{-3}\,\text{mol Pa}^{-1}\,\text{m}^{-3} \times 4.878 \times 10^{-4}\,\text{Pa} = 4.063 \times 10^{-6}\,\text{mol m}^{-3}$$

Once again, the hydrophobic nature of hypothene results in relatively high concentrations in all phases that are comprised of or contain OM.

The actual amounts of chemical in each subcompartment are determined by the product of the concentration in the component times the component volume, the latter again given by the product of the total phase volume and the volume fraction of the phase component.

Using the input area and water depth, we can calculate the total bulk water volume as

$$V_W^{Bulk} = A_{W-Sed} \times Y_W = 10^3\,\text{m}^2 \times 20\,\text{m} = 2.0 \times 10^4\,\text{m}^3$$

The bulk sediment volume is given by

$$V_{Sed}^{Bulk} = A_{W-Sed} \times Y_{Sed} = 10^3\,\text{m}^2 \times 3.0 \times 10^{-2}\,\text{m} = 30\,\text{m}^3$$

Therefore, for each subcompartment of the bulk water as well as bulk water itself, the total amount of hypothene in moles is given by

$$m_W^{Bulk} = V_W^{Bulk} C_W^{Bulk} = 2.0 \times 10^4\,\text{m}^3 \times 3.33 \times 10^{-7}\,\text{mol m}^{-3} = 6.67 \times 10^{-3}\,\text{mol}$$

$$m_W^W = v_W^{f-W} V_W^{\text{Bulk}} C_W^W = 1.0 \times 2.0 \times 10^4 \text{ m}^3 \times 2.2038 \times 10^{-7} \text{mol m}^{-3} = 4.408 \times 10^{-3} \text{mol}$$

$$m_W^{TSP} = v_W^{f-TSP} V_W^{\text{Bulk}} C_W^{TSP} = 3.339 \times 10^{-6} \times 2.0 \times 10^4 \text{ m}^3 \times 3.382 \times 10^{-2} \text{ mol m}^{-3} = 2.258 \times 10^{-3} \text{mol}$$

$$m_W^{OM} = v_W^{f-OM} V_W^{\text{Bulk}} C_W^{OM} = 2.232 \times 10^{-6} \times 2.0 \times 10^4 \text{ m}^3 \times 5.060 \times 10^{-2} \text{mol m}^{-3} = 2.259 \times 10^{-3} \text{mol}$$

$$m_W^{MM} = v_W^{f-MM} V_W^{\text{Bulk}} C_W^{MM} = 1.107 \times 10^{-6} \times 2.0 \times 10^4 \text{ m}^3 \times 5.508 \times 10^{-7} \text{mol m}^{-3} = 1.219 \times 10^{-8} \text{ mol}$$

Similarly for the sediment phase:

$$m_{\text{Sed}}^{\text{Bulk}} = V_{\text{SedDry}} C_{\text{Sed}}^{\text{SedDry}} = 30 \text{ m}^3 \times 1.470 \times 10^{-2} \text{mol m}^{-3} = 4.41 \times 10^{-1} \text{ mol}$$

$$m_W^W = v_{\text{Sed}}^{f-W} V_{\text{Sed}}^{\text{Bulk}} C_{\text{Sed}}^W = 0.8 \times 30 \text{ m}^3 \times 1.626 \times 10^{-6} \text{mol m}^{-3} = 3.9024 \times 10^{-5} \text{mol}$$

$$m_{\text{Sed}}^{\text{SedDry}} = v_{\text{Sed}}^{f-\text{SedDry}} V_{\text{Sed}}^{\text{Bulk}} C_{\text{SedDry}}^{\text{Bulk}} = 0.2 \times 30 \text{ m}^3 \times 7.341 \times 10^{-2} \text{ mol m}^{-3} = 4.405 \times 10^{-1} \text{mol}$$

$$m_{\text{Sed}}^{OM} = v_{\text{Sed}}^{f-OM} V_{\text{Sed}}^{\text{Bulk}} C_{\text{Sed}}^{OM} = 0.03936_8 \times 30 \text{ m}^3 \times 3.733 \times 10^{-1} \text{mol m}^{-3} = 4.4088 \times 10^{-1} \text{mol}$$

$$m_{MM}^{\text{Sed}} = v_{\text{Sed}}^{f-MM} V_{\text{Sed}}^{\text{Bulk}} C_{MM}^{\text{Sed}} = 0.1606_{32} \times 30 \text{ m}^3 \times 4.063 \times 10^{-6} \text{mol m}^{-3} = 1.958 \times 10^{-5} \text{ mol}$$

The percent mass of chemical in each subcompartment is given by

$$\% \text{Mass} = \left(\frac{\text{mass in compartment}}{\text{total mass in phase}} \right) \times 100\%$$

to the following percent mass contents for each subcompartment:

	% Water Phase	% Sediment Phase
Water	66.116	0.008849
Sediment solids/TSP	33.884	99.991
OM	33.884	99.987
MM	0.000183	0.004442

Note that the OM in the sediment is the primary site for the concentration of hypothene. In the water, the fact that there is relatively little sediment present results in the water holding the bulk of the hypothene, about twice as much as partitions to the total suspended particles.

vi. Calculate the transport and persistence of hypothene in all components of both water and sediment phases:

We now need to determine the various D-values that describe the transport processes under consideration. As always, some D-values will be combined as parallel processes, or as "partners" in the context of the Whitman two-resistance approach to diffusion across phase interfaces.

The processes are:
Diffusion from the water phase to the sediment phase
Diffusion from sediment phase to water phase
Sediment deposition from the water phase into the sediment phase
Resuspension of sediment from the sediment phase to the water phase
Burial of active sediment into the inactive sediment zone
Chemical decomposition or reaction in the sediment phase

For the diffusion from water to sediment, we need to determine the corresponding mass-transfer coefficient from the molecular diffusivity B, the void volume fraction for the sediment, and the diffusion path length

$$
\begin{aligned}
D_{W-\text{Sed}}^{\text{Diff}} &= \left(\frac{B_{\text{Sed}}^{\text{Diff}} A_{\text{Sed}-W} Z_W}{Y_{\text{Sed}}} \right) \left(v_{\text{Sed}}^{f-\text{Void}} \right)^{1.5} \\
&= \left(\frac{2.0 \times 10^{-6}\, \text{m}^2\, \text{h}^{-1} \times 1000\, \text{m}^2 \times 3.33 \times 10^{-3}\, \text{mol Pa}^{-1}\, \text{m}^{-3}}{0.015\, \text{m}} \right) 0.8^{1.5} \\
&= 3.18 \times 10^{-4}\, \text{mol Pa}^{-1}\, \text{h}^{-1}
\end{aligned}
$$

Diffusion from sediment to water is assumed to be by the same mechanism, i.e., both transfer processes assume that transport through the active sediment layer of 1.5×10^{-2} m depth controls diffusion in both directions.

Deposition, resuspension, and burial are all advective processes. For such processes, all D-values are of the form GZ, where G is an equivalent bulk volume flow in $\text{m}^3\, \text{h}^{-1}$. As various environmentally relevant processes are reported as fluxes or rates with various favored units, some conversion is necessary to bring them onto equal footing from an equivalent flow in $\text{m}^3\, \text{h}^{-1}$ viewpoint.

For example, consider the sedimentation flux, input as $3\, \text{g}\, \text{m}^{-2}\, \text{day}^{-1}$. This is the assumed average flux of particles that collected in a square meter each day. To consider this from the point of view of a flow of particles across the water–sediment interface, we need to consider how much water contains 3.0 g of sediment. Once we know that volume, we can then determine the length of a column of water with a 1 m^2 cross-sectional area that contains a day's worth of particles. This volume is then considered to bring its sediment content to the 1 m^2 interfacial area each day, and is analogous to a flow rate of sediment-containing water across that threshold.

Thus, the G-value or equivalent bulk water volume flow rate for sediment deposition is given by

$$
\begin{aligned}
G_{TSP}^{\text{Dep}} &= \frac{L_{\text{Sed}} A_{\text{Sed}-W}}{\rho_{TSP} \times 24\, \text{h}\, \text{day}^{-1} \times 1 \times 10^3\, \text{g}\, \text{kg}^{-1}} \\
&= \frac{3\, \text{g}\, \text{m}^{-2}\, \text{day}^{-1} \times 1.00 \times 10^3\, \text{m}^2}{1497.4\, \text{kg}\, \text{m}^{-3} \times 24\, \text{h}\, \text{day}^{-1} \times 1 \times 10^3\, \text{g}\, \text{kg}^{-1}} \\
&= 8.348 \times 10^{-5}\, \text{m}^3\, \text{h}^{-1}
\end{aligned}
$$

The corresponding D-value for sedimentation is then

$$
\begin{aligned}
D_{TSP}^{\text{Dep}} &= G_{TSP}^{\text{Dep}} Z_{TSP} \\
&= 8.348 \times 10^{-5}\, \text{m}^3\, \text{h}^{-1} \times 511.53\, \text{mol Pa}^{-1}\, \text{m}^{-3} \\
&= 4.271 \times 10^{-2}\, \text{mol Pa}^{-1}\, \text{h}^{-1}
\end{aligned}
$$

The analogous resuspension D-value differs only in the density of the particles resuspended and the Z-value for sediment particles:

$$
\begin{aligned}
G_{\text{Sed}}^{\text{Resusp}} &= \frac{L_{\text{Resusp}} A_{\text{Sed}-W}}{\rho_{TSP} \times 24\, \text{h}\, \text{day}^{-1} \times 1 \times 10^3\, \text{g}\, \text{kg}^{-1}} \\
&= \frac{1\, \text{g}\, \text{m}^{-2}\, \text{day}^{-1} \times 1.00 \times 10^3\, \text{m}^2}{2204.74\, \text{kg}\, \text{m}^{-3} \times 24\, \text{h}\, \text{day}^{-1} \times 1 \times 10^3\, \text{g}\, \text{kg}^{-1}} \\
&= 1.890 \times 10^{-5}\, \text{m}^3\, \text{h}^{-1}
\end{aligned}
$$

The corresponding D-value for resuspension is then

$$D_{Sed}^{Resusp} = G_{Sed}^{Resusp} Z_{SedDry}^{Sed}$$

$$= 1.890 \times 10^{-5} \, m^3 \, h^{-1} \times 150.65 \, mol \, Pa^{-1} m^{-3}$$

$$= 2.847 \times 10^{-3} \, mol \, Pa^{-1} h^{-1}$$

For the process of burial, the calculation is similar to that of resuspension:

$$G_{Sed}^{Burial} = \frac{L_{Burial} A_{Sed-W}}{\rho_{TSP} \times 24 \, h \, day^{-1} \times 1 \times 10^3 \, g \, kg^{-1}}$$

$$= \frac{1.5 \, g \, m^{-2} \, day^{-1} \times 1.00 \times 10^3 \, m^2}{2204.74 \, kg \, m^{-3} \times 24 \, h \, day^{-1} \times 1 \times 10^3 \, g \, kg^{-1}}$$

$$= 2.835 \times 10^{-5} \, m^3 \, h^{-1}$$

The corresponding D-value for burial is then given by

$$D_{Sed}^{Burial} = G_{Sed}^{Burial} Z_{SedDry}^{Sed}$$

$$= 2.835 \times 10^{-5} \, m^3 \, h^{-1} \times 150.65 \, mol \, Pa^{-1} \, m^{-3}$$

$$= 4.271 \times 10^{-3} \, mol \, Pa^{-1} h^{-1}$$

Since burial and resuspension involve all the same properties except for the transfer rate, we could also simply determine the "net burial" D-value by appropriate scaling of the resuspension D-value:

$$D_{Sed}^{Net \, Burial} = \left(\frac{L_{Burial}}{L_{Resusp}} \right) D_{Sed}^{Resusp} = \left(\frac{1.5}{1} \right) \times 2.847 \times 10^{-3} \, mol \, Pa^{-1} h^{-1}$$

$$= 4.271 \times 10^{-3} \, mol \, Pa^{-1} h^{-1}$$

The D-value for decomposition or reaction is given by the product of the rate constant, the compartment volume, and the bulk sediment Z-value. We are given the degradation half-time of 1.2×10^5 h, from which we can determine the rate constant:

$$k_{Sed}^{Deg} = \frac{-\ln(0.5)}{\tau_{1/2}} = \frac{0.693}{1.2 \times 10^5 \, h} = 5.776 \times 10^{-6} \, h^{-1}$$

$$D_{Sed}^{Deg} = k_{Sed}^{Deg} V_{Sed} Z_{Sed}^{Bulk}$$

$$= 5.776 \times 10^{-6} \, h^{-1} \times 1000 \, m^2 \times 0.03 \, m \times 30.13 \, mol \, Pa^{-1} \, m^{-3}$$

$$= 5.221 \times 10^{-3} \, mol \, Pa^{-1} h^{-1}$$

Since decomposition and burial are irreversible and parallel processes, their D-values sum to give the corresponding D-value for total irreversible transfer from the sediment:

$$D_{Sed}^{Irr} = D_{Sed}^{Burial} + D_{Sed}^{Deg}$$

$$= 4.271 \times 10^{-3} \, mol \, Pa^{-1} h^{-1} + 5.221 \times 10^{-3} \, mol \, Pa^{-1} h^{-1}$$

$$= 9.492 \times 10^{-3} \, mol \, Pa^{-1} h^{-1}$$

Similarly, D-values for the total reversible transfer from sediment to water, and from water to sediment may also be calculated as simple sums:

$$D_{W-Sed}^{Rev} = D_{TSP}^{Dep} + D_{Sed-W}^{Diff}$$

$$= 4.271 \times 10^{-2} \, \text{mol Pa}^{-1} \text{h}^{-1} + 3.18 \times 10^{-4} \, \text{mol Pa}^{-1} \text{h}^{-1}$$

$$= 4.303 \times 10^{-3} \, \text{mol Pa}^{-1} \text{h}^{-1}$$

and

$$D_{Sed-W}^{Rev} = D_{Sed}^{Resusp} + D_{W-Sed}^{Diff}$$

$$= 2.847 \times 10^{-3} \, \text{mol Pa}^{-1} \text{h}^{-1} + 3.18 \times 10^{-4} \, \text{mol Pa}^{-1} \text{h}^{-1}$$

$$= 3.165 \times 10^{-3} \, \text{mol Pa}^{-1} \text{h}^{-1}$$

The rates associated with each process are determined by the product of the corresponding D-value and compartment fugacity, and may be converted into the more convenient units of kg year^{-1} with the molar mass. In tabular form, we then have

Transport Process	D-Value (mol Pa^{-1} h^{-1})	Fugacity (Pa)	Transfer Rate (mol h^{-1})	Transfer Rate (kg year^{-1})
Diffusion (W – Sed)	3.18×10^{-4}	6.615×10^{-5}	2.103×10^{-8}	5.526×10^{-5}
Diffusion (Sed – W)	3.18×10^{-4}	4.878×10^{-4}	1.551×10^{-7}	4.077×10^{-4}
TSP deposition	4.27×10^{-2}	6.615×10^{-5}	2.824×10^{-6}	7.421×10^{-3}
Resuspension	2.85×10^{-3}	4.878×10^{-4}	1.389×10^{-6}	3.650×10^{-3}
Burial	4.27×10^{-3}	4.878×10^{-4}	2.083×10^{-6}	5.475×10^{-3}
Decomposition	5.22×10^{-3}	4.878×10^{-4}	2.547×10^{-6}	6.693×10^{-3}
Total (W – Sed)	4.30×10^{-2}	6.615×10^{-5}	2.845×10^{-6}	7.476×10^{-3}
Total (Sed – W)	3.17×10^{-3}	4.878×10^{-4}	1.544×10^{-6}	4.058×10^{-3}
Total burial and decomposition	9.49×10^{-3}	4.878×10^{-4}	4.630×10^{-6}	1.217×10^{-2}

The characteristic times for different removal processes may be determined by the ratio of the total amount of chemical in the compartment divided by the rate of removal by the process in question. For example, for sediment burial and reaction (i.e., irreversible losses), we have the total amount of hypothene in the sediment given by the product of the concentration of hypothene in the sediment (as Zf) and the sediment volume:

$$m_{Sed} = C_{Sed} V_{Sed}$$

$$= Z_{Sed}^{Bulk} f_{Sed} A_{Sed-W} Y_{Sed} = 30.14 \, \text{mol Pa}^{-1} \text{m}^{-3} \times 4.878 \times 10^{-4} \, \text{Pa} \times 1 \times 10^3 \, \text{m}^2 \times 3.0 \times 10^{-2} \, \text{m}$$

$$= 4.411 \times 10^{-1} \, \text{mol}$$

The characteristic time is therefore

$$\tau_{IrrLoss} = \frac{m_{Sed}}{r_{IrrLoss}} = \frac{m_{Sed}}{\left(r_{Sed}^{Burial} + r_{Sed}^{Deg} \right)}$$

$$= \frac{4.411 \times 10^{-1} \, \text{mol}}{(2.083 \times 10^{-6} + 2.547 \times 10^{-6}) \, \text{mol h}^{-1}} = 9.526 \times 10^4 \, \text{h}$$

$$= \frac{9.526 \times 10^4 \, \text{h}}{24 \, \text{h day}^{-1} \times 365.25 \, \text{days year}^{-1}} = 10.867 \, \text{year}$$

The corresponding half-time for loss is simply the product of $^-\ln(0.5)$ or 0.6931 and the characteristic time:

$$\tau_{\text{IrrLoss}}^{1/2} = 0.6931 \times \tau_{\text{IrrLoss}}$$

$$= 0.6931 \times 9.526 \times 10^4 \, \text{h} = 6.603 \times 10^4 \, \text{h}$$

$$= 0.6931 \times 10.867 \, \text{year} = 7.53 \, \text{year}$$

Bringing these calculations together into one equation, we have for the half-time in years:

$$\tau_i^{1/2} = \left(\frac{0.6931}{24 \times 365.25} \right) \frac{Z_i^{\text{Bulk}} f_i V_i}{\displaystyle\sum_{i=1}^{\text{All losses}} r_i}$$

Thus, for water-to-sediment transfer, we can write

$$\tau_{\text{Sed}-W}^{1/2} = \frac{\ln(2)}{24 \, \text{h day}^{-1} \times 365.25 \, \text{day year}^{-1}} \left(\frac{Z_W^{\text{Bulk}} f_W V_W}{r_{W-\text{Sed}}} \right)$$

$$= \frac{0.6931}{24 \times 365.25} \times \left(\frac{5.042 \times 10^{-3} \, \text{mol Pa}^{-1} \text{m}^{-3} \times 6.612 \times 10^{-5} \, \text{Pa} \times \left[1 \times 10^3 \, \text{m}^2 \times 20 \, \text{m} \right]}{2.845 \times 10^{-6} \, \text{mol h}^{-1}} \right)$$

$$= 0.185 \, \text{year}$$

Similarly, for sediment-to-water transfer, we can write

$$\tau_{W-\text{Sed}}^{1/2} = \frac{\ln(2)}{24 \, \text{h day}^{-1} \times 365.25 \, \text{day year}^{-1}} \left(\frac{Z_{\text{Sed}}^{\text{Bulk}} f_{\text{Sed}} V_{\text{Sed}}}{r_{\text{Sed}-W}} \right)$$

$$= \frac{0.6931}{24 \times 365.25} \times \left(\frac{30.14 \, \text{mol Pa}^{-1} \text{m}^{-3} \times 4.878 \times 10^{-4} \, \text{Pa} \times \left[1 \times 10^3 \, \text{m}^2 \times 0.030 \, \text{m} \right]}{1.544 \times 10^{-6} \, \text{mol h}^{-1}} \right)$$

$$= 22.58 \, \text{year}$$

Finally, we may ask the question: What is the half-time for all processes that cause removal from the sediment. This value is relevant for remediation, for example, in situations where the water fugacity is reduced to zero and the sediment "recovers" with this half-time. The calculation is similar, only requiring that the rates for all removal processes from the sediment be combined, whether irreversible or not. This is conveniently done by summing the appropriate rates:

$$r_{\text{Sed}}^{\text{TotLoss}} = r_{\text{Sed}}^{\text{IrrLoss}} + r_{\text{Sed}-W}^{\text{Tot}} = 4.630 \times 10^{-6} \, \text{mol h}^{-1} + 1.544 \times 10^{-6} \, \text{mol h}^{-1}$$

$$= 6.174 \times 10^{-6} \, \text{mol h}^{-1}$$

Now, we proceed in the same manner:

$$\tau_{\text{Sed}-\text{TotLoss}}^{1/2} = \frac{\ln(2)}{24 \, \text{h day}^{-1} \times 365.25 \, \text{day year}^{-1}} \left(\frac{Z_{\text{Sed}}^{\text{Bulk}} f_{\text{Sed}} V_{\text{Sed}}}{r_{\text{Sed}}^{\text{TotLoss}}} \right)$$

$$= \frac{0.6931}{24 \times 365.25} \times \left(\frac{30.14 \, \text{mol Pa}^{-1} \text{m}^{-3} \times 4.878 \times 10^{-4} \, \text{Pa} \times \left[1 \times 10^3 \, \text{m}^2 \times 0.030 \, \text{m} \right]}{6.174 \times 10^{-6} \, \text{mol h}^{-1}} \right)$$

$$= 5.65 \, \text{year}$$

Therefore, we can see that, if free from any water-to-sediment transfer, the concentration of hypothene in the sediment would drop to half its initial concentration after 5.65 years.

vii. Determine the steady-state fugacity ratio and predict steady-state fugacities:

It may be of interest to ask the question: "What would the fugacity in the sediment be after a long time if the water fugacity remained unchanged?" Such a question would be appropriate for a situation where a contaminated water stream comes in contact with uncontaminated sediment, for example. Alternatively, we might ask the same question about the water, on the assumption that the sediment fugacity remains constant. To answer this question, we must invoke the steady-state condition introduced above, in which the transfer processes between the two phases are equal:

$$f_W \left(D_W^{\text{Dep}} + D_{\text{Sed}-W}^{Ov} \right) = f_S \left(D_{\text{Sed}}^{\text{Resusp}} + D_{\text{Sed}-W}^{Ov} + D_{\text{Sed}}^{\text{Burial}} + D_{\text{Sed}}^{\text{Deg}} \right)$$

The corresponding ratio of the two fugacities of interest at steady state is

$$\frac{f_S}{f_W} = \frac{\left(D_W^{\text{Dep}} + D_{\text{Sed}-W}^{Ov} \right)}{\left(D_{\text{Sed}}^{\text{Resusp}} + D_{\text{Sed}-W}^{Ov} + D_{\text{Sed}}^{\text{Burial}} + D_{\text{Sed}}^{\text{Deg}} \right)}$$

Under these conditions, the rate of transport out of the water is equal to that of transport out of the sediment, i.e., the system is at steady state. For our case:

$$\frac{f_S}{f_W} = \frac{\left(4.27 \times 10^{-2} + 3.18 \times 10^{-4} \right)}{\left(2.85 \times 10^{-3} + 3.18 \times 10^{-4} + 4.27 \times 10^{-3} + 5.22 \times 10^{-3} \right)}$$

$$= \frac{\left(4.3028 \times 10^{-2} \right)}{\left(1.2658 \times 10^{-2} \right)} = 3.40$$

Using this steady-state fugacity ratio, we can solve for a given phase's fugacity f_i, by assuming that the other is fixed at the initial value f_j^{Fixed}:

$$f_{\text{Sed}}^{SS} = 3.40 f_W^{\text{Fixed}} = 3.40 \times 6.612 \times 10^{-5} \text{Pa} = 2.247 \times 10^{-4} \text{Pa}$$

$$f_W^{SS} = \frac{f_{\text{Sed}}^{\text{Fixed}}}{3.40} = \frac{4.878 \times 10^{-4} \text{Pa}}{3.40} = 1.435 \times 10^{-4} \text{Pa}$$

Although organisms living in the water or sediment are not explicitly included in the mass and transport balances in the *Sediment* model, it is possible to estimate levels of contamination and BCFs for both. Such estimates are made on the assumption that these organisms are at equilibrium with their immediate environment, so that fish exist at the same fugacity as the water, and benthic organisms at the same fugacity as the sediment. As introduced above, we can estimate the BCFs as

$$K_{\text{Biota}-W} = v_L^{f-W\text{Biota}} K_{OW} = 0.05 \times 10^6 = 5.0 \times 10^4$$

$$K_{\text{Biota}-\text{Sed}} = v_L^{f-\text{SedBiota}} K_{OW} = 0.03 \times 10^6 = 3.0 \times 10^4$$

Since the BCF is simply the ratio of chemical concentration in the organism with respect to the surrounding medium, the biotic concentrations follow immediately:

$$C_{\text{Biota}}^W = K_{\text{Biota}-W} C_W = 5.0 \times 10^4 \times 2.204 \times 10^{-7} \text{mol m}^{-3} = 1.102 \times 10^{-2} \text{mol m}^{-3}$$

$$C_{\text{Biota}}^{Sed} = K_{\text{Biota}}^{Sed} C_{Sed} = 3.0 x 10^4 \times 1.626 x 10^{-6} \text{mol m}^{-3} = 4.878 x 10^{-2} \text{mol m}^{-3}$$

Note that the appropriate concentration for the medium in which the organism exists is that of the "pure" water only in either the water or the sediment phase. That is, we consider the organism to be at the same fugacity as the water fraction of the compartment in question, and ignore fugacity contributions from any of the "objects" in the water, such as suspended particles or sediment solids. This is required as a result of the way in which the BCF is defined.

The output from the spreadsheet *Sediment* model is as follows for this problem:

Sediment Results

Chemical Name	Hypothene
Environment Name	Hypothene

Total Chemcial Conc. in Water	1.00E-04	g/m³
Chemical Conc. in Sed. Soldis (dry)	1.00E+01	µg/g

	mol	g
Chemical in System	4.48E-01	1.34E+02

Fugacity

Water	6.612E-05	Pa
Sediment	4.878E-04	Pa

Sediment to Water Ratio	7.38E+00

Prediction of Fugacities at Steady-State

Sediment to Water Ratio at Steady-State	3.40

Water Fugacity	1.43E-04	Pa
Sediment Fugaicty	2.25E-04	Pa

Biota

Biotic Concentrations	mol/m³	µg/g
Water Organisms	1.10E-02	3.31E+00
Sediment Organisms	4.88E-02	1.46E+01

Bioconcentration Factor (BCF)	unitless
Water Organisms	5.00E+04
Sediment Organisms	3.00E+04

Phase Properties

	Z Value	Concentration			Amount			
	mol/m³·Pa	mol/m³	g/m³	µg/g	mol	g	% in System	% in Water
Bulk Water	5.042E-03	3.333E-07	1.000E-04	1.000E-04	6.667E-03	2.000E+00	1.489	100.000
Water	3.333E-03	2.204E-07	6.612E-05	6.612E-05	4.408E-03	1.322E+00	0.985	66.116
Total Susp. Solids	5.116E+02	3.382E-02	1.015E+01	6.777E+00	2.259E-03	6.777E-01	0.505	33.884
Organic Matter	7.653E+02	5.060E-02	1.518E+01	1.518E+01	2.259E-03	6.777E-01	0.505	33.884
Mineral Matter	8.333E-03	5.510E-07	1.653E-04	6.612E-05	1.220E-08	3.660E-06	2.725E-06	1.830E-04

	Z Value	Concentration			Amount			
	mol/m³·Pa	mol/m³	g/m³	µg/g	mol	g	% in System	% in Sed
Bulk Sediment	3.014E+01	1.470E-02	4.410E+00	3.554E+00	4.410E-01	1.323E+02	98.511	100.000
Water	3.333E-03	1.626E-06	4.878E-04	4.878E-04	3.902E-05	1.171E-02	8.717E-03	8.849E-03
Total Sed. Solids	1.507E+02	7.349E-02	2.205E+01	2.778E-01	4.409E-01	1.323E+02	98.502	99.991
Organic Matter	7.653E+02	3.733E-01	1.120E+02	1.120E+02	4.409E-01	1.323E+02	98.498	99.987
Mineral Matter	8.333E-03	4.065E-06	1.219E-03	4.878E-04	1.959E-05	5.876E-03	4.376E-03	4.442E-03

Transfer Rates

| | D Value | G Value | Rates | | % of Net |
	mol/Pa*h	m³/h	mol/h	kg/year	Deposition
Diffusion (Water-Sed)	3.180E-04	-	2.103E-08	5.526E-05	0.74
Diffusion (Sed-Water)	3.180E-04	-	1.551E-07	4.077E-04	5.45
Deposition	4.271E-02	8.348E-05	2.824E-06	7.421E-03	99.26
Resuspension	2.847E-03	1.890E-05	1.389E-06	3.650E-03	48.82
Burial	4.271E-03	2.835E-05	2.083E-06	5.475E-03	73.23
Reaction	5.221E-03	-	2.547E-06	6.693E-03	89.52
Water-Sediment Transfer	4.303E-02	-	2.845E-06	7.476E-03	100.00
Sediment-Water Transfer	3.165E-03	-	1.544E-06	4.058E-03	54.28
Sediment Burial and Reaction	9.492E-03	-	4.630E-06	1.217E-02	162.76

Half-time

| | Half-time | |
	hours	years
Water-Sediment Transfer	1624.1	0.185
Sediment-Water Transfer	197927.0	22.594
Sediment Burial and Reaction	66004.5	7.535
Sediment Losses and Transfer	49498.0	5.650

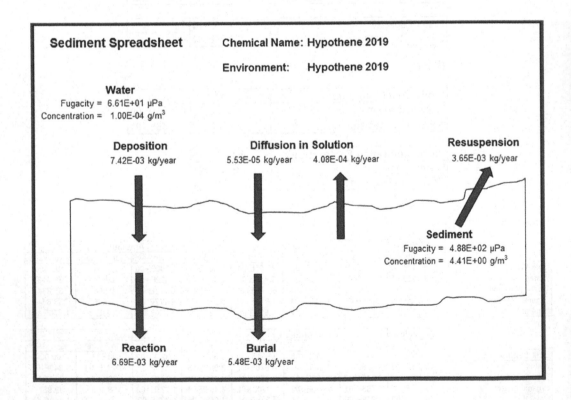

Sediment Spreadsheet   Chemical Name: Hypothene 2019

Environment: Hypothene 2019

Water
Fugacity = 6.61E+01 µPa
Concentration = 1.00E-04 g/m³

Deposition   **Diffusion in Solution**   **Resuspension**
7.42E-03 kg/year   5.53E-05 kg/year   4.08E-04 kg/year   3.65E-03 kg/year

Sediment
Fugacity = 4.88E+02 µPa
Concentration = 4.41E+00 g/m³

Reaction   **Burial**
6.69E-03 kg/year   5.48E-03 kg/year

6.5 QUANTITATIVE WATER-AIR–SEDIMENT INTERACTION: THE *QWASI* MODEL FOR LAKES

Having established air–water and sediment–water exchange models, it is relatively straightforward to combine them in a lake model by adding reaction and advective inflow and outflow terms. The result is the *QWASI* model, the first version of which was published by Mackay et al. in 1983, and was subsequently applied to Lake Ontario (Mackay, 1989). Other reports include an application to a variety of chemicals by Mackay and Diamond (1989), to organochlorine chemicals produced by the pulp and paper industry by Mackay and Southwood (1992), the use of spreadsheets to aid fitting parameter values to the model (Southwood et al., 1989), to situations in which surface microlayers are important (Southwood et al., 1999), and to metals by Woodfine et al. (2000). In principle, the *QWASI* model can be applied to any well-mixed body of water for which the hydraulic and particulate flows are defined.

The original QWASI model was updated and published in spreadsheet format in 2014 (Mackay et al., 2014), and included improved methods for obtaining input parameters such as partition ratios. This version is available from the CEMC website. The model enables the user to define the emission rate and advective inputs to the water and the air concentration. It then calculates fugacities, concentrations and masses in water and sediment, and all fluxes.

Figure 6.4 shows the transport and transformation processes treated in the QWASI model. Table 6.3 lists the D-values and the corresponding fugacity in the rate expressions. Subscripts refer to water (W), air (A), sediment (Sed), suspended particles in water (TSP), and aerosol particles in air (Q) and rain (Rain). G-values are flows ($m^3 \ h^{-1}$) of a phase, e.g., G_{Sed}^{Burial} is the rate in $m^3 \ h^{-1}$ of sediment burial. Superscripts refer to burial (Burial), resuspension (Resusp), decomposition (Deg), and

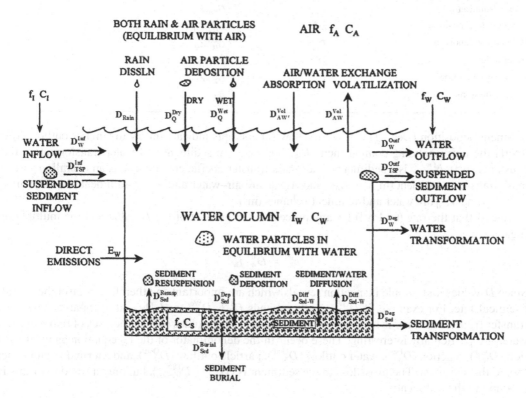

FIGURE 6.4 Transport and transformation processes treated in the QWASI model, consisting of a defined atmosphere with water and sediment compartments.

TABLE 6.3
D-Values in the QWASI Model and Their Corresponding Fugacities

Process	D-Value	Definition of D-Value	Multiplying Fugacity
Sediment burial	D_{Sed}^{Burial}	$G_{Sed}^{Burial} Z_{Sed}$	f_{Sed}
Sediment transformation	D_{Sed}^{Deg}	$V_{Sed} k_{Sed}^{Deg} Z_{Sed}$	f_{Sed}
Sediment resuspension	D_{Sed}^{Resusp}	$G_{Sed}^{Resusp} Z_{Sed}$	f_{Sed}
Sediment to water diffusion	D_{Sed-W}^{Diff}	$k_{Sed-W} A_{Sed-W} Z_{Sed}$	f_{Sed}
Water to sediment diffusion	D_{W-Sed}^{Diff}	$k_{W-Sed} A_{Sed-W} Z_W$	f_W
Net sediment–water diffusion	D_{Sed-W}^{Ov}	$\dfrac{1}{\left(\dfrac{1}{D_{Sed-W}^{Diff}} + \dfrac{1}{D_{W-Sed}^{Diff}} \right)}$	$(f_{Sed} - f_W)$
Sediment deposition	D_W^{Dep}	$G_W^{Dep} Z_{TSP}$	f_W
Water transformation	D_W^{Deg}	$V_W k_W^{Deg} Z_W$	f_W
Volatilization	D_{AW}^{Vol}	$k_{AW}^{Ov} A_{AW} Z_W$	f_W
Absorption	D_{AW}^{Abs}	$k_{AW}^{Ov} A_{AW} Z_A$	f_A
Net air–water diffusion	D_{AW}^{Ov}	$\dfrac{1}{\left(\dfrac{1}{D_{AW}^{Abs}} + \dfrac{1}{D_{WA}^{Vol}} \right)}$	$(f_A - f_W)$
Water outflow	D_W^{Outf}	$G_{Outf}^{W} Z_W$	f_W
Water particle outflow	D_{TSP}^{Outf}	$G_{TSP}^{Outf} Z_{TSP}$	f_W
Rain dissolution	D_{Rain}	$G_{Rain} Z_W$	f_A
Wet particle deposition	D_Q^{Wet}	$G_Q^{Wet} Z_Q$	f_A
Dry particle deposition	D_Q^{Dry}	$G_Q^{Dry} Z_Q$	f_A
Water inflow	D_W^{Inf}	$G_W^{Inf} Z_W$	f_W^{Inf}
Water particle inflow	D_{TSP}^{Inf}	$G_{TSP}^{Inf} Z_{TSP}$	f_W^{Inf}
Direct emissions	—	E_W	

sediment deposition (Dep). The advective flow superscripts indicate inflow to (Inf) or outflow from (Outf) the corresponding compartment. k_{Sed}^{Deg} and k_W^{Deg} are sediment and water transformation rate constants (h⁻¹). k_{Sed-W}^{Ov} is a sediment–water mass-transfer coefficient and k_{AW}^{Ov} is an overall air–water mass-transfer coefficient (m h⁻¹). A_{AW} and A_{Sed-W} are air–water and water–sediment interface areas (m²). V_W and V_{Sed} are water and sediment volumes (m³).

Recall that the rate (mol h⁻¹) for any process is the product of the D-value and the multiplying fugacity, e.g.,

$$r_W^{Inf} = D_W^{Inf} f_W^{Inf}$$

Since D-values add, simple inspection reveals which are important and therefore control the overall chemical fate. For example, if D_{AW}^{Ov} greatly exceeds D_Q^{Dry}, D_Q^{Wet}, and D_{Rain}, it is apparent that most transfer from air to water is by absorption. The relative magnitudes of the processes of removal from water are particularly interesting. These occur in the denominator of the f_W equation as volatilization (D_W^{Vol}), reaction (D_W^{Deg}), water outflow (D_W^{Outf}), particle outflow (D_Q^{Outf}), and a term describing net loss to the sediment. The gross loss to the sediment is ($D_{Sed}^{Dep} + D_{W-Sed}^{Diff}$), but only a fraction of this is retained in the sediment:

$$\text{Sed. fraction retained:} \left(\frac{D_{Sed}^{Deg} + D_{Sed}^{Burial}}{D_{Sed}^{Resusp} + D_{Sed-W}^{Diff} + D_{Sed}^{Deg} + D_{Sed}^{Burial}} \right)$$

The remaining fraction being returned to the water:

$$\text{Sed. fraction returned:} \left(\frac{D_{Sed}^{Resusp} + D_{Sed-W}^{Diff}}{D_{Sed}^{Resusp} + D_{Sed-W}^{Diff} + D_{Sed}^{Deg} + D_{Sed}^{Burial}} \right)$$

The details of the equations in the steady-state QWSAI model are as follows:

1. Sediment mass-balance and fugacity:
   The rates of input to the sediment equal those of output under steady-state conditions:

$$\overset{\text{All sources}}{\sum r_{Sed}^{In}} = \overset{\text{All losses}}{\sum r_{Sed}^{Out}}$$

or

$$r_{W-Sed}^{Diff} + r_W^{Dep} = r_{Sed-W}^{Diff} + r_{Sed}^{Resusp} + r_{Sed}^{Deg} + r_{Sed}^{Burial}$$

or

$$f_W \left(D_{Sed-W}^{Ov} + D_W^{Dep} \right) = f_{Sed} \left(D_{Sed-W}^{Ov} + D_{Sed}^{Resusp} + D_{Sed}^{Deg} + D_{Sed}^{Burial} \right)$$

The corresponding sediment fugacity may then be expressed as

$$f_{Sed} = \frac{f_W \left(D_{Sed-W}^{Ov} + D_W^{Dep} \right)}{\left(D_{Sed-W}^{Ov} + D_{Sed}^{Resusp} + D_{Sed}^{Deg} + D_{Sed}^{Burial} \right)}$$

2. Water mass-balance and fugacity:
   In the same manner as with the sediment, the rates of input to the water equal those of the output under steady-state conditions:

$$r_W^{Em} + r_W^{Inf} + r_{TSP}^{Inf} + r_{AW}^{Diff} + r_Q^{Dry} + r_Q^{Wet} + r_{Rain} + r_{Sed}^{Resusp} + r_{Sed-W}^{Diff}$$

$$= r_{WA}^{Diff} + r_W^{Outf} + r_{TSP}^{Outf} + r_{W-Sed}^{Diff} + r_W^{Dep} + r_W^{Deg}$$

or

$$E_W + f_W^{Inf} \left(D_W^{Inf} + D_{TSP}^{Inf} \right) + f_A \left(D_{AW}^{Ov} + D_Q^{Dry} + D_Q^{Wet} + D_{Rain} \right) + f_{Sed} \left(D_{Sed}^{Resusp} + D_{Sed-W}^{Ov} \right)$$

$$= f_W \left(D_{AW}^{Ov} + D_W^{Deg} + D_W^{Outf} + D_{TSP}^{Outf} + D_W^{Dep} + D_{W-Sed}^{Ov} \right)$$

The corresponding water fugacity may then be expressed as

$$f_W = \frac{E_W + f_W^{Inf} \left(D_W^{Inf} + D_{TSP}^{Inf} \right) + f_A \left(D_{AW}^{Ov} + D_Q^{Dry} + D_Q^{Wet} + D_{Rain} \right) + f_{Sed} \left(D_{Sed}^{Resusp} + D_{Sed-W}^{Ov} \right)}{\left(D_{AW}^{Ov} + D_W^{Deg} + D_W^{Outf} + D_{TSP}^{Outf} + D_W^{Dep} + D_{W-Sed}^{Ov} \right)}$$

Note that the QWASI model assumes a constant air-phase fugacity, such that f_A is a parameter set by the initial conditions and does not vary. As well, it is assumed that the inflow water fugacity is constant. We therefore have two equations with two unknowns, f_W and f_{Sed}.

3. Solve for water fugacity:

Solving by substitution for f_{Sed} leads to

$$f_W = \frac{E_W + f_W^{Inf}\left(D_W^{Inf} + D_{TSP}^{Inf}\right) + f_A\left(D_{AW}^{Ov} + D_Q^{Dry} + D_Q^{Wet} + D_{Rain}\right) + \left[\dfrac{f_W\left(D_{Sed-W}^{Ov} + D_W^{Dep}\right)}{\left(D_{Sed-W}^{Ov} + D_{Sed}^{Resusp} + D_{Sed}^{Deg} + D_{Sed}^{Burial}\right)}\right]\left(D_{Sed}^{Resusp} + D_{Sed-W}^{Ov}\right)}{\left(D_{AW}^{Ov} + D_W^{Deg} + D_W^{Outf} + D_{TSP}^{Outf} + D_W^{Dep} + D_{W-Sed}^{Ov}\right)}$$

or

$$f_W = \frac{E_W + f_W^{Inf}\left(D_W^{Inf} + D_{TSP}^{Inf}\right) + f_A\left(D_{AW}^{Ov} + D_Q^{Dry} + D_Q^{Wet} + D_{Rain}\right)}{\left(D_{AW}^{Ov} + D_W^{Deg} + D_W^{Outf} + D_{TSP}^{Outf}\right) + \left[\dfrac{\left(D_{Sed-W}^{Ov} + D_W^{Dep}\right)\left(D_{Sed}^{Deg} + D_{Sed-W}^{Burial}\right)}{\left(D_{Sed-W}^{Ov} + D_{Sed}^{Resusp} + D_{Sed}^{Deg} + D_{Sed}^{Burial}\right)}\right]}$$

Once the water fugacity is known, the sediment fugacity follows from the initial expression for f_{Sed} given above. This gives the entire set of fugacities, concentrations, masses, and fluxes at steady state for the defined inputs from emission, water inflow, and atmospheric inputs.

The steady-state solution describes conditions that will be reached after prolonged exposure of the lake to constant input conditions, i.e., emissions, air fugacity, and inflow water fugacity.

When the equations are solved, the concentrations, amounts, and fluxes can be calculated. An illustration of such an output is given in Figure 6.5 for PCBs in Lake Ontario (Mackay, 1989). Such mass-balance diagrams clearly show which processes are most important for the chemical of interest.

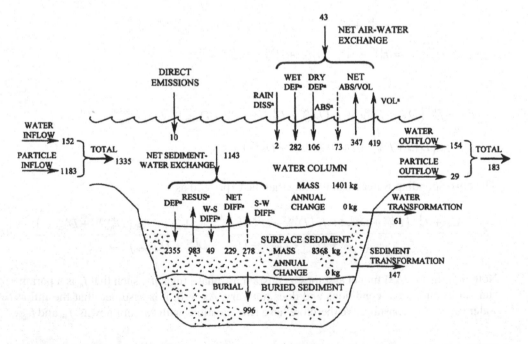

FIGURE 6.5 Illustrative results from a QWASI model calculation of steady-state behavior of PCBs in Lake Ontario (Mackay, 1989).

6.5.1 QWASI Dynamic Model

The details of the equations in the dynamic or unsteady-state QWSAI model are developed as follows:

1. Define the rate of change of matter in each compartment:

$$VZ\frac{df}{dt} = \left(\text{total input rate} - \text{total output rate}\right)$$

The sediment differential equation is

$$V_{\text{Sed}} Z_{\text{Sed}}^{\text{Bulk}} \left(\frac{df_{\text{Sed}}}{dt}\right) = \left(r_{W-\text{Sed}}^{\text{Diff}} + r_W^{\text{Dep}}\right) - \left(r_{\text{Sed}-W}^{\text{Diff}} + r_{\text{Sed}}^{\text{Resusp}} + r_{\text{Sed}}^{\text{Deg}} + r_{\text{Sed}}^{\text{Buria}}\right)$$

or

$$V_{\text{Sed}} Z_{\text{Sed}}^{\text{Bulk}} \left(\frac{df_{\text{Sed}}}{dt}\right) = f_W \left(D_{\text{Sed}-W}^{\text{Ov}} + D_W^{\text{Dep}}\right) - f_{\text{Sed}} \left(D_{\text{Sed}-W}^{\text{Ov}} + D_{\text{Sed}}^{\text{Resusp}} + D_{\text{Sed}}^{\text{Dec}} + D_{\text{Sed}}^{\text{Burial}}\right)$$

The water differential equation is

$$V_W Z_W^{\text{Bulk}} \left(\frac{df_W}{dt}\right) = \left(r_W^{\text{Em}} + r_W^{\text{Inf}} + r_{TSP}^{\text{Inf}} + r_{AW}^{\text{Diff}} + r_Q^{\text{Dry}} + r_Q^{\text{Wet}} + r_{\text{Rain}} + r_{\text{Sed}}^{\text{Resusp}} + r_{\text{Sed}-W}^{\text{Diff}}\right)$$

$$- \left(r_{WA}^{\text{Diff}} + r_W^{\text{Outf}} + r_{TSP}^{\text{Outf}} + r_{W-\text{Sed}}^{\text{Diff}} + r_W^{\text{Dep}} + r_W^{\text{Deg}}\right)$$

or

$$V_W Z_W^{\text{Bulk}} \left(\frac{df_W}{dt}\right) = E_W + f_W^{\text{Inf}} \left(D_W^{\text{Inf}} + D_{TSP}^{\text{Inf}}\right) + f_A \left(D_{AW}^{\text{Ov}} + D_Q^{\text{Dry}} + D_Q^{\text{Wet}} + D_{\text{Rain}}\right)$$

$$+ f_{\text{Sed}} \left(D_{\text{Sed}}^{\text{Resusp}} + D_{\text{Sed}-W}^{\text{Ov}}\right) - f_W \left(D_{AW}^{\text{Ov}} + D_W^{\text{Outf}} + D_{TSP}^{\text{Outf}} + D_{\text{Sed}-W}^{\text{Ov}} + D_W^{\text{Dep}} + D_W^{\text{Deg}}\right)$$

Here, superscript "Bulk" refers to the bulk or total phase including dissolved and sorbed material.

This pair of equations can be written more compactly by defining symbols for terms that appear as a group:

$$\frac{df_W}{dt} = I_1 + I_2 f_{\text{Sed}} - I_3 f_W$$

$$\frac{df_S}{dt} = I_4 f_W - I_{\text{Sed}} f_{\text{Sed}}$$

where

$$I_1 = \left[\frac{E_W + f_W^{\text{Inf}} \left(D_W^{\text{Inf}} + D_{TSP}^{\text{Inf}}\right) + f_A \left(D_{AW}^{\text{Ov}} + D_Q^{\text{Dry}} + D_Q^{\text{Wet}} + D_{\text{Rain}}\right)}{V_W Z_W^{\text{Bulk}}}\right]$$

$$I_2 = \frac{\left(D_{Sed}^{Resusp} + D_{Sed-W}^{Ov}\right)}{V_W Z_W^{Bulk}}$$

$$I_3 = \frac{\left(D_{AW}^{Ov} + D_W^{Outf} + D_{TSP}^{Outf} + D_{Sed-W}^{Ov} + D_W^{Dep} + D_W^{Deg}\right)}{V_W Z_W^{Bulk}}$$

$$I_4 = \frac{\left(D_{Sed-W}^{Ov} + D_W^{Dep}\right)}{V_{Sed} Z_{Sed}^{Bulk}}$$

$$I_5 = \frac{\left(D_{Sed-W}^{Ov} + D_{Sed}^{Resusp} + D_{Sed}^{Deg} + D_{Sed}^{Burial}\right)}{V_{Sed} Z_{Sed}^{Bulk}}$$

These differential equations may be solved by numerical techniques, and generate a time-dependent picture of the changing emission fugacities, concentrations, and phase loads with time.

A useful but restricted solution is shown below for defined initial conditions, assuming that the input terms remain constant with time. Assuming initial fugacities of f_W^0 and f_{Sed}^0 and using the labels f_W^∞ and f_{Sed}^∞ for the steady-state fugacities at very long times, the solution is

$$f_W^t = f_W^\infty + I_8 \exp\left(-[I_6 - I_7]t\right) + I_9 \exp\left(-[I_6 + I_7]t\right)$$

$$f_{Sed}^t = f_{Sed}^\infty + \frac{I_8\left[(I_3 - I_6 + I_7)\exp\left(-[I_6 - I_7]t\right)\right]}{I_2} + \frac{I_9\left[(I_3 - I_6 - I_7)\exp\left(-[I_6 + I_7]t\right)\right]}{I_2}$$

where

$$f_W^\infty = \frac{I_1 I_5}{\left(I_3 I_5 - I_2 I_4\right)}$$

and

$$f_{Sed}^\infty = \frac{I_1 I_4}{\left(I_3 I_5 - I_2 I_4\right)}$$

as in the steady-state solution above, and

$$I_6 = \frac{(I_3 + I_5)}{2}$$

$$I_7 = \frac{\sqrt{(I_3 - I_5)^2 + 4 I_2 I_4}}{2}$$

$$I_8 = \frac{\left[-I_2\left(f_{Sed}^\infty - f_{Sed}^0\right) + (I_3 - I_6 - I_7)\left(f_W^\infty - f_W^0\right)\right]}{2 I_7}$$

and

$$I_9 = \frac{\left[+I_2\left(f_{Sed}^{\infty} - f_{Sed}^0\right) - \left(I_3 - I_6 + I_7\right)\left(f_W^{\infty} - f_W^0\right)\right]}{2I_7}$$

The full details of this solution may be found in the original QWASI paper by Mackay et al. (1983).

Worked Example 6.4: Model QWASI calculation.

Consider a constant emission of 2190 kg year⁻¹ (1 mol h⁻¹) of hypothene into the waters of Lake Ontario.

A. Use the chemical and environmental data given below to determine, for both water and sediment, the D-values, total inputs and outputs, fugacities at steady state, concentrations, total amounts, and residence times. Assume a temperature of 15°C in the system.

B. Using these results, assume that organisms within the water or sediment are at equal fugacity with the medium to estimate the concentration of hypothene in fish and benthic (sediment-dwelling) organisms. Assume values of $Z = 1200$ for fish and benthic organisms.

Reproduced below are the chemical and environmental properties of hypothene and Lake Ontario used as model input data for this Worked Example, as they appear entered into the "Inputs—Chemical Properties" and "Inputs–Environment" of the QWASI program in the spreadsheet version:

Chemical Name	Hypothene
CAS	

Molar Mass (g/mol)	250
Data Temperature (°C)	9
Melting Point (°C)	0
Vapor Pressure (Pa)	1.00E+00
Solubility in Water (g/m³)	20
Henry's Law Constant (Pa·m³/mol)	10

Reaction Half-Lives (hours)

In Water	6.93E+02
In Sediment	6.93E+03

Partition Coefficients

	unitless
logK$_{OW}$	4
Air-Water (K$_{AW}$)	
Aerosol-Air (K$_{QA}$)	
OR	
Aerosol-Water (K$_{QW}$)	

	L/kg
K$_{OC}$	4.10E+03
Suspended Particles-Water	
Suspended Particles-Water (Inflow)	
Sediment-Water	
Resuspended Sediment-Water	

Environment Name	Lake Ontario (2012 D5 SI)

Dimensions

Water Surface Area (m²)	1.91E+10
Water Volume (m³)	1.64E+12
Sediment Active Layer Depth (m)	0.02

Concentration of Solids

Aerosols in Air (µg/m³)	30
in Water Column (mg/L)	0.64
in Inflow Water (mg/L)	24
in Sediment (vol/vol)	0.15

Density (kg/m³)

Aerosols	1500
Particles in Water	2400
Sediment Solids	2400

Organic Carbon (OC) Content (mass/mass)

Particles in Water	0.25
Particles in Water (inflow)	0.25
Sediment Solids	0.04
Resuspended Sediment	0.035

Rates

Water Inflow (m³/h)	22000000
Water Outflow (m³/h)	26000000
Sedimentation (g/m²·day)	1.4
Sediment Burial (g/m²·day)	0.6
Sediment Resuspension (g/m²·day)	0.6
Aerosol Deposition (m/h)	7.2
Scavenging Ratio (vol air/ vol rain)	200000
Rain Rate (m/year)	0.96

Mass Transfer Coefficients (m/h)

Volatilization - Air Side	1
Volatilization - Water Side	0.01
Sediment-Water Diffusion	0.0004

A. Using this data, we proceed to calculate the necessary Z- and D-values required to determine the prevailing fugacities at steady state.

i. Calculate or estimate the various Z-values:

Air:

$$Z_A = \frac{1}{RT} = \frac{1}{8.3145\,\text{Pa}\,\text{mol}^{-1}\,\text{m}^3\,\text{K}^{-1} \times (273.15 + 9.00)\text{K}} = 4.26 \times 10^{-4}\,\text{mol}\,\text{Pa}^{-1}\,\text{m}^{-3}$$

Water, rainwater:

$$Z_W = \frac{1}{H} = \frac{1}{10\,\text{Pa}\,\text{mol}^{-1}\,\text{m}^3} = 1.0 \times 10^{-1}\,\text{mol}\,\text{Pa}^{-1}\,\text{m}^{-3}$$

Sediment:

$$Z_{Sed} = Z_W K_{Sed-w} = Z_W K_{OC} m_{Sed}^{f-OC} \left(\rho_{Sed}/1000 \right)$$

$$= 1.00 \times 10^{-1} \, \text{mol Pa}^{-1} \, \text{m}^{-3} \times 4.1 \times 10^3 \times 0.04 \times 2.4$$

$$= 39.36 \, \text{mol Pa}^{-1} \, \text{m}^{-3}$$

The assumption is made that the Z-value of the pore water in the sediment is the same as that of the pure lake water itself, and that partitioning in the sediment is only to the OC fraction. As this ratio is a mass–mass ratio, conversion to a volume–volume ratio requires multiplication by the ratio of densities of the solid sediment to the water. The same approach is taken for suspended particles in lake water and inflow water, as well as for resuspended sediment particles, for which the calculation is same as for the sediment above.

Suspended particles in water:

$$Z_{TSP} = Z_W K_{TSP-w} = Z_W K_{OC} m_{Sed}^{f-TSP} \left(\rho_{TSP}/1000 \right)$$

$$= 1.00 \times 10^{-1} \, \text{mol Pa}^{-1} \, \text{m}^{-3} \times 4.1 \times 10^3 \times 0.25 \times 2.4$$

$$= 246 \, \text{mol Pa}^{-1} \, \text{m}^{-3}$$

Resuspended sediment particles:

$$Z_{Sed}^{Resusp} = Z_W K_{Sed-W}^{Resusp} = Z_W K_{OC} m_{Resusp}^{f-OC} \left(\rho_{Resusp}/1000 \right)$$

$$= 1.00 \times 10^{-1} \, \text{mol Pa}^{-1} \text{m}^{-3} \times 4.1 \times 10^3 \times 0.035 \times 2.4$$

$$= 34.4 \, \text{mol Pa}^{-1} \, \text{m}^{-3}$$

As the value for K_{QA} is already inputted, the aerosol Z-value is easy to determine.

Aerosol:

$$Z_Q = Z_A K_{QA} = 4.26 \times 10^{-4} \, \text{mol Pa}^{-1} \, \text{m}^{-3} \times 1.00 \times 10^7$$

$$= 4.26 \times 10^3 \, \text{mol Pa}^{-1} \, \text{m}^{-3}$$

ii. Calculate bulk Z-values:

Recall from Chapter 2 that the Z-value for a medium comprised of two or more components is merely the sum of the individual component Z-values, weighted by their corresponding volume fractions:

$$Z_T = \sum_{i=1}^{n} v_i^f Z_i$$

Thus, for each of air, water, and sediment, we can calculate bulk Z-values:

Bulk air:

$$Z_A^{Bulk} = v_A^f Z_A + v_Q^f Z_Q$$

$$= 1.00 \times 4.26 \times 10^{-4} \, \text{Pa mol}^{-1} \, \text{m}^{-3} + 2.00 \times 10^{-11} \times 4.26 \times 10^3 \, \text{Pa mol}^{-1} \, \text{m}^{-3}$$

$$= 4.26 \times 10^{-4} \, \text{Pa mol}^{-1} \, \text{m}^{-3}$$

Bulk water:

$$Z_W^{\text{Bulk}} = v_W^f Z_W + v_{TSP}^f Z_{TSP}$$

$$= 1.00 \times 1.00 \times 10^{-1} \text{Pa mol}^{-1} \text{m}^{-3} + 2.67 \times 10^{-7} \times 246 \text{Pa mol}^{-1} \text{m}^{-3}$$

$$= 1.00 \times 10^{-1} \text{Pa mol}^{-1} \text{m}^{-3}$$

Bulk inflow water:

$$Z_W^{\text{Infl-Bulk}} = v_{\text{InflW}}^f Z_W + v_{TSP}^f Z_{TSP}$$

$$= 1.00 \times 1.00 \times 10^{-1} \text{Pa mol}^{-1} \text{m}^{-3} + 1.00 \times 10^{-5} \times 246 \text{Pa mol}^{-1} \text{m}^{-3}$$

$$= 1.02 \times 10^{-1} \text{Pa mol}^{-1} \text{m}^{-3}$$

Bulk sediment:

$$Z_{\text{Sed}}^{\text{Bulk}} = v_W^f Z_W + v_{\text{Sed}}^f Z_{\text{Sed}}$$

$$= 0.85 \times 1.00 \times 10^{-1} \text{Pa mol}^{-1} \text{m}^{-3} + 0.15 \times 39.4 \text{Pa mol}^{-1} \text{m}^{-3}$$

$$= 5.99_5 \text{Pa mol}^{-1} \text{m}^{-3}$$

Here it is worth noting that the impact on the bulk Z-values due to aerosols in air and total suspended solids in water is insignificant, due to their low volume fractions. We could have assumed that they would have no impact and simply used the pure water and air Z-values in this step. This is not always the case, so caution is advised with such an assumption.

iii. Calculate or estimate the various D-values:

As with the sediment model, transport D-values are of the form GZ, where G is an equivalent bulk volume flow in $\text{m}^3 \text{ h}^{-1}$. The G-value or equivalent bulk water volume flow rate for sediment deposition is given by

$$G_{\text{Sed}}^{\text{Dep}} = \frac{L_{\text{Sed}} \times A_{\text{Sed}-W}}{\rho_{\text{Sed}} \times 24 \text{h day}^{-1} \times 1 \times 10^3 \text{ g kg}^{-1}}$$

$$= \frac{1.4 \text{ g m}^{-2} \text{day}^{-1} \times 1.91 \times 10^{10} \text{m}^2}{2400 \text{ kg m}^{-3} \times 24 \text{h day}^{-1} \times 1 \times 10^3 \text{ g kg}^{-1}}$$

$$= 4.64 \times 10^2 \text{ m}^3 \text{ h}^{-1}$$

The following table gives the actual and equivalent G-values, the corresponding Z-value from above, and the product, which is the D-value for the process in question:

Process	Symbol	Flow Rate ($\text{m}^3 \text{ h}^{-1}$)	Z-value for Medium	D-Value for Process ($\text{Pa mol}^{-1} \text{ m}^{-3}$)
Sediment burial	$G_{\text{Sed}}^{\text{Burial}}$	1.99×10^2	Z_{Sed}	$D_W^{\text{Burial}} = 7830$
Sediment resuspension	$G_{\text{Sed}}^{\text{Resusp}}$	1.99×10^2	Z_{Sed}	$D_{\text{Sed}}^{\text{Resusp}} = 6850$
Water–sediment diffusion	$G_{\text{Sed}-W}^{\text{Ov}}$	7.64×10^6	Z_W	$D_{\text{Sed}-W}^{\text{Ov}} = 7.64 \times 10^5$
Sediment deposition	G_W^{Dep}	4.64×10^2	Z_{TSP}	$D_W^{\text{Dep}} = 1.14 \times 10^5$
Volatilization	G_{AW}^{Ov}	1.91×10^8	Z_W	$D_W^{\text{Vol}} = 1.91 \times 10^7$

(*Continued*)

Process	Symbol	Flow Rate (m³ h⁻¹)	Z-value for Medium	D-Value for Process (Pa mol⁻¹ m⁻³)
Water outflow	G_W^{Outf}	2.60×10^7	Z_W	$D_W^{Outf} = 2.60 \times 10^6$
Water particle outflow	G_{TSP}^{Outf}	6.93	Z_{TSP}	$D_{TSP}^{Outf} = 1.71 \times 10^3$
Rain dissolution	G_{Rain}	2.09×10^6	Z_W	$D_{Rain} = 2.09 \times 10^5$
Wet particle deposition	G_Q^{Wet}	8.37	Z_Q	$D_Q^{Wet} = 3.57 \times 10^4$
Dry particle deposition	G_Q^{Dry}	2.75	Z_Q	$D_Q^{Dry} = 1.17 \times 10^4$
Water inflow	G_W^{Inf}	2.2×10^7	Z_W	$D_W^{Inf} = 2.20 \times 10^6$
Water particle inflow	G_{TSP}^{Inf}	220	Z_{TSP}	$D_{TSP}^{Inf} = 5.41 \times 10^4$

iv. Determine fugacity of water and sediment under steady-state conditions:
With D-values in hand, we may now use the fugacity expressions derived above:

$$f_W = \frac{E_W + f_W^{Inf}\left(D_W^{Inf} + D_{TSP}^{Inf}\right) + f_A\left(D_{AW}^{Ov} + D_Q^{Dry} + D_Q^{Wet} + D_{Rain}\right)}{\left(D_{AW}^{Ov} + D_W^{Deg} + D_W^{Outf} + D_{TSP}^{Outf} + D_W^{Dep} + D_{W-Sed}^{Ov}\right) - \left[\dfrac{\left(D_{Sed-W}^{Ov} + D_W^{Dep}\right)\left(D_{Sed}^{Deg} + D_{Sed}^{Burial}\right)}{\left(D_{Sed-W}^{Ov} + D_{Sed}^{Resusp} + D_{Sed}^{Deg} + D_{Sed}^{Burial}\right)}\right]}$$

or

$$f_W = \frac{\begin{bmatrix}1.00 + 0\left(2.20 \times 10^6 + 5.41 \times 10^4\right) + \\ 0\left(1.91 \times 10^7 + 1.17 \times 10^4 + 3.57 \times 10^4 + 2.09 \times 10^5\right)\end{bmatrix} \text{mol h}^{-1}}{\begin{bmatrix}\left(1.91 \times 10^7 + 1.62 \times 10^8 + 2.60 \times 10^6 + 1.71 \times 10^3 + 1.14 \times 10^5 + 7.64 \times 10^5\right) \\ -\dfrac{\left(7.64 \times 10^5 + 1.14 \times 10^5\right)\left(6850 + 7.64 \times 10^5\right)}{\left(7.64 \times 10^5 + 6850 + 2.29 \times 10^5 + 7830\right)}\end{bmatrix} \text{mol Pa}^{-1}\,\text{h}^{-1}}$$

or

$$f_W = \frac{1}{\left[1.85 \times 10^8 - \left(\dfrac{6.77 \times 10^{11}}{1.1 \times 10^6}\right)\right]} = 5.44 \times 10^{-9}\,\text{Pa}$$

With the steady-state fugacity of the water in hand, the fugacity of the sediment follows directly:

$$f_{Sed} = \frac{f_W\left(D_{Sed-W}^{Ov} + D_W^{Dep}\right)}{\left(D_{Sed-W}^{Ov} + D_{Sed}^{Resusp} + D_{Sed}^{Deg} + D_{Sed}^{Burial}\right)}$$

or

$$f_{Sed} = \frac{5.44 \times 10^{-9}\,\text{Pa}\left(7.64 \times 10^5 + 1.14 \times 10^5\right)}{\left(7.64 \times 10^5 + 6.85 \times 10^3 + 2.29 \times 10^5 + 7.83 \times 10^3\right)}$$

$$= \frac{7.48 \times 10^{-3}\,\text{Pa}}{1.1 \times 10^6} = 4.74 \times 10^{-9}\,\text{Pa}$$

Note that in this case f_W is 15% greater than f_s, whereas often f_W will be less than f_s.

v. Determine steady-state concentrations in the water and sediment:
Concentrations in the various phases both bulk and constituents follow directly from the corresponding products of fugacity and fugacity capacities, Zf:

$$C_W^{\text{Bulk}} = Z_W^{\text{Bulk}} f_W = 1.00 \times 10^{-1} \text{mol Pa}^{-1} \text{m}^{-3} \times 5.44 \times 10^{-9} \text{Pa} = 5.44 \times 10^{-10} \text{mol m}^{-3}$$

$$C_{\text{Dissolved}}^W = Z_W f_W = 1.00 \times 10^{-1} \text{mol Pa}^{-1} \text{m}^{-3} \times 5.44 \times 10^{-9} \text{Pa} = 5.44 \times 10^{-10} \text{mol m}^{-3}$$

$$C_{TSP}^W = Z_{TSP}^W f_W = 246 \text{mol Pa}^{-1} \text{m}^{-3} \times 5.44 \times 10^{-9} \text{Pa} = 1.34 \times 10^{-6} \text{mol m}^{-3}$$

$$C_{\text{Sed}}^{\text{Bulk}} = Z_{\text{Sed}}^{\text{Bulk}} f_{\text{Sed}} = 5.99 \text{mol Pa}^{-1} \text{m}^{-3} \times 4.74 \times 10^{-9} \text{Pa} = 2.84 \times 10^{-8} \text{mol m}^{-3}$$

$$C_W^{\text{Sed}} = Z_W^{\text{Sed}} f_{\text{Sed}} = 1.0 \times 10^{-1} \text{mol Pa}^{-1} \text{m}^{-3} \times 4.74 \times 10^{-9} \text{Pa} = 4.74 \times 10^{-10} \text{mol m}^{-3}$$

$$C_{\text{Solid}}^{\text{Sed}} = Z_{\text{Solid}}^{\text{Sed}} f_{\text{Sed}} = 39.4 \text{mol Pa}^{-1} \text{m}^{-3} \times 4.74 \times 10^{-9} \text{Pa} = 1.87 \times 10^{-7} \text{mol m}^{-3}$$

vi. Determine total molar quantities in various compartments and components at steady state:
Total amounts follow from the product of compartment volume and concentration (VC), or the product VZf:

$$m_{\text{Dissolved}}^W = V_W C_{\text{Dissolved}}^W = 1.62 \times 10^{12} \text{m}^3 \times 5.44 \times 10^{-10} \text{mol m}^{-3} = 881 \text{mol}$$

$$m_{TSP}^W = V_{TSP}^W C_{TSP}^W = 4.32 \times 10^5 \text{mol Pa}^{-1} \text{m}^{-3} \times 1.34 \times 10^{-6} \text{mol m}^{-3} = 0.578 \text{mol}$$

$$m_{\text{Solid}}^{\text{Sed}} = V_{\text{Sed}} C_{\text{Sed}}^{\text{Bulk}} = 3.82 \times 10^8 \text{m}^3 \times 2.84 \times 10^{-8} \text{Pa} = 10.8 \text{mol}$$

$$m = 892 \text{mol}$$

Recall that we began the calculation with the assumption of an emission to the water compartment of 1 mol h^{-1}.

vii. Determine residence times in water and sediment at steady state:
Recall that the residence times are given by the ratio of total molar content in a compartment divided by the rate of removal:

$$\tau = \frac{\sum_{i=1}^{n} m_i}{\sum_{i=1}^{n} r_i}$$

For water, relevant loss mechanisms are diffusion to air, outflow losses, and decomposition:

$$\tau_W = \frac{m_W}{\left(r_{AW}^{\text{Ov}} + r_W^{\text{Dec}} + r_W^{\text{Outf}} \right)}$$

$$= \frac{8.81 \times 10^2 \text{mol}}{\left(0.104 + 0.881 + 1.41 \times 10^{-2} \right) \text{mol h}^{-1}}$$

$$= 8.82 \times 10^2 \text{h} = 36.7 \text{days}$$

For sediment, relevant losses are burial and decomposition:

$$\tau_{Sed} = \frac{m_{Sed}}{\left(r_{Sed}^{Dec} + r_{Sed}^{Burial}\right)}$$

$$= \frac{10.8\,mol}{\left(1.08\times10^{-3} + 3.71\times10^{-5}\right)mol\,h^{-1}}$$

$$= 9.67\times10^{3}\,h = 403\,days$$

B. Estimate the concentration of hypothene in fish and benthic (sediment-dwelling) organisms. Assume values of $Z = 1200$ for fish and benthic organisms.

Given the assumed Z-value of 1200 mol Pa^{-1} m^{-3} for fish and benthic organisms as stated, the concentrations in each is obtained by assuming equifugacity with the lake water or sediment, respectively:

$$C_{Fish}^{W} = Z_{Fish}^{W}f_{W} = 1200\,mol\,Pa^{-1}m^{-3} \times 5.43\times10^{-9}\,Pa = 6.53\times10^{-6}\,mol\,m^{-3}$$

$$= 6.53\times10^{-6}\,mol\,m^{-3} \times 250\,g\,mol^{-1} = 1.63\times10^{-3}\,mol\,m^{-3}$$

$$C_{Benthos}^{Sed} = Z_{Benthos}^{Sed}f_{Sed} = 1200\,mol\,Pa^{-1}\,m^{-3} \times 4.74\times10^{-9}\,Pa = 5.69\times10^{-6}\,mol\,m^{-3}$$

$$= 5.69\times10^{-6}\,mol\,m^{-3} \times 250\,g\,mol^{-1} = 1.42\times10^{-3}\,mol\,m^{-3}$$

This example, while tedious to do by hand, is readily implemented on a spreadsheet, or the software available on the internet can be used. It gives a clear quantification of all the fluxes and demonstrates which processes are most important. A mass-balance diagram in Figure 6.6 illustrates the process rates clearly.

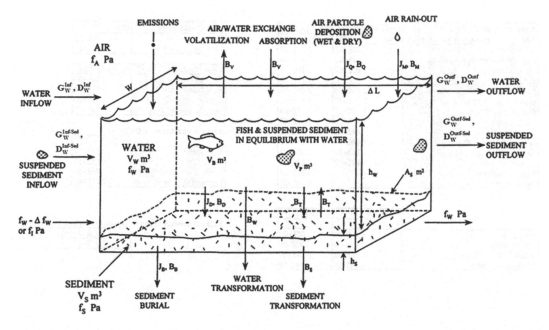

FIGURE 6.6 Mass-balance diagram for a QWASI model calculation, showing media and transport processes.

The results of Worked Example 6.4 in the QWASI spreadsheet program are reproduced below for reference:

QWASI RESULTS

Chemical Name	Hypothene QWASI
Environment Name	Lake Ontario Mackay 2012

Fugacity Ratio	1

Subcooled Liq. Vap. Press.	1	Pa

Partition Coefficients	dimensionless	L/kg	
Suspended Particles-Water	2.46E+03	1.03E+03	(estimated using Koc and OC content of suspended particles)
Suspended Particles-Water (Inflow)	2.46E+03	1.03E+03	(estimated using Koc and OC content of inflow particles)
Sediment-Water	3.94E+02	1.64E+02	(estimated using Koc and OC content of sediment solids)
Resuspended Sediment-Water	3.44E+02	1.44E+02	(estimated using Koc and OC content of resuspended sediment)

	Volume Fraction	Volume	Z Value	VZ
		m³	mol/m³·Pa	mol/Pa
Air: Bulk	-	-	4.26E-04	-
Gas Phase	1.00E+00	-	4.26E-04	-
Aerosols	2.00E-11	-	4.26E+03	-
Water: Bulk	-	1.62E+12	1.00E-01	1.62E+11
Liquid Phase	1.00E+00	1.62E+12	1.00E-01	1.62E+11
Suspended Particles	2.67E-07	4.32E+05	2.46E+02	1.06E+08
Inflow Water: Bulk	-	-	1.02E-01	-
Liquid Phase	1.00E+00	-	1.00E-01	-
Suspended Particles	1.00E-05	-	2.46E+02	-
Sediment: Bulk	-	3.82E+08	5.99E+00	2.29E+09
Pore Water	8.50E-01	3.25E+08	1.00E-01	3.25E+07
Solids	1.50E-01	5.73E+07	3.94E+01	2.26E+09
Resuspended Sediment	-	-	3.44E+01	-
Rain	-	-	1.00E-01	-

Advection, Reaction and Intercompartmental Transfers	Flow Rate (G)	D Value
	m³/h	mol/Pa·h
Water Inflow	2.20E+07	2.20E+06
Suspended Particle Inflow	2.20E+02	5.41E+04
Water Outflow	2.60E+07	2.60E+06
Suspended Particle Outflow	6.93E+00	1.71E+03
Rain Dissolution and Deposition	2.09E+06	2.09E+05
Dry Aerosol Deposition	2.75E+00	1.17E+04
Wet Aerosol Deposition	8.37E+00	3.57E+04
Volatilization	1.91E+08	1.91E+07
Sedimentation	4.64E+02	1.14E+05
Water-Sediment Diffusion	7.64E+06	7.64E+05
Sediment Resuspension	1.99E+02	6.85E+03
Sediment Burial	1.99E+02	7.83E+03
Reaction in Water	-	1.62E+08
Reaction in Sediment	-	2.29E+05

Rate Constants	h⁻¹
Reaction in Water	1.00E-03
Reaction in Sediment	1.00E-04

Total D values	mol/Pa·h
From Air	1.94E+07
From Water	1.85E+08
Inflow	2.25E+06
From Sediment	1.01E+06

Intercompartmental D Values

	mol/Pa·h
air to water	1.94E+07
water to air	1.91E+07
water to sediment	8.78E+05
sediment to water	7.71E+05

	Fugacity	Activity
	Pa	
Air: Bulk	0.00E+00	0.00E+00
Water: Bulk	5.43E-09	5.43E-09
Inflow Water: Bulk	0.00E+00	0.00E+00
Sediment: Bulk	4.74E-09	4.74E-09

Total Chemical Mass in Water-Sediment System

8.92E+02	mol
2.23E+02	kg

	Concentrations			Amounts		
	mol/m³	g/m³	µg/g dry wgt	mol	kg	%
Air: Bulk	0.00E+00	0.00E+00	-	-	-	-
Gas Phase	0.00E+00	0.00E+00	-	-	-	-
Aerosols	0.00E+00	0.00E+00	0.00E+00	-	-	-
Water: Bulk	5.44E-10	1.36E-07	-	8.81E+02	2.20E+02	9.88E+01
Liquid Phase	5.43E-10	1.36E-07	-	8.80E+02	2.20E+02	9.87E+01
Suspended Particles	1.34E-06	3.34E-04	1.39E-04	5.77E-01	1.44E-01	6.48E-02
Inflow Water: Bulk	0.00E+00	0.00E+00	-	-	-	-
Liquid Phase	0.00E+00	0.00E+00	-	-	-	-
Suspended Particles	0.00E+00	0.00E+00	0.00E+00	-	-	-
Sediment: Bulk	2.84E-08	7.09E-06	-	1.08E+01	2.71E+00	1.22E+00
Pore Water	4.74E-10	1.18E-07	-	1.54E-01	3.84E-02	1.72E-02
Solids	1.86E-07	4.66E-05	1.94E-05	1.07E+01	2.67E+00	1.20E+00
Rain	0.00E+00	0.00E+00	-	-	-	-

MASS BALANCES

System	kg/year	mol/h
Total Chemical Inputs	2.19E+03	1.00E+00
Emission	2.19E+03	1.00E+00
Inflow	0.00E+00	0.00E+00
Air to water transfer	0.00E+00	0.00E+00
Total Chemical Losses	2.19E+03	1.00E+00
Outflow	3.10E+01	1.41E-02
Water to air transfer	2.27E+02	1.04E-01
Total Transformation	1.93E+03	8.82E-01
Sediment Burial	8.12E-02	3.71E-05

Residence Time	h	d
Water	8.82E+02	3.67E+01
Sediment	9.67E+03	4.03E+02
System	8.92E+02	3.72E+01

Water	kg/year	mol/h
Total Chemical Inputs	2.20E+03	1.00E+00
Emission	2.19E+03	1.00E+00
Inflow	0.00E+00	0.00E+00
Air to water transfer	0.00E+00	0.00E+00
Sediment to water transfer	8.00E+00	3.65E-03
Total Chemical Losses	2.20E+03	1.00E+00
Outflow	3.10E+01	1.41E-02
Water to air transfer	2.27E+02	1.04E-01
Water to sediment transfer	1.04E+01	4.77E-03
Transformation in water	1.93E+03	8.81E-01

	h	d
Response Time	8.78E+02	3.66E+01

Sediment	kg/year	mol/h
Total Chemical Inputs	1.04E+01	4.77E-03
Water to sediment transfer	1.04E+01	4.77E-03
Total Chemical Losses	1.04E+01	4.77E-03
Sediment to water transfer	8.00E+00	3.65E-03
Transformation in sediment	2.37E+00	1.08E-03
Sediment Burial	8.12E-02	3.71E-05

	h	d
Response Time	2.27E+03	9.46E+01

Rate Details	kg/year	mol/h
Emission to Water	2.19E+03	1.00E+00
Water Inflow	0.00E+00	0.00E+00
Particle Inflow	0.00E+00	0.00E+00
Water Outflow	3.09E+01	1.41E-02
Particle Outflow	2.03E-02	9.27E-06
Rain Dissolution	0.00E+00	0.00E+00
Dry Aerosol Deposition	0.00E+00	0.00E+00
Wet Aerosol Deposition	0.00E+00	0.00E+00
Absorption	0.00E+00	0.00E+00
Volatilization	2.27E+02	1.04E-01
Sedimentation	1.36E+00	6.20E-04
Water-Sediment Diffusion	9.09E+00	4.15E-03
Sediment-Water Diffusion	7.92E+00	3.62E-03
Sediment Resuspension	7.11E-02	3.25E-05
Sediment Burial	8.12E-02	3.71E-05
Water Transformation	1.93E+03	8.81E-01
Sediment Transformation	2.37E+00	1.08E-03

Water Response Times

	h	d
Water Inflow	7.37E+04	3.07E+03
Particle Inflow	3.00E+06	1.25E+05
Water Outflow	6.23E+04	2.60E+03
Particle Outflow	9.50E+07	3.96E+06
Rain Dissolution	7.74E+05	3.23E+04
Dry Aerosol Deposition	1.38E+07	5.76E+05
Wet Aerosol Deposition	4.54E+06	1.89E+05
Volatilization/Absorption	8.49E+03	3.54E+02
Sedimentation	1.42E+06	5.91E+04
Water-Sediment Diffusion	2.12E+05	8.84E+03
Sediment Resuspension	2.37E+07	9.86E+05
Transformation	1.00E+03	4.17E+01

Sediment Response Times

	h	d
Sedimentation	2.00E+04	8.35E+02
Water-Sediment Diffusion	2.99E+03	1.25E+02
Sediment Resuspension	3.34E+05	1.39E+04
Sediment Burial	2.92E+05	1.22E+04
Transformation	1.00E+04	4.17E+02

The diagrammatical output that corresponds to these data looks like this:

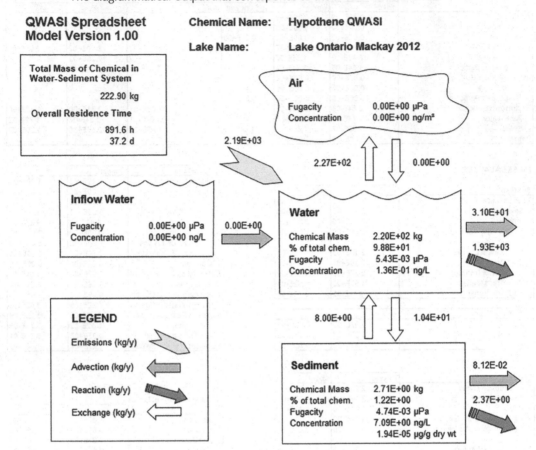

Examination of the diagram demonstrates that, despite the fact that hypothene is released to the water compartment, its fugacity and concentration are higher in the sediment, where it "likes to be," as reflected the magnitude of $K_{OW} = 1.0 \times 10^4$. Nevertheless, the amount of chemical remains highest in the water phase due to the slow transfer rate from water to sediment, which ensures that losses due to outflow are significant, as are volatilization losses to air.

6.6 MULTI-SEGMENT MODELS: THE *QWASI* MODEL FOR RIVERS

6.6.1 EULERIAN AND LAGRANGIAN COORDINATE SYSTEMS

In most models described and developed above, we have assumed a compartment or mass-balance envelope as being fixed in space. This can be called the Eulerian coordinate system. When there is appreciable flow through the envelope, it may be better to define the envelope as being around a certain amount of material and allow that envelope of material to change position with time. This "fix a parcel of material then follow it in time as it moves" approach is often applied to rivers when we wish to examine the changing condition of a volume of water as it flows downstream and undergoes various reactions. This approach can be called the Lagrangian coordinate system. It is also applied to "parcels" of air emitted from a stack and subject to wind drift. Both systems must give the same results, but it may be easier to write the equations in one system than the other. The following Worked Example is an illustration of the Eulerian and Lagrangian approaches.

Worked Example 6.5: Concentrations downstream of a discharge into a river by the Eulerian and Lagrangian approaches.

Consider a river into which the 1.8 million population of a city discharges a detergent at a rate of 1 pound per capita per year, i.e., the discharge is 1.8 million pounds per year. The aim is to calculate the concentrations at distances of 1 and 10 miles downstream, from knowledge of the degradation rate of the detergent and the constant downstream flow conditions (given below). This can be done in Eulerian or Lagrangian coordinate systems.

The input data are first converted to SI units:

Discharge rate (E): 1.8×10^6 lb year^{-1} = 93300 g h^{-1}
River flow velocity (U): 1 ft s^{-1} = 1097 m h^{-1}
River depth (h): 3 ft = 0.91 m
River width (w): 20 yards = 18.3 m
Degradation half-life ($t_{1/2}$): 0.3 days = 7.2 h

i. Calculate initial parameters:
   River flow rate: Uhw = 18,270 m^3 h^{-1}
   Degradation rate constant $k = 0.693/t_{1/2} = 0.096$ h^{-1}.
   Mixed detergent concentration: $C_0 = E/(Uhw) = 5.1$ g m^{-3}

ii. Lagrangian solution:
   A given parcel of chemical-containing water that maintains its integrity, i.e., it does not diffuse or disperse, will decay according to the equation:

$$C_t = C_0 \exp(-kt)$$

where t is the time from discharge. At 1 mile (1609 m), the time t will be 1609/U or 1.47 h, and at 10 miles, it will be 14.7 h.

   Substituting shows that, after 1 and 10 miles, the concentrations will be 4.4 and 1.24 g m^{-3}. The chemical will reach half its input concentration when t is 0.693/k or 7.2 h, which corresponds to 7900 m or 4.92 miles. This Lagrangian solution is straightforward, but it is valid only if conditions in the river remain constant and negligible upstream–downstream dispersion occurs.

iii. Eulerian solution:
   We now simulate the river as a series of connected reaches, segments, or well-mixed lakes, each being L or 200 m long. Each reach thus has a volume V of Lhw or 3330 m^3. A steady-state mass balance on the first reach gives

$$r^{\text{In}} = r^{\text{Out}}$$

or

$$UhwC_0 = UhwC_1 + VkC_1$$

where C_0 and C_1 are the input and output concentrations, respectively. C_1 is also assumed to be the concentration throughout the entire segment. It follows that

$$C_1 = \frac{C_0}{\left(1 + \left(\dfrac{Vk}{Uhw}\right)\right)} = \frac{C_0}{\left(1 + \left(\dfrac{kL}{U}\right)\right)}$$

Note the consistency of the dimensions, kL/U being dimensionless. The group $(1 + kL/U)$ has a value of 1.0175, thus C_1 is 0.983C_0, and 1.7% of the chemical is lost in each segment. The same equation applies to the second reach, thus C_2 is 0.983C_1 or $(0.983)^2C_0$. In general, for the nth reach

$$C_n = (0.983)^n C_0 = \frac{C_0}{(1 + kL/U)^n}$$

One mile is reached when n is 8, and 10 miles correspond to n of 80; thus, C_8 is $(0.983)^8 C_0$ or 4.45, and C_{80} is 1.29. The half distance will occur when 0.983^n is 0.5, i.e., when $n = \log (0.5)/\log (0.983)$ or 40, corresponding to 8000 m or 5 miles.

The Eulerian answer is thus slightly different. It could be made closer to the Lagrangian result by carrying more significant figures or by decreasing L and increasing n. An advantage of the Eulerian system is that it is possible to have segments with different properties such as depth, width, velocity, volume, and temperature. There can be additional inputs. The general equation employing the group $(1 + kL/U)^n$ will not then apply, each segment having a specific value of this factor. The mathematics enthusiast will note that L/U is t/n, where t is the flow time to a distance nL m. The Lagrangian factor is thus $(1 + kt/n)^n$, which approaches $\exp(kt)$ when n is large. It is good practice to do the calculation in both systems (even approximately) and check that the results are reasonable. Some water quality models of rivers and estuaries can have several hundred segments, thus it is difficult to grasp the entirety of the results, and mistakes can go undetected.

The QWASI lake equations can be modified to describe the chemical fate in rivers by either of these two methods. The river can be treated as a series of connected lakes or reaches (Eulerian approach), each of which is assumed to be well mixed, with unique water and sediment concentrations. There can be varying discharges into each reach, and tributaries can be introduced as desired. The larger the number of reaches, the more closely simulated is the true "plug-flow" condition of the river. The second approach is to set up and solve the Lagrangian differential equation for water concentration as a function of river length. This has been discussed by Mackay et al. (1983), and an application to surfactant decay in a river has been described by Holysh et al. (1985).

A differential equation is set up for the water column as a function of flow distance or time, and steady-state exchange with the sediment is included. The equation can then be solved from an initial condition with zero or constant inputs of chemical. The practical difficulty is that changes in flow volume, velocity, or river width or depth cannot be easily included; therefore, the equation necessarily applies to idealized conditions.

This equation may be useful for calculating a half-time or half-distance of a substance in a river as the concentration decays as a result of volatilization or degradation. A version is the oxygen sag equation. This contains an additional term for oxygen consumption by OM added to the river. This model was first developed by Streeter and Phelps in 1925 and is described in texts such as that of Thibodeaux (1996). This is historically significant as being among the first successful applications of mathematical models to the fate of a chemical (oxygen) in the aquatic environment.

A lake, river, or estuary rarely can be treated as a single well-mixed "box" of water, and a more accurate simulation is obtained if the system is divided into a series of connected boxes. In the case of a river, it may be acceptable to use the output from one box as input to the next downstream box and treat any downstream–upstream flow as being negligible compared with upstream–downstream flow. In slowly moving water, this will be invalid if there are significant flows in both directions. In principle, if there are n water and n sediment boxes, there are $2n$ mass balance equations containing $2n$ fugacities, and the equations can be solved algebraically for steady-state conditions or numerically for dynamic conditions.

In the case of steady-state conditions, a major simplification is possible if it is assumed that there is no direct sediment–sediment transfer between reaches; i.e., all transfer is via the water column. Each sediment mass-balance equation then can be written to express the sediment fugacity as a function of the fugacity in the overlying water. The fugacity in sediment then can be eliminated entirely, leaving only n water fugacities to be solved. This is equivalent to calculating the water fugacity as in the *QWASI* model by including loss to the sediment in the denominator. A total loss D-value can thus be calculated for the water and sediment.

Algebraic solution of the equations is straightforward, provided the number of boxes is small and there is minimal branching. For a set of boxes connected in series with both "upstream and downstream" flows, the solution becomes simple, elegant, general, and intuitively satisfying because of its transparency. If the number of boxes is very large (above about 8) and there is appreciable branching, it may be easier to set up the equations in differential form and solve them numerically in time, with constant inputs to reach a steady-state solution.

The dynamic version allows the user to observe changes in concentrations with time. The steady-state version gives the concentrations when the system has reached a condition where all concentrations remain constant with respect to time. If a long enough integration period is used for the dynamic version, the concentrations approach those in the steady-state version. It is best to use the steady-state version if the user is concerned only with the end result, and the dynamic version if it is desired to track changes in the system over time or how long it will take for the system to approach a steady state.

It is good practice to check the consistency between steady-state and dynamic solutions by comparing the steady-state output with the dynamic output obtained after integrating for a prolonged period at constant input rates, such that a steady state has been achieved.

The following papers are examples of multi-QWASI model applications: Lun et al. (1998) describe the fate of PAHs in the Saguenay River in Quebec. Ling et al. (1993) treat the fate of chemicals in a harbor including vertical segmentation. Hickie and Mackay (2000) describe the fate of PAHs from atmospheric sources to Lac Saint Louis in the St. Lawrence River. Diamond et al. (1994) treat the fate of a variety of organic chemicals and metals in a highly segmented model of the Bay of Quinte which is connected to Lake Ontario.

6.7 CHEMICAL FATE IN SLUDGE AMENDED SOILS AND BIOPILES. THE *BASL4* MODEL

Sewage treatment plants (STPs) generate considerable quantities of sludges or biosolids with high OM contents as well as valuable nutrients such as nitrogen and phosphorus. They may also contain metals and organic chemicals that are present in the influent water. These biosolids are frequently used as soil conditioners in an agricultural setting, thus contributing to the fertility of soils. Typical biosolids application rates are in the range of 2–10 tonnes per hectare per year, however, if the metal content is high, there may be restrictions on the quantity and duration of the application. Land reclamation projects may also generate contaminated soils from municipal and industrial sites. Notable are soils removed from sites subject to spills of hydrocarbons and other materials that require proper disposal and monitoring.

It proves useful to assess the fate of organic chemicals in these biosolids and soils to ensure that excessive concentrations do not arise. This can be accomplished by the use of mass-balance models with an aim to describe the fate of organic chemicals in the amended soils to ensure that crops are not contaminated. In the case of biopiles, it is desirable to assess the rate of decontamination over time and to indicate the need for incorporating engineered interventions such as ventilation and nutrient addition. The models may be applied in advance to gain a perspective on the decontamination rates, to contribute to restrictions on amendment rates, and to assist system design. They may be detailed or evaluative in nature.

An example of an evaluative chemical feed model is the Biosolid Amended Soil Level IV fugacity model (*BASL4*) described by Hughes et al. (2008, 2011). It is based on the Soil model described earlier, and is only briefly outlined here. The model input includes the physical chemical properties of the organic chemical, especially the octanol–water partition ratio, solubility, and vapor pressure and the biodegradation rate in the amended soil expressed as an estimated half-life. Soil properties include the volume fractions of air, water, and OC. The biosolids may be applied directly at a defined rate to the soil surface and if desired incorporated into the soil to a defied depth by plowing. Multiple layers of soil may be treated. Conventional Level IV differential equations are set up in fugacity format and solved by numerical integration to describe the fate of the chemical in the soil over time as the chemical is subject to evaporation, degrading reactions, and transport by leaching. Diffusion and bioturbation vertically and biodegradation by bacteria and uptake into by resident organisms such as worms and vegetation may also be considered. The fate of the chemical in the amended soil may be computed over a period of months and years as necessary.

Coulon et al. (2010) have described fugacity models of chemical fate, especially hydrocarbons in engineered biopiles, including the effect of temperature and extended to treatment of hydrocarbon contaminated sounds in cold climates. A number of studies involving non-fugacity models of chemical fate in soils and biopiles have been reported by US government agencies, notably the US Environmental Protection Agency.

The BASL4 model is available for download from the Trent website and is fully described in the study by Hughes et al. (2008).

6.8 LONG-RANGE TRANSPORT: THE *TAPL3* MODEL AND CHARACTERISTIC TRAVEL DISTANCES

Chemical persistence leads to one of the most vexing international environmental issues, that of long-range transport (LRT) to areas remote from the chemical source. When a chemical persists for long periods of time, transport by air or water advection becomes a significant dispersion mechanism. LRT presents an interesting challenge because, like persistence, LRT cannot be measured in the environment. Its extent can only be estimated using models. Most interest is in atmospheric LRT; however, in some cases, oceans and rivers can also play a significant role. Even migrating biota can contribute to LRT.

The most promising approach is to consider the fate of chemical in a parcel of "Lagrangian" air passing over soil or water and subject to degradation, deposition, and re-evaporation. Such systems have been suggested by Van Pul et al. (1998), Bennett et al. (1998), and Beyer et al. (2000). Beyer et al. (2000) showed that an LRT distance in air can be deduced from a simple Level III model as the product of the wind velocity, the overall persistence or residence time of the chemical, and the fraction of the chemical in the atmosphere.

In this section we give an overview of the salient aspects of such an LRT modeling calculation. It is based on the work of Beyer et al. (2000) and employs the *TaPL3* (Transport and Persistence Level 3) program, available from the CEMC website. A similar approach has been taken by Wegman et al. (2009) under the name of the OECD model. TaPL3 is essentially a standard Level III calculation, but without any advection contributions. TaPL3 takes advantage of the fact that a Level III calculation that does not involve any modeled air movement can be used to determine all of the necessary quantity estimates needed to determine the Characteristic Transport Distance (CTD). The CTD is the distance traveled by a fluid medium such as air or water at which the initial concentration in the medium drops to 1/e (about 37%) of its initial concentration. Thus, as demonstrated in the work of Beyer et al., the Level IV problem of a Lagrangian air mass movement is reduced to a Level III steady-state "box" problem.

The 2000 version of TaPL3 was updated in 2003 by Webster et al. to include provision for temperature variation, improved treatment of vegetation, and an alternative "hopping" descriptor for the potential for LRT.

The TaPL3 model is based on the movement of a Lagrangian air parcel over a stationary landscape phase. The parcel moves over a stationary medium such as soil and is considered at all times to be at steady state in terms of transfer to and from the air to the soil. In the formulation of the theory of characteristic travel distance by Bennett et al. (1998), the air (and hence soil) concentration undergoes a continuous decrease with distance travel. The transfer from air to soil and the reverse must be in mass balance. Recasting the original derivation by Bennett et al., we can write

$$C_A U_A (h_A w) + C_S k_{SA}^M (w\Delta x) = (C_A + \Delta C_A) U_A (h_A w) + C_A k_A^{Deg} (h_A w\Delta x) + C_A k_{AS}^M (w\Delta x)$$

where the various symbols represent concentration (C_A and C_S), the change in air concentration along a distance Δx (ΔC_A), air–body height (h_A), air–soil interface width (w), air–soil contact distance along direction of travel (Δx), air-travel velocity (U_A), mass-transport coefficients for air–soil and soil–air interfacial transport (k_{AS}^M and k_{SA}^M), and the air degradation first-order rate constant (k_A^{Deg}).

Rearrangement of this equation and conversion to a differential equation leads to

$$\frac{dC_A}{dx} = \frac{\left(-C_A\left(\left[k_A^{\text{Deg}} \right] + \left[\frac{k_{AS}^M}{h_A} \right] \right) + \left[\frac{C_S k_{SA}^M}{h_A} \right] \right)}{U_A}$$

A mass balance in the soil compartment gives rise to a second expression that allows for the elimination of the soil concentration as a variable:

$$C_A k_{AS}^M \left(w\Delta x \right) = C_S k_{SA}^M \left(w\Delta x \right) + C_S k_S^{\text{Deg}} \left(h_S w\Delta x \right)$$

Here we also introduce the first-order rate constant for decomposition in the soil (k_S^{Deg}) and the soil depth as the height of the soil compartment (h_S).

Now:

$$C_S = \frac{C_A k_{AS}^M}{\left(k_{SA}^M + k_S^{\text{Deg}} h_S \right)}$$

So that substitution in the differential equation for air concentration change above leads to

$$\frac{dC_A}{dx} = \frac{-C_A\left(\left[k_A^{\text{Deg}} \right] + \left[\frac{F k_{AS}^M}{h_A} \right] \right)}{U_A} \quad \text{where } F = \left[\frac{k_S^{\text{Deg}}}{k_{SA}^M / h_S + k_S^{\text{Deg}}} \right]$$

F is equal to the fraction of the transported chemical that does not return to the air parcel, i.e., it degrades in the soil. This equation is easily rearranged into first-order decay form:

$$\frac{dC_A}{C_A} = \left[\frac{-\left(\left[k_A^{\text{Deg}} \right] + \left[\frac{F k_{AS}^M}{h_A} \right] \right)}{U_A} \right] dx$$

Consequently, it may be integrated by inspection to yield

$$C_A = C_A^0 \exp(-kx) \quad k = \frac{-\left(\left[k_A^{\text{Deg}} \right] + \left[\frac{F k_{AS}^M}{h_A} \right] \right)}{U_A}$$

Of key interest is the characteristic travel distance (CTD), defined as $1/k$, at which point ($1/e$) or 37% of the chemical remains in the air body. Here, the CTD is given by

$$L_A^{CTD} = \frac{1}{k} = \frac{U_A}{\left(\left[k_A^{\text{Deg}}\right] + \left[\frac{Fk_{AS}^M}{h_A}\right]\right)} = \frac{U_A}{k_{\text{eff}}}$$

The term k_{eff} is the total or effective loss rate constant, as defined by Bennett et al. (1998).

The work of Beyer et al. has demonstrated that a Level III model can be used to determine L_A^{CTD} without having to resort to the solution to the differential equation derived above. This result is the basis of the TaPL3 model approach, and accounts for the "3" in its name. We may demonstrate this fact by rewriting the above expression for the CTD in terms of the total mass of chemical in the air, m_A:

$$L_A^{CTD} = \frac{U_A m_A}{\left(\left[k_A^{\text{Deg}} m_A\right] + \left[\frac{Fk_{AS}^M m_A}{h_A}\right]\right)} = \frac{U_A m_A}{\left(r_A^{\text{Deg}} + r_{\text{Dep}}^{\text{Net}}\right)} = \frac{U_A m_A}{r_A^{\text{Tot}}}$$

Here r_A^{Tot} is the total rate of loss from the air, excluding advection. We have substituted the following rate expressions for clarity:

$$r_A^{\text{Deg}} = k_A^{\text{Deg}} m_A \ \text{ and } \ r_{\text{Dep}}^{\text{Net}} = \frac{Fk_{AS}^M m_A}{h_A}$$

The persistence time for the chemical is simply the characteristic time, which is given by the ratio of total mass over total rate of loss:

$$\tau_R = \frac{m_T}{r_A^{\text{Tot}}} \ \text{ or } \ r_A^{\text{Tot}} = \frac{m_T}{\tau_R}$$

Bringing these results together, one may express the characteristic travel distance in terms of the fraction of chemical in the air in a steady-state situation:

$$L_A^{CTD} = \frac{U_A m_A}{r_A^{\text{Tot}}} = \frac{U_A m_A}{(m_T / \tau_R)} = U_A \tau_R \left(\frac{m_A}{m_T}\right) = U_A \tau_R m_A^f$$

where $m_A^f = \dfrac{m_A}{m_T}$ is the fractional concentration of chemical in the air. This simpler Level III calcula-

tion is sufficient to determine the values of the non-advective characteristic time for degradation, as well as the fraction of chemical in the air. The speed of the moving medium, in this case air, must be provided by the user for the particular situation in hand.

It may be demonstrated that the CTD for a moving medium in steady-state contact with "j" non-mobile phases is given by

$$L_i^{CTD} = \frac{U_i m_i}{r_i^{\text{Tot}}} = U_i \tau_R m_i^f$$

The essence of a TaPL3 model calculation is therefore to assess the necessary parameters needed to estimate the characteristic travel time for a chemical. A model environment with a surface of 10% water and 90% soil, and air column height of 1000 m, water and soil depths of 20 cm, wind velocity of 4 m s^{-1} (14.4 km h^{-1}, as used by Bennett et al. (1998)), and water velocity of 1 m s^{-1}.

Worked Example 6.6

Application of the TaPL3 model to hypothene.

Apply the TaPL3 model to hypothene, with properties similar to DDT and determine the characteristic travel time for transport when released to (i) air or (ii) water.

The input properties for hypothene with properties similar to DDT appear as follows:

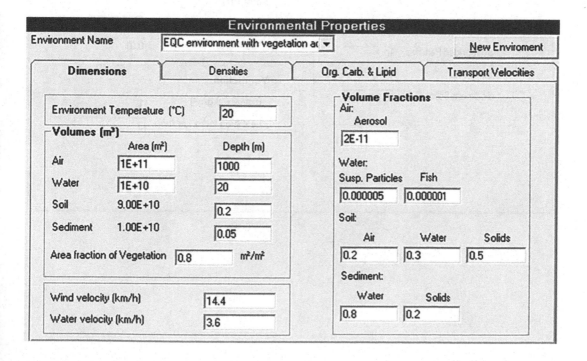

Environmental Properties

Environment Name   EQC environment with vegetation ac ▼   <u>N</u>ew Enviroment

| Dimensions | **Densities** | Org. Carb. & Lipid | Transport Velocities |

Densities for subcompartments (kg/m³)

Note that 'air vapour' is assumed to have a density of 1.185 kg/m³ and 'pure' water has a density of 1000 kg/m³

Density of fish refers to the mass of fish / volume of fish

Air:	Air	Aerosol		
	1.185	2000		
Water:	Water	Susp. Particles	Fish	
	1000	1500	1000	
Soil:	Air	Water	Solid	
	1.185	1000	2400	
Sediment:	Water	Solid		
	1000	2400		
Vegetation:		900		

Environmental Properties

Environment Name   EQC environment with vegetation ac ▼   <u>N</u>ew Enviroment

| Dimensions | Densities | **Org. Carb. & Lipid** | Transport Velocities |

Organic Carbon (g/g)

Suspended Particles	0.2
Soil Solids	0.02
Sediment Solids	0.04

Lipid (g/g)

Fish Lipid	0.05
Vegetation Lipid	0.01

Vegetation

Biomass (kg/m²)	1
Leaf Area Index (m²/m²)	3

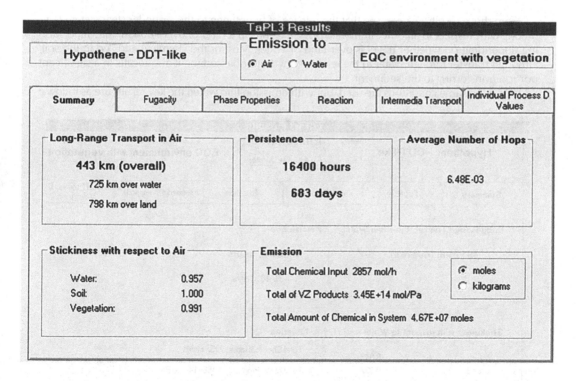

The summary of results from the TaPL3 computation is shown here, with an estimated overall LRT distance:

The results are also presented by TaPL3 in diagrammatic form are

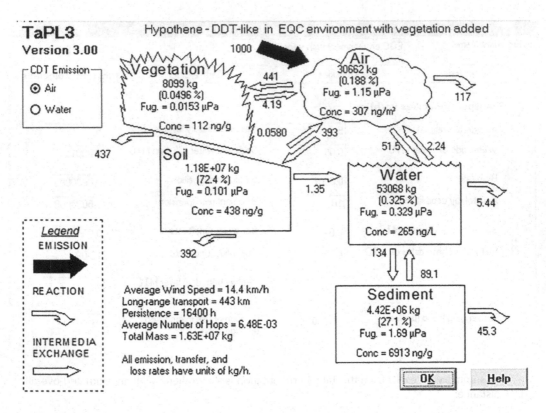

From these results, we can see that the release of hypothene to the air under steady-state conditions with respect to all other immobile phases results in a CTD of over 700 km, the value varying depending on whether travel is over land or water. Most of the chemical ends up in the soil, although transfer through water to the relatively hydrophobic sediment also results in a significant portion transferring to the sediment.

When the same conditions are used, but with emission into water, the key results are as follows:

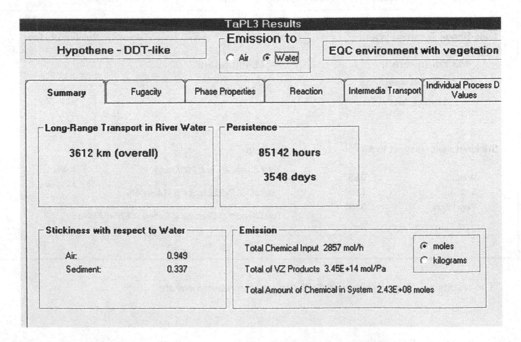

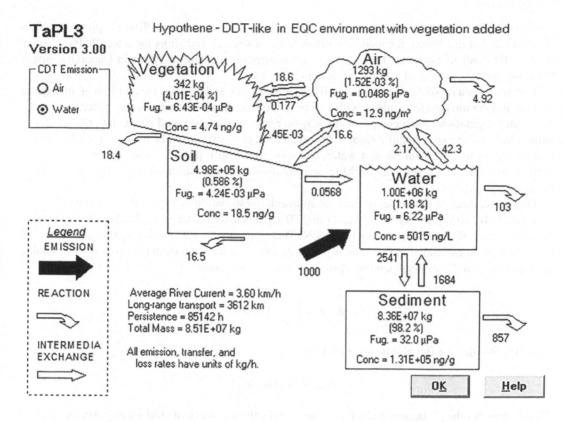

Virtually all the chemical ends up in the sediment, due to its very high fugacity capacity. The characteristic travel time is much longer at over 3500 km, primarily driven by the slow indirect transfer from water to soil via air, a "kinetic delay" effect (Mackay et al., 2019). The air capacity for carrying hypothene does not allow the concentration in the soil to reach high values, despite the high fugacity capacity of soil compared with water.

6.9 URBAN ENVIRONMENTAL MODELS: THE MULTIMEDIA URBAN MODEL (*MUM*)

Most multimedia environmental fate and exposure models are designed to handle typical "natural" environmental situations such as lakes, forests, fields, soils, and air bodies. However, most humans do not spend most of their time in such environments. Rather, humans are by-and-large urban creatures, living in environments that are comprised of concrete, asphalt, glass, and other building surfaces, as well as highly altered biosphere components such as lawns, isolated city trees and shrubs, and relatively few managed water bodies. For this reason, consideration of human health and environmental exposure must take these "non-traditional" aspects of the human environment into account. Exposure due to human activity may be highly localized, such as with industrial point-source emissions and WWTP discharge, or diffuse, such as with agricultural application of pesticides. Additionally, there is evidence that chemical contamination levels and human exposures are higher in urban regions, i.e. "people pollute."

Models designed to account for the environments in which most humans live are called "urban" models. The "Multimedia Urban Model" or MUM (Diamond et al., 2001) was developed by Diamond and co-workers in 2001 and is representative of this type of model. MUM was first applied to evaluative urban environments and specifically to the lower Don River watershed in Toronto, Canada, to consider the emissions and fate of several classes of semivolatile organic compounds (SVOCs) such as PAHs, PCBs, and organophosphate esters. Since that time it has been used to look at the role of indoor dust as a pathway for polybrominated diphenyl ether (PBDE) exposure in 2005 (Jones-Otazo et al., 2005), for modeling urban films as sources and sinks for select PCBs (Csiszar et al., 2012) and to estimate the emissions and environmental fate of PAHs in Catalonia, Spain (Dominguez-Morueco et al., 2016), for example.

The model is a modification of a Level III fugacity model, and features the addition of an organic film on impervious surfaces along with five more familiar compartments of air, water, soil, sediment, and vegetation. Each compartment may be comprised of pure and particulate phases, these being assumed to exist at equilibrium within each bulk compartment. The model allows for advective transport to and from air and water, emission to all bulk phases, losses due to irreversible vertical airborne transport, soil leaching to groundwater, sediment burial, and export/removal of vegetative litter.

Unique features of the model include the approach to estimation of the Z-values for the bulk film, assumed to be 30% pure "organic" matter and 70% particulate matter. The pure film is assumed to have the same OC fraction as octanol, i.e., 0.74. Therefore, the Z-value for the organic component of the film may be estimated in the same way that Z_O may be estimated from the product of a suitable partition ratio and the corresponding Z-value of the partner phase:

$$K_{Oi}Z_i = \left(\frac{Z_O}{Z_i}\right)Z_i = Z_O$$

Here, the Z-value for the film is thus given by

$$Z_{\text{Film}} = v_{OC}^f\left(K_{OA}Z_A\right)$$

The Z-value for the particulate matter is assumed to be the same as that used for airborne particulate matter, based on the reasoning that the film particulates originate from the atmosphere. The model uses the relationship of Bidleman and Harner (2000) correlating the particle–air partition ratio K_P (m^3 µg^{-1}) with K_{OA}, assuming an OM content in the aerosol of 20%. Thus,

$$Z_{\text{Film}}^Q = Z_A^Q = \left[10^{\left(\log K_{OA} + \log v_{OC}^f - 11.91\right)}\right]Z_A\rho_Q \times 10^9$$

The bulk film Z-value is determined as a volume fraction-weighted sum of these two Z-values.

The total area of impervious film is calculated based on aerial estimates and an impervious surface index (ISI) to account for the three-dimensional surface area of buildings, similar to a leaf area index (LAI) (Monteith and Unsworth, 2013). D-values for various inter-compartment transfer processes are then adjusted to account for unique aspects of the urban environment, such as the high adhesion probability for particles on the impervious "greasy" films and rapid conveyance of water from built-up areas. Mass-balance equations are then set up for the six compartments and solved by substitution for the bulk compartment fugacities.

The MUM was updated in 2018 (Rodgers et al., 2018). The single-parameter Linear Free Energy Relationships (sp-LFERs) in the original model were replaced by the more recently developed and more accurate polyparameter LFERs (pp-LFERs). The latter were taken from the UFZ-LSER database of Goss and co-workers (Ulrich et al., UFZ-LSER database; http://www.ufz.de/lserd). This modified model was applied to the transport, fate, and emission of organophosphate esters in Toronto, Canada.

6.10　INDOOR AIR MODELS: THE *INDOOR* MODEL

We present here a very simple model of chemical fate in indoor air. Numerous studies have shown that humans are exposed to much higher concentrations of certain chemicals indoors than outdoors.

Notable are radon, CO, CO_2, formaldehyde, pesticides, fragrances, and volatile solvents present in glues, paints, and a variety of consumer products.

The key issue is that, whereas advective flow rates in air are large outdoors, they are constrained to much smaller values indoors. Attempts to reduce heating costs often result in reduced air exchange, leading to increased chemical "entrapment." A submarine or a space vehicle is an extreme example of reduced advection. Bennett and Furtaw (2004) developed a fugacity-based indoor residential pesticide fate model that includes air, aerosols, carpet, smooth flooring, walls in both treated and adjacent indoor chambers. Fairly complicated models of chemical emission, sorption, reaction, and exhaust in multichamber buildings have been compiled [e.g., Nazaroff and Cass (1986, 1989) and Thompson et al., (1986)], but we treat here only the simple model developed by Mackay and Paterson (1983), which shows how *D*-values can be used to estimate indoor concentrations caused by evaporating pools or spills of chemicals.

An example of the effective use of fugacity for compiling mass-balances indoors is the *INPEST* model, developed in Japan by Matoba et al. (1995, 1998a, 1998b). This model successfully describes the changing concentration of pyrethroid pesticides applied indoors in the hours and days following their application. Because of the reduced advection, there is a potential for high concentrations and exposures immediately following pesticide use, and it may be desirable to evacuate the room or building for a number of hours to allow the initial peak concentration in air to dissipate. The *INPEST* fugacity model, which is in the form of a spreadsheet, can be used to suggest effective strategies for avoiding excessive exposure. It can be used to compare pesticides and explore the effects of different application practices.

The diagrammatic output of the *Indoor* model available from the CEMC website for infiltration of toluene from outside air into a model house is shown in Figure 6.7.

One may note that the transfer rates are driven by fugacity differences in the various compartments.

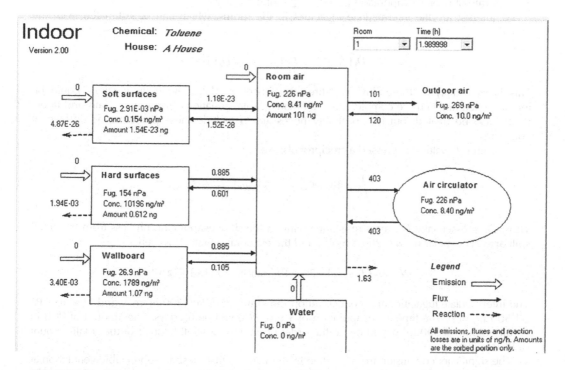

FIGURE 6.7 Diagrammatic output of the *Indoor* model for a calculation of toluene infiltration from outside air into an initially "clean" house.

Worked Example 6.7: Indoor evaporation of a chemical.

An example of a "spill" of a small quantity (1 g) of PCB over 0.01 m² (e.g., from a fluorescent ballast) was considered by Mackay and Paterson (1983). The PCB fugacity was 0.12 Pa and the outdoor concentration was taken as 4 ng m⁻³ or 3.7×10^{-8} Pa. Determine the evaporation rate, the intermediate fugacities, and the time for evaporation.

We treat a situation in which a pool of chemical is evaporating into the basement air space of a two-room (basement plus ground level) building with air circulation. If the building were entirely sealed and the chemical were non-reactive, evaporation would continue until the fugacity throughout the entire building equaled that of the pool (f_P). Of course, it is possible that the pool would have been completely evaporated by that time.

The evaporation rate can be characterized by a D-value D_1 corresponding to kAZ, the product of the mass-transfer coefficient, pool area, and air phase Z-value. If the chemical were in solution, it would be necessary to invoke liquid and gas phase D-values in series, i.e., the two-film theory, as discussed in Chapter 4.

The evaporated chemical may then be advected from one room to another with a D-value D_2 defined as GZ, the product of the air flow or exchange rate and the air phase Z-value. From this second room, it may be advected to the outdoors with another D-value D_3. These advection rates are normally characterized as "air changes per hour" or ACH, which is the advection rate G divided by the room or building volume and is the reciprocal of the air residence time. Typical ACHs for houses range from 0.25 to 1.5 per hour. The outdoor air has a defined background fugacity f_A.

It is apparent that the chemical experiences three D-values in series in its journey from spill to outdoors, thus the total D-value will be given by

$$\frac{1}{D_T} = \frac{1}{D_1} + \frac{1}{D_2} + \frac{1}{D_3}$$

and the rate of overall evaporation is $D_T(f_P - f_A)$ mol h⁻¹.

Of interest are the intermediate fugacities in the rooms, which can be estimated from the equations:

$$N = D_1 \left(f_P - f_1 \right) = D_2 \left(f_1 - f_2 \right) = D_3 \left(f_2 - f_A \right)$$

This is essentially a "three-film" or "three-resistance" model. Degrading reactions could be included, leading to more complex, but still manageable, equations. Sorption to walls and floors could also be treated, but it is probably necessary to include these processes as differential equations.

The three D-values (expressed as reciprocals) are

$$\frac{1}{D_1} = 49,000, \ \frac{1}{D_2} = 30, \text{ and } \frac{1}{D_3} = 15$$

Thus, D_T is essentially D_1, most resistance lying in the slow evaporation process from the small spill area. The molar mass is 260 g mol⁻¹, and the evaporation rate N is then

$$N = D_T \left(f_P - f_A \right) = 2.4 \times 10^{-6} \text{ mol h}^{-1} = 6.4 \times 10^{-4} \text{ g h}^{-1}$$

The intermediate fugacities and concentrations are 3.6×10^{-5} Pa (3800 ng m⁻³) and 11×10^{-5} Pa (11,500 ng m⁻³). The time for evaporation of 1 g of PCB will be 65 days. The amount of PCB in an air volume of 500 m³ would be on the order of 0.004 g, a small fraction of the small amount of PCB spilled.

The significant conclusion from Worked Example 6.7 is that, despite appreciable ventilation at an ACH of 0.5 h⁻¹, the indoor air concentrations are over 1000 times those outdoors. Fortunately, the indoor fugacity is still very much lower than the pool fugacity. Similar behavior applies to other solvents, pesticides, and chemicals that may be used and released indoors. Although the amounts spilled or released are small, the restricted advective dilution results in concentrations that are much higher than are normally encountered outdoors. In many cases, this phenomenon is suspected to be the cause of the "sick building" problem in which residents complain repeatedly about headache, nausea, and tiredness. The cure is to eliminate the source or increase the ventilation rate. Fugacity calculations can contribute to understanding such problems.

6.11   CONTINENTAL AND GLOBAL MODELS

The ultimate mass-balance model of chemical fate is the one that describes the dynamic behavior of the substance in the entire global environment. At present, only relatively simplistic treatments of chemical fate at this scale have been accomplished, but it is likely that more complex and accurate models will be produced in the future. Meteorologists can now describe the dynamic behavior of the atmosphere in considerable detail. Similarly, oceanographers are able to describe ocean currents with acceptable precision. Ultimately, there may be linked meteorological/oceanographic/terrestrial models in which the ultimate fate of 100 kg of DDT applied in Mexico can be predicted over the decades during which it migrates globally.

The obvious ethical implication is that a nation should not use a substance in such a way that other nations suffer significant exposure and adverse effects. These situations have already occurred with acid rain and Arctic and Antarctic contamination by persistent organic substances.

The construction of global-scale models opens up many new and interesting prospects. It appears that there is a global fractionation phenomenon as a result of chemicals migrating at different rates and tending to condense at lower temperatures. Chemicals that do reach cold regions may be better preserved there because of the reduced degradation rates. Chemicals appear to be subject to "grasshopping" (or "kangarooing" in the Southern Hemisphere) as they journey, deposit, evaporate, and continue hopping from place to place during annual seasons until they are ultimately degraded as shown in Figure 6.8.

Accounts of these phenomena, and models that attempt to quantify them, are given in a series of papers by Wania and Mackay (1993, 1995, 1996, 1999a) and Wania et al. (1998). The most successful modeling to date has been of a-HCH, which was produced as an impurity in the insecticide, technical lindane (Wania and Mackay, 1999a) but is no longer produced. An interesting insight from that study is an assertion that, despite a-HCH never having been used in the Arctic, about half the remaining mass on this planet now resides in the Arctic Ocean regions. This *GloboPOP* model is

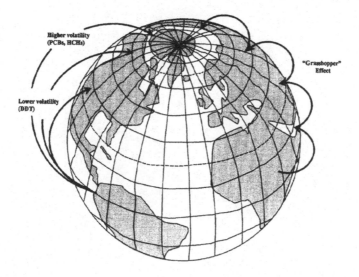

FIGURE 6.8   Schematic diagram of chemical "grasshopping" on a global scale.

available from the University of Toronto from the website https://www.utsc.utoronto.ca/labs/wania/ downloads/. A link is maintained from the Trent CEMC website, from which other models from groups other than the CEMC are also available.

As better models of global fate become available, they will provide invaluable tools with which humanity can design, select, and use chemicals on our planet with little or no fear of adverse consequences. Whether we will be sufficiently enlightened to achieve this is a question only time will tell.

SYMBOLS AND DEFINITIONS INTRODUCED IN CHAPTER 6

Symbol	Common Units	Description
B_i	$m^2\,h^{-1}$	Porous medium "i" molecular diffusivity
B_{Ei}	$m^2\,h^{-1}$	Effective porous medium "i" diffusivity, corrected for tortuosity and void volume
F_Y	—	Tortuosity factor
F_A or v	—	Void volume fraction
h	m	River depth
L_A^{CTD}	m	Characteristic travel distance
S	—	Sensitivity parameter
$t_i^{1/2}$	H	Half-time for loss in phase "i"
$U*$	$m\,h^{-1}$	Friction velocity
W	m	River width

7 Chemical Uptake by Organisms

Bioavailable Solute Fraction (BSF) in water: $\text{BSF} = \dfrac{1}{\left(1 + \left(\dfrac{K_{OW}C_W^{OC}\,(\text{kg L}^{-1})}{\rho_{OM}\,(\text{kg L}^{-1})}\right)\right)}$

Approximate Z-value in organisms such as fish: $Z_F = v_L^{f-\text{Fish}} Z_O = v_L^{f-\text{Fish}} K_{OW} Z_W$

Concentration Format	Fugacity Format
Fish Bioconcentration Time-Dependence:	
$\dfrac{dC_F}{dt} = k_1 C_W - k_2 C_F$	$(V_F Z_F)\dfrac{df_W}{dt} = D_V(f_W - f_F)$
Uptake:	
$C_F = \left(\dfrac{k_1}{k_2}\right)C_W\left[1 - e^{-k_2 t}\right]w$	$f_F = f_W\left(1 - \exp\left(-(D_V/V_F Z_F)t\right)\right)$
Clearance:	
$C_F = C_F^0 e^{-k_2 t}$	$f_F = f_F^0 \exp\left(-(D_V/V_F Z_F)t\right)$
Fish Bioaccumulation Time-Dependence:	
$\dfrac{dC_F}{dt} = \dfrac{(k_1 C_W + k_D C_D)}{(k_V + k_E + k_C + k_W)}$	$\left(\dfrac{df_F}{dt}\right) = \left(\dfrac{1}{V_f Z_f}\right)D_V f_W + D_D f_D$
$C_F = \dfrac{(k_1 C_W + k_D C_D)}{(k_V + k_E + k_C + k_W)}$	$- f_F(D_V + D_M + D_E)$
	$f_F = \dfrac{(D_V f_W + D_D f_D)}{\left(D_V + D_M + (D_D/Q)\right)}$
Biomagnification Factor (BMF):	
$\text{BMF} = \dfrac{C_F}{C_{\text{Diet}}}$	$\text{BMF} = \dfrac{Z_F}{Z_{\text{Diet}}}$

Trophic Magnification Factor: TMF is an "averaged" BMF obtained from the slope of a plot of log concentration versus trophic level for a food web

7.1 BIOACCUMULATION, BIOCONCENTRATION, BIOMAGNIFICATION, TMFs: THE *FISH* MODEL

The phenomenon of bioaccumulation is very important as a means by which chemicals present at low concentration in water become concentrated in aquatic organisms, often by many orders of magnitude, thus causing a potential hazard to aquatic species such as fish, and especially to birds and humans who consume them. For example, DDT may be found in fish at concentrations a million times that of the water in which they live. The primary cause of this effect is simply the difference in Z-values between water and fish lipids as characterized by K_{OW}. However, there are also other, more subtle effects at work. The kinetics of uptake are also important, because a fish may never reach thermodynamic equilibrium. There is also a fascinating biomagnification phenomenon that is not yet fully understood in which concentrations increase progressively through food chains. An additional incentive for quantifying bioaccumulation is that it is often easier to sample and analyze fish or other organisms in preference to water since water concentrations are much lower and more variable.

It is useful to define some terminology, although opinions differ on the correct usage. *Bioconcentration* refers here to uptake by respiration from water, usually under laboratory conditions when the fish are not fed. *Bioaccumulation* results from the total (water plus food) uptake process and can occur in the laboratory or field. *Biomagnification* is a special case of bioaccumulation in which there is an increase in concentration or fugacity from food to fish. This situation may occur for non-metabolizing chemicals of log K_{OW} exceeding approximately 5.

A comprehensive review of methods of estimating bioaccumulation is that of Gobas and Morrison (2000). Other reviews are the texts by Connell (1990) and Hamelink et al. (1994) and the paper by Mackay and Fraser (2000). The models described here are based on those of Clark et al. (1990), Gobas (1993), and Campfens and Mackay (1997). Figure 7.1 shows the processes of uptake and clearance by a fish. Armitage et al. (2018) have developed a comprehensive Bioaccumulation Assessment Tool (BAT) that addresses this issue and provides estimates of relevant properties of chemicals and organisms.

The approach taken here is to set out the basic mass-balance equations in conventional rate constant form, then show that they are equivalent to the fugacity versions using D-values. The final model gives both rate constants and D-values.

7.1.1 Concentration Format

Here, we treat the fish as one compartment or "box." The conventional concentration expression for uptake of chemical by fish from water, through the gills only under laboratory conditions, was first written by Neely et al. (1974) as

$$\frac{dC_F}{dt} = k_1 C_W - k_2 C_F$$

where C_F and C_W are concentrations in fish and water, k_1 is an uptake rate constant, and k_2 is the clearance rate constant that may include several loss processes as depicted in Figure 7.1. In a simple bioconcentration test, k_2 is primarily the respiratory loss.

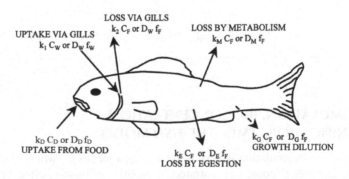

FIGURE 7.1 Fish bioaccumulation processes expressed as concentration/rate constants and fugacity/ D-values.

Apparently, the chemical passively diffuses into the fish via the gills into the circulating blood and then into other tissues, taking the same route as oxygen. In the laboratory, it is usual to expose a fish to a constant water concentration for a period of time during which the concentration in the fish should rise from zero to C_F according to the integrated version of the differential equation with C_F initially zero and C_W constant:

$$C_F = \left(\frac{k_1}{k_2}\right) C_W \left[1 - \exp(-k_2 t)\right]$$

This solution gives rise to the following well-known "exponential rise to an asymptotic limit" form, illustrated in Figure 7.2, for a chemical that has a log K_{OW} of 4.7 and a bioconcentration factor (BCF) of 5000, corresponding to a lipid content of 5% with a k_1 of 50 h^{-1}.

7.1.2 Bioconcentration of a Persistent (Non-Biotransforming) Chemical

There are several constraints on the design of a successful experimental bioconcentration test. The test organism is not fed, so that it cannot survive indefinitely. This fact constrains the test duration to a typical maximum of 96 h or 4 days. In some cases, a shorter duration of 48 h is used. To achieve appreciable uptake, the uptake rate constant must be relatively high, implying that the organism is small, e.g., 1 g or less in body weight. Typical organisms are fathead minnows, guppies, daphnids, or even algae. The concentrations of the chemical in the water must be constant and selected to be well below lethal levels. This implies that the chemical activity be less than 10^{-3} or even 10^{-4}, i.e., the concentration should be a factor of 1000–10,000 below the solubility limit.

If equilibrium cannot be achieved feasibly for the chemical of high K_{OW}, the BCF can be estimated as k_1/k_2, from separate measurements of k_1 (from an initial uptake rate) or k_2 (from a separate depuration test).

Aquatic toxicity tests are similar in principle, but the aim is to determine the chemical concentration in water that causes a 50% mortality as at prescribed times of 48 or 96 h. As equilibrium

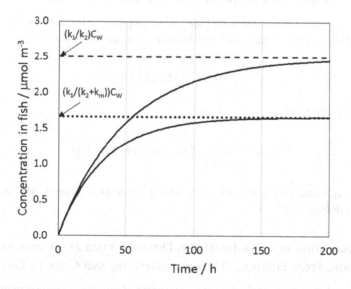

FIGURE 7.2 Time-dependence of a persistent chemical of log K_{OW} = 4.7 in 1.0 g fish of 5% lipid content exposed to water of 10^{-9} mol m^{-3} chemical concentration, with rate constants of k_1 = 50.0 and k_2 = 0.02 h^{-1}. The upper curve corresponds to a fish with no metabolic losses. The lower curve corresponds to a biotransforming chemical with a k_M of 0.01 h^{-1}. The upper dashed and lower dotted lines indicate the corresponding asymptotic limits of 2.5 and 1.67 μmol m^{-3}.

approached between water and the fish, equifugacity applies with Z-values of Z_W in water and approximately $Z_W v_L^f K_{OW}$ in the fish, where v_L^f is the volume fraction of lipid in the fish. The BCF C_F/C_W is then $v_L^f K_{OW}$ or Z_F/Z_W.

A useful rule of thumb is that for a 1 g fish, k_1 is approximately 1000 days^{-1} or 40 h^{-1}. The implication is that the fish absorbs chemicals from a volume of water 1000 times its own volume per day or 40 times its volume per hour. The uptake rate constant can be scaled to larger or smaller fish by raising the volume to the power -0.67. For example, for a larger 10 g fish, k_1 falls by a factor of 2.13 to $1000 \times 10^{-0.67}$ or 214 days^{-1}.

After prolonged exposure, when the product $k_2 t$ is large, i.e., >4, C_F approaches $(k_1/k_2)C_W$ or $K_{FW}C_W$, where K_{FW} is the equilibrium BCF.

If the fish is then placed in clean water, and loss or clearance or depuration follows, the corresponding depuration equation being:

$$C_F = C_F^0 \exp(-k_2 t)$$

where C_F^0 is the concentration in the fish at the start of clearance. Such an equation predicts simple exponential decay to a limit of zero, i.e., the fish depurates exponentially.

It is apparent that there are three parameters, k_1, k_2, and K_{FW} or k_1/k_2, thus only two can be defined independently. The most fundamental are k_1 (which is a kinetic rate constant term quantifying the volume of water that the fish respires per unit of time and from which it removes chemical, divided by the volume of the fish) and K_{FW}, which is a thermodynamic term reflecting equilibrium partitioning, equal to Z_F/Z_W. If the fish mass is used instead of its volume, k_1 and K_{FW} are numerically the same, but differ in their units.

The loss rate constant k_2 is best regarded as k_1/K_{FW}. The uptake and clearance half-times are both

$$\tau_{1/2} = \frac{0.693 K_{FW}}{k_1} = \frac{0.693}{k_2}$$

7.1.3 Bioconcentration of a Biotransforming Chemical in a Fish

In this case, the loss rate constant now includes both respiratory loss and biotransformation with rate constants k_2 and k_M, respectively, and the differential equation becomes

$$\frac{dC_F}{dt} = k_1 C_W - C_F (k_2 + k_M)$$

This equation integrates to give

$$C_F = \frac{C_W k_1}{(k_2 + k_M)} \left(1 - \exp(-(k_2 + k_M)t)\right)$$

At long times C_F approaches $C_W \times (k_1/(k_2+k_M))$ and a lower BCF applies with a ratio $k_1/(k_2+k_M)$. This is illustrated in Figure 7.2.

7.1.4 Bioaccumulation in a Fish Including Dietary Uptake and Losses by Respiration, Fecal Egestion, Biotransformation, and Growth Dilution

Bioaccumulation is the most ecologically relevant metric of biouptake since it includes uptake from bulk water and the diet and additional loss processes. The earlier uptake equation can be readily expanded to include uptake from food (diet) with a rate constant k_D and food concentration C_D, loss by respiration with a rate constant k_2, loss by metabolism with a rate constant k_M, and loss by egestion in feces with rate constant k_E, namely:

$$\frac{dC_F}{dt} = k_1 C_W + k_D C_D - C_F \left(k_2 + k_M + k_E \right)$$

If the fish is growing, there will be growth dilution, which can be included as an additional loss rate constant k_G, which is the fractional increase in fish volume per hour, i.e., k_G is $(dC_F/dt)/C_F$, and therefore

$$\frac{dC_F}{dt} = k_1 C_W + k_D C_D - C_F \left(k_2 + k_M + k_E + k_G \right)$$

At steady-state conditions, the left side is zero, and we have

$$0 = k_1 C_W + k_D C_D - C_F \left(k_2 + k_M + k_E + k_G \right)$$

or

$$C_F = \frac{k_1 C_W + k_D C_D}{\left(k_2 + k_M + k_E + k_G \right)}$$

Gobas (1993) and Arnot and Gobas (2004) have suggested correlations for these rate constants as a function of fish size, temperature, and dissolved oxygen content in the water. The mass balance around the fish can be deduced and the important processes identified. The rate constant k_1 is much larger than k_D, typically by a factor of 10,000. Thus, uptake from water and food becomes equal when C_D is about 1000–10,000 C_W, which corresponds to a K_{OW} of 10^5–10^6 and 10% lipid. For lower K_{OW} chemicals, uptake from water dominates, whereas for higher K_{OW} chemicals, uptake from food dominates.

7.1.4 Fugacity Format

We can rewrite the bioconcentration equation in the equivalent fugacity form as

$$(V_F Z_F)\frac{df_W}{dt} = D_V \left(f_W - f_F \right)$$

where D_V is a gill ventilation D-value, analogous to k_1, and applies to both uptake and loss. This form implies that the fish is merely seeking to establish equilibrium with its surrounding water.

The corresponding uptake and clearance equations are

$$f_F = f_W \left(1 - \exp\left(-\left(D_V/V_F Z_F \right) t \right) \right]$$

$$f_F = f_F^0 \exp\left(-\left(D_V/V_F Z_F \right) t \right)$$

The following expressions relate the rate constants and D-values, showing that the two approaches are ultimately identical algebraically.

$$k_1 = \frac{D_V}{V_F Z_W} \qquad k_2 = \frac{D_V}{V_F Z_F}$$

and

$$\frac{Z_F}{Z_W} = K_{FW} = \frac{\left(k_1 V_F/D_V \right)}{\left(k_2 V_F/D_V \right)} = \frac{k_1}{k_2}$$

As was discussed earlier, Z_F can be approximated as $v_L^f Z_O$, where v_L^f is the volume fraction lipid content of the fish, and Z_O is the Z-value for octanol or lipid. K_{FW} is then $v_L^f K_{OW}$, where v_L^f is typically 0.05 or 5%.

From an examination of uptake data, Mackay and Hughes (1984) suggested that D_V is controlled by two resistances in series, a water resistance term D_W and an organic resistance term D_O. Since the resistances are in series,

$$\frac{1}{D_V} = \frac{1}{D_W} + \frac{1}{D_O}$$

The nature of the processes controlling D_W and D_O is not precisely known, but it is suspected that they are a combination of flow (GZ) and mass-transfer (kAZ) resistances. If we substitute GZ for each D, recognizing that G may be fictitious, we obtain

$$\frac{1}{k_2} = \frac{V_F Z_F}{D_V} = V_F v_L^f Z_O \left(\frac{1}{G_W Z_W} + \frac{1}{G_O Z_O} \right) = \left(\frac{V_F v_L^f}{G_W} \right) K_{OW} + \left(\frac{V_F v_L^f}{G_O} \right)$$

The characteristic time for chemical depuration in the fish is given by the total amount present divided by the rate of depuration or loss:

$$\tau_i = \frac{V_F v_L^f}{G_i}$$

Upon substitution, we find that the loss rate constant is related to the characteristic times for depuration in the water or octanol phases as

$$\frac{1}{k_2} = \frac{V_F v_L^f}{G_W} K_{OW} + \frac{V_F v_L^f}{G_O} = \tau_W K_{OW} + \tau_O$$

By plotting $1/k_2$ against K_{OW} for a series of chemicals taken up by goldfish, Mackay and Hughes (1984) estimated that τ_W was about 0.001 h and τ_O was approximately 300 h. This is another example of probing the nature of series or "two-film" resistances using chemicals of different partition ratio as discussed in Chapter 4.

The times τ_W and τ_O are characteristic of the fish species and vary with fish size and their metabolic or respiration rate, as discussed by Gobas et al. (1989) and Arnot and Gobas (2004). The uptake and clearance equilibria and kinetics, i.e., bioconcentration phenomena, of a persistent chemical in a fish are thus entirely described by K_{OW}, v_L^f, τ_W, and τ_O.

The bioconcentration equation can be expanded as before to include uptake from food (diet) with a D-value D_D, loss by egestion (including urine, D_E), loss by metabolism (D_M), and growth dilution (D_G) giving

$$V_F Z_F \left(\frac{df_F}{dt} \right) = D_V f_W + D_D f_D - f_F (D_V + D_M + D_E + D_G)$$

The growth dilution term can become very important for hydrophobic chemicals for which the D_V and D_M terms are small. The primary determinant of concentration is then how fast the fish can grow and thus dilute the chemical. It should be noted that this treatment of growth is simplistic in that growth is assumed to be first order and does not affect other D-values.

Female fish may experience an additional reproductive loss by egg laying, but this is necessarily an episodic event and requires special treatment that is beyond our scope here.

A mass-balance envelope problem arises when treating the food uptake by digestive processes. The entire fish, including gut contents, can be treated as a single compartment. In this case, the food uptake D-value is simply the GZ product of the food consumption rate and its Z-value, i.e., $G_D Z_D$, subscript D applying to food (diet). Z_D can be estimated as $v_D^f Z_O$, where v_D^f is the lipid content of the diet. Often, the fugacity of a chemical in the food f_D will approximate the fugacity in water. The rate of chemical uptake into the body of the fish is then $E_D D_D f_D$ or $D_{DE} f_D$, where E_D is the uptake efficiency of the chemical. In reality, the gut is "outside" the epithelial tissue of the fish, and it is better to treat the fish as only the volume inside the epithelium. In this case, the uptake D-value is $G_D Z_D E_D$. To avoid confusion, we can define two uptake D-values, D_{DE}, which includes the efficiency, and D_D, which does not. The same problem applies to egestion where we define D_{EE} as including a transport efficiency and D_E, which does not.

The digestive process that controls E_D and thus D_{DE} is more complicated and less understood than gill uptake. The first problem is quantifying the uptake efficiency, i.e., the ratio of quantity of chemical absorbed by the fish to the chemical consumed. It is generally about 50%–90%. Gobas et al. (1989) have suggested that the uptake efficiency E_D from food in the gastrointestinal tract of a "clean" fish can also be described by a two-film approach, yielding

$$\frac{1}{E_D} = A_W K_{OW} + A_O$$

where A_W and A_O are water and organic resistance terms similar in principle to τ_W and τ_O, but are dimensionless. A_O appears to have a magnitude of about 2, and A_W a magnitude of about 10^{-7}; thus, for all but the most hydrophobic chemicals, E_D is about 50%. When K_{OW} exceeds 10^7, the efficiency drops off because of a high water phase resistance in the gut.

A major difficulty is encountered when describing the loss of chemical in feces and urine. In principle, D-values can be defined, but it is quite difficult and messy to measure G and Z; therefore, neither are usually known accurately. It is probable that the digestion process, which removes both mass and lipids to provide matter and energy to the fish, reduces both G_D and Z_D so that D_E for egestion is smaller than D_D. The simplest expedient is to postulate that D_D is reduced by a factor Q_D, thus we estimate D_E for loss by egestion as D_D/Q_D or D_{DE}/Q_D, i.e., D_E or D_{EE}. The resistances causing E_D for uptake are assumed to apply to loss by egestion. This assumption is probably erroneous, but it is acceptable for many purposes, especially because there is usually an insufficiency of data to justify different values for uptake and loss.

The steady-state solution to the differential equation for the entire fish becomes

$$f_F = \frac{\left(D_V f_W + D_D f_D\right)}{\left(D_V + D_M + \left(D_D/Q_D\right) + D_G\right)}$$

For the fish inside the epithelium, D_D is replaced by D_{DE}.

Assuming that f_D equals f_W, it is clear that f_F will approach f_W only when the D_V term dominates in both the numerator and denominator. If K_{OW} is large, e.g., 10^6, the term D_D will exceed D_V (because Z_D will greatly exceed Z_W), and the uptake of chemical in food becomes most important. This results in the following simplification of the expression for the chemical fugacity in the fish:

$$f_F \simeq \frac{\left(D_D f_D\right)}{\left(D_D/Q_D\right)} = Q_D f_D \quad \text{or} \quad Q_D \simeq \frac{f_F}{f_D}$$

Under such circumstances, the fish fugacity then tends toward $Q_D f_D$, i.e., the fish achieves a fugacity BMF of Q_D. Q_D is thus a maximum BMF as well as being a ratio of D-values.

This biomagnification behavior was first clearly documented in terms of fugacity by Connolly and Pedersen (1988), Q_D typically having a value of typically 3–8. Biomagnification is not immediately obvious until the fugacities or lipid-normalized concentrations are examined instead of the total concentrations. At each step in the food chain, or at each trophic level, there is a possibility of a fugacity multiple being necessary. It is thus apparent that fish fugacities and concentrations are a reflection of a complex combination of kinetic and equilibrium terms that can in principle be described by D-values.

The detailed physiology of the factors controlling Q has been investigated in a series of elegant experiments by Gobas and colleagues (1993, 2000). The fugacity change in the gut contents as they journey through the gastrointestinal tract was followed by headspace analysis. These experiments showed convincingly that the hydrolysis and absorption of lipids reduce the Z-value, causing the fugacity to increase as a result of loss of the lipid "solvent." Additionally, the mass of food is reduced, thus G also decreases. The net effect is a decrease in GZ by about a factor Q_D of 4–8. It is noteworthy that Q_D for mammals and birds is much larger, e.g., 30, thus biomagnification is more significant for these animals, rendering them more vulnerable to toxic effects of persistent chemicals. Biomagnification in mammals has been reviewed by deBruyn and Gobas (2006).

Worked Example 7.1: Determine the fugacity of a fish under steady-state exposure to water and food of known fugacity, and determine the extent of biomagnification.

Calculate the fugacity of a fish at steady state when exposed to uptake of a hydrophobic chemical hypothene-6 (log K_{OW} = 6.0) from water and food. Deduce the fluxes and determine if biomagnification occurs.

Input data: $f_W = 10^{-6}$ Pa $f_D = 2 \times 10^{-6}$ Pa. D-values (mol Pa^{-1} h^{-1}): $D_V = 10^{-4}$, $D_D = 10^{-3}$, $D_M = 10^{-4}$, $D_C = 10^{-4}$, $Q_D = 4$

$$f_F = \frac{\left(D_V f_W + D_D f_D\right)}{\left(D_V + D_M + \left(D_E = D_D/Q_D\right) + D_G\right)}$$

$$= \frac{(10^{-4} \times 10^{-6} + 10^{-3} \times 2 \times 10^{-6})\,\text{mol h}^{-1}}{\left(10^{-4} + 10^{-4} + \left(10^{-3}/4\right) + 10^{-4}\right)\text{mol Pa}^{-1}\,\text{h}^{-1}}$$

$$= \frac{(10^{-10} + 2 \times 10^{-9})}{5.5 \times 10^{-4}}\,\text{Pa} = 3.82 \times 10^{-6}\,\text{Pa}$$

Therefore, we have

$$\frac{f_F}{f_D} = \frac{3.82 \times 10^{-6}\,\text{Pa}}{2 \times 10^{-6}\,\text{Pa}} = 1.91$$

This BMF value indicates an increase in fugacity from diet to fish by a factor of 1.91, demonstrating that biomagnification occurs. Note that there is a slight biomagnification in the food to begin with, when compared with the ambient water, reflecting some biomagnification in the diet itself. The fugacities increase from water (1 μPa) to diet (2 μPa) to fish (3.82 μPa).

It is noteworthy that the "concentration-based" BMF for this situation can be shown to be 2.41, which is close to the fugacity-based BMF but not equal. This difference arises as a result of the difference in Z-values for the diet and the fish. When these Z-values are equal, the two versions of BMF are identical. Such a situation will arise when the lipid content of the diet and the fish is equal.

7.1.5   Application of the *Fish* Bioaccumulation Model

In 1993, Gobas published a pioneering paper in which he described and modeled the bioaccumulation of a series of organic chemicals in an eight-member food chain in Lake Ontario.

Here we exploit this model applied to PCBs in Lake Ontario in a 2.4 kg salmonid. A version of the model in spreadsheet form is available from the CEMC website. It has been modified somewhat from the original Gobas version. Time units of days are employed.

There are three sets of input data. First are the data for the chemical properties including the name, molar mass, K_{OW}, Henry's law constant, and Z_W.

Chemical Properties

Parameter	Value	Units
Chemical name	Hypothene-6	—
Molar mass	300	g mol^{-1}
Log K_{OW}	6	—
K_{OW}	1.00E+06	—
H	5	Pa m^3 mol^{-1}
Z_W	0.2	mol m^{-3} Pa

Second are the properties of the fish including species name, mass and volume, lipid content and parameters relating to uptake by respiration, food consumption, biotransformation half-life, and growth. These data are sufficient to enable the BCF to be defined as a product of lipid content and K_{OW} for both the fish and its diet.

Organism Properties

Parameter	Value	Units
Species	Salmonid	—
Weight (W)	2.41	kg
Lipid content (L)	0.16	fraction
Fish volume (V)	0.00241	m^3
Lipid density (kg L^{-1})	0.9	kg L^{-1}
ED	0.42	—
Metabolic half-life	1000	days
Q_w (Regression parameter)	149.68	—
Q_L (Regression parameter)	1.50E+00	—

Third are the exposure data for the chemical including temperature, total concentration in water, and organic carbon concentration in the water. These data enable a BSF to be calculated and thus the dissolved concentration in water. The concentration in the diet can be selected or estimated and is similar to the product of the BCF of the diet and concentration in water. A BMF for the diet can be included if desired.

The model calculates values of the six rate constants and the corresponding D-values. These are k_1 (respiratory input), k_D (dietary input), k_2, (loss by respiration), k_E (fecal egestion), k_M (metabolic or biotransformation), and k_G (growth dilution). These rate constants are readily converted to D-values as described earlier and as shown below. Allometric relationships may be used to estimate the respiration rate, the food uptake rate, and the growth rate of the fish as shown below in the details of Worked Example 7.2.

Exposure Properties

Parameter	Value	Units
Temperature	8	°C
Water concentration (C_w)	1.1	ng L^{-1}
Bio available solute factor (BSF)	7.83E–01	—
Dissolved water concentration (CWD)	8.61E–01	ng L^{-1}
OC density (kg L^{-1})	0.9	kg L^{-1}
OC in water (kg L^{-1})	2.50E–07	kg L^{-1}
F_w	1.43E–08	Pa
C_diet_g	1	mg kg
C_diet	3.33E–03	mol m^{-3}
L_diet	0.05	—
Z_diet	1.00E+04	mol Pa^{-1} m^{-3}
Weight of diet (W_diet)	3.20E–02	Kg
V_diet	3.20E–05	m^3
Fugacity_diet	3.33E–07	
Z-value for fish (Z_fish)	3.20E+04	mol Pa^{-1} m^{-3}

By assuming a steady-state condition, the concentration and fugacity in the fish can be calculated as shown in detail in Worked Example 7.2 that follows. The overall mass balance of the chemical content in the fish can then be calculated, and the relative importance of the various input and output processes can be evaluated. The residence time of the chemical in the fish can be calculated as the amount in the fish divided by the input or output rate. Finally, the biouptake metrics of the bioaccumulation factor (BAF) and biomagnification factor (BMF) can be determined. The result is a complete, quantitative description of the chemical uptake by the fish. Identical results are obtained using concentrations and rate constants and fugacity D-values.

The *FISH* model may be used to calculate the various flux terms from input data on the chemical's properties, concentration in water, and various physiological constants. It includes a "bioavailability" calculation in the water by estimating sorption to organic matter in particles. The model is particularly useful for exploring how variation in K_{OW} and metabolic half-life affect bioaccumulation, and shows the relative importance of food and water as sources of chemical. An overall residence time is calculated that indicates the time required for contamination or decontamination to take place. The following Worked Example gives the full detail of such a computation, as carried out within the *FISH* model spreadsheet program.

Worked Example 7.2: Application of the FISH model computational approach to a salmonid.

Determine the concentration of a hydrophobic chemical hypothene-6 (log K_{OW} = 6.0) in a salmonid (salmon-like fish) under conditions similar to those published for Lake Ontario in Gobas (1993), using the data given above.

Time units of hours are general used for fugacity equations but, in the case of bioaccumulation, it is conventional to use "days" as the time unit. Either may be used, but the selected unit must be clearly defined to avoid confusion.

Here follow the details of the calculations necessary for this determination of the concentration of hypothene-6 in the fish, and the various rates and half-times for the various transport processes. Rationalizations for most steps may be found in the material that has already been presented in this and earlier chapters, with the remainder available from Gobas (1993).

i. Calculate the bioavailable solute fraction (BSF) and dissolved water concentration:

The effective concentration of hypothene-6 in the water experienced by the fish is calculated with the BSF, which is the fraction of chemical in the water that is present as a solute in the water, after accounting for the fraction that is strongly associated with particulate matter. The BSF, as defined by Gobas (1993), is given by

$$BSF = \frac{1}{\left(1 + \left(\frac{K_{OW} C_W^{OC} \,(kg\,L^{-1})}{\rho_{OC} \,(kg\,L^{-1})}\right)\right)}$$

Here we find

$$BSF = \frac{1}{\left(1 + \left(\frac{1 \times 10^6 \times 2.5 \times 10^{-7} \,kg\,L^{-1}}{0.9 \,kg\,L^{-1}}\right)\right)} = 0.783$$

The product of the measured water concentration and the BSF gives the dissolved water concentration, the actual concentration experienced by the fish via gill and dermal uptake:

$$C_W^W = C_W^{Bulk} BSF = 1.1 \,ng\,L^{-1} \times 0.783 = 0.861 \,ng\,L^{-1}$$

ii. Determine Z-for hypothene-6 in the water, food, and fish:

The Z-value for water is obtained in the usual way from the reciprocal of the Henry's law constant:

$$Z_W = \frac{1}{H} = \frac{1}{5 \,Pa\,mol^{-1}\,m^3} = 0.2 \,mol\,Pa^{-1}\,m^{-3}$$

The food is assumed to have its predominant capacity for holding hypothene-6 (and any other hydrophobic chemical) due to its lipid content. Therefore, it is reasonable to assign a food (diet) Z-value based on the overall food lipid content of 5% by volume $\left(v_L^{f-D}\right)$ and the Z-value of octanol:

$$Z_D = v_L^{f-D} Z_O = v_L^{f-D} K_{OW} Z_W$$

$$= 0.05 \times 10^6 \times 0.2 \,mol\,Pa^{-1}\,m^{-3}$$

$$= 1.0 \times 10^4 \,mol\,Pa^{-1}\,m^{-3}$$

The Z-value for hypothene-6 in the fish is also estimated by assuming that all the fugacity capacity in the fish are determined by the lipid content, and that the Z-value for that lipid is the same as n-octanol. Thus, using K_{OW} and the calculated Z_W to obtain Z_O as with the food, we have

$$Z_F = v_L^{f-F} Z_O = v_L^{f-F} K_{OW} Z_W$$

$$= 0.16 \times 10^6 \times 0.2 \,mol\,Pa^{-1}\,m^{-3}$$

$$= 3.2 \times 10^4 \,mol\,Pa^{-1}\,m^{-3}$$

iii. Determine the fugacity of hypothene-6 in the water and the diet:

The fugacity of hypothene-6 in water is easily determined using the dissolved hypothene-6 concentration and the corresponding Z-value:

$$f_W = \frac{C_W^W}{Z_W} = \left(\frac{0.860_{87}\,\text{ng}\,\text{L}^{-1}}{0.2\,\text{mol}\,\text{Pa}^{-1}\,\text{m}^{-3}}\right) \times \left(\frac{10^{-9}\,\text{g}\,\text{ng}^{-1} \times 10^3\,\text{L}\,\text{m}^{-3}}{300\,\text{g}\,\text{mol}^{-1}}\right) = 1.435 \times 10^{-8}\,\text{Pa}$$

The dietary fugacity is determined in the same way:

$$f_D = \frac{C_D}{Z_D} = \left(\frac{1.0\,\text{mg}\,\text{kg}^{-1}}{1.0 \times 10^{-4}\,\text{mol}\,\text{Pa}^{-1}\,\text{m}^{-3}}\right) \times \left(\frac{10^{-3}\,\text{g}\,\text{mg}^{-1} \times 1 \times 10^3\,\text{kg}\,\text{m}^{-3}}{300\,\text{g}\,\text{mol}^{-1}}\right) = 8.57 \times 10^{-5}\,\text{Pa}$$

iv. Calculate rate constants for the various transport processes:

The rate constants and D-values for uptake from food and gills, as well as loss due to gill transfer, metabolism, egestion, metabolism, and growth dilution, are calculated as in Gobas (1993) and Mackay (2001):

The uptake by the gills has been theorized to involve both lipid and hydrophilic gill uptake, with associated transport parameters Q_L and Q_W for these respective processes, in L day^{-1} units. The latter is defined by Gobas and Mackay (1987) as

$$Q_W = 88.3(M_F)^{0.6(\pm 0.2)}\,\text{L}\,\text{day}^{-1}$$

Here M_F is the mass of the fish in kg.

No such expression for the lipid uptake has been determined, but it is generally believed that it is about 1% of Q_W, hence

$$Q_L \approx 0.01 \times Q_W = 0.01 \times 88.3 \times (M_F)^{0.6(\pm 0.2)}\,\text{L}\,\text{day}^{-1}$$

These two parameters are then combined in a two-resistance manner to yield the rate constant for uptake:

$$\frac{1}{k_1} = M_F\left(\frac{1}{Q_W} + \frac{1}{Q_L K_{OW}}\right)$$

With these equations on hand we have

$$Q_W = 88.3 \times (2.41)^{0.6} = 149.68\,\text{L}\,\text{day}^{-1}$$

$$Q_L = 0.01 \times 149.68 = 1.4968\,\text{L}\,\text{day}^{-1}$$

$$\frac{1}{k_1} = 2.41\,\text{kg}\left(\frac{1}{149.68\,\text{L}\,\text{day}^{-1}} + \frac{1}{1.4968\,\text{L}\,\text{day}^{-1} \times 1 \times 10^6}\right) = 1.610 \times 10^{-2}\,\text{kg}\,\text{day}\,\text{L}^{-1}$$

$$k_1 = \frac{1}{1.610\,\text{kg}\,\text{day}} = 62.10\,\text{L}\,\text{kg}^{-1}\,\text{day}^{-1} \approx 62.1\,\text{day}^{-1}$$

The determination of the reverse gill loss rate constant is obtained by substituting the definition of k_1 given above into the expression for K_{OW}, which results in

$$\frac{1}{k_2} = v_L^{f-\text{Fish}} M_F\left(\frac{K_{OW}}{Q_W} + \frac{1}{Q_L}\right)$$

where the lipid volume fraction of the fish times its mass, $v_L^{f-Fish} M_F$, is an estimate of the mass of lipid in the fish. Such an approach yields

$$\frac{1}{k_2} = 0.16 \times 2.41 \left(\frac{1 \times 10^6}{149.69 \, day^{-1}} + \frac{1}{1.4969 \, day^{-1}} \right) = 2.576 \times 10^3 \, day$$

$$k_2 = \frac{1}{2.576 \times 10^3 \, day} = 3.881 \times 10^{-4} \, day^{-1}$$

Note here that it is possible to derive a simpler expression for k_2 more directly from k_1:

$$\left(\frac{1}{k_2} \right) = v_L^{f-F} M_F \left(\frac{K_{OW}}{Q_W} + \frac{1}{Q_L} \right)$$

$$= v_L^{f-F} M_F K_{OW} \left(\frac{1}{Q_W} + \frac{1}{Q_L K_{OW}} \right)$$

$$= v_L^{f-F} K_{OW} \left[M_F \left(\frac{1}{Q_W} + \frac{1}{Q_L K_{OW}} \right) \right]$$

$$= v_L^{f-F} K_{OW} \left(\frac{1}{k_1} \right)$$

Therefore, we have

$$k_2 = \left(\frac{k_1}{v_L^{f-F} K_{OW}} \right)$$

Using this simpler expression, we arrive at the same value for k_2:

$$k_2 = \left(\frac{k_1}{v_L^{f-F} K_{OW}} \right) = \left(\frac{62.10 \, kg^{-1} \, day^{-1}}{0.16 \, kg \times 1 \times 10^6} \right) = 3.881 \times 10^{-4} \, day^{-1}$$

The food ingestion rate in kg day^{-1} is estimated with a temperature-dependent expression based on a bioenergetics model that is dependent on the fish mass and ambient temperature T in degrees Celsius:

$$F_D = 0.022 \left(M_F / kg \right)^{0.85} \exp(0.06 \times T)$$

Thus we have

$$F_D = 0.022(2.41)^{0.85} \exp(0.06 \times 8) = 7.509 \times 10^{-2} \, kg \, day^{-1}$$

In terms of the relative mass of food consumed in a day, we have

$$\left(\frac{7.509 \times 10^{-2} \, kg \, day^{-1}}{2.41 \, kg} \right) \times 100\% = 3.1\% \, day^{-1}$$

Not all food that is eaten results in uptake of the chemical. The model accounts for the extent to which a chemical is absorbed from the diet by way of a dietary uptake efficiency E_D. This factor is estimated as

$$E_D = \frac{1}{\left(5.3 \times 10^{-8} \times K_{OW} + 2.3\right)}$$

Here, we have

$$E_D = \frac{1}{\left(5.3 \times 10^{-8} \times 1 \times 10^6 + 2.3\right)} = 0.425$$

The rate constant for dietary uptake is then given by

$$k_D = \frac{E_D F_D}{M_F} = \frac{0.425 \times 7.509 \times 10^{-2}\ \text{kg day}^{-1}}{2.41\ \text{kg}} = 0.01324\ \text{day}^{-1}$$

The fish consumes 3.1% of its body volume or mass per day; however, only 42% of the chemical is absorbed, such that the net chemical uptake corresponds to 1.3% of its body volume per day.

The rate constant for egestion is estimated based on the assumption of a 75% efficiency in transfer from food to fish, i.e., 25% of the hypothene-6 leaves with the feces. This corresponds to a dietary loss parameter value of $Q_D = 0.25$ (discussed in detail earlier):

$$k_E = Q_D k_D = 0.25 \left(\frac{E_D F_D}{M_F}\right) = 0.25 \times 1.324 \times 10^{-2}\ \text{day}^{-1} = 3.311 \times 10^{-3}\ \text{day}^{-1}$$

For a highly persistent and unreactive chemical that is known to bioaccumulate, it would be reasonable to assume a rate constant for metabolism k_M of zero. However, to preserve generality, we assume here a metabolic half-life of 1000 days or about 3 years, for which the first-order rate constant is

$$k_M = 6.93 \times 10^{-4}\ \text{day}^{-1}$$

Growth dilution is assumed to follow this relationship:

$$k_G = 1.33 \times 10^{-3} \left(M_F\ \text{kg}^{-1}\right)^{-0.2}$$

Thus, we have

$$k_G = 1.33 \times 10^{-3} \times 2.41^{-0.2} = 1.115 \times 10^{-3}\ \text{day}^{-1}$$

The rate constant for all loss processes is the sum of all individual loss rate constants, so that

$$k_T^{\text{Loss}} = k_2 + k_E + k_M + k_G$$

or

$$k_T^{\text{Loss}} = \left(3.88 \times 10^{-4} + 3.31 \times 10^{-3} + 6.93 \times 10^{-4} + 1.12 \times 10^{-3}\right) \text{day}^{-1}$$

$$= 5.507 \times 10^{-3}\ \text{day}^{-1}$$

v. Calculate all D-values for transfer processes (in units of mol Pa^{-1} day^{-1}):
   All D-values require the volume of the fish (V_F) to be calculated. This value is obtained by dividing the mass of the fish by its assumed density of 1000 kg m^{-3}.
   Thus,

$$V_F = \frac{2.41\ \text{kg}}{1000\ \text{kg m}^{-3}} = 2.41 \times 10^{-3}\ \text{m}^{-3}$$

For the uptake from the gills, D_V, we have

$$D_V = k_1 Z_W V_F = \left(62.1_0 \ \text{day}^{-1} \times 0.2 \ \text{mol Pa}^{-1} \ \text{m}^{-3} \times 2.41 \times 10^{-3} \ \text{m}^3\right)$$

$$= 2.993 \times 10^{-2} \ \text{mol Pa}^{-1} \ \text{day}^{-1}$$

Losses through the gills are also described by D_V:

$$D_V = k_2 Z_F V_F = \left(3.88_{14} \times 10^{-4} \ \text{day}^{-1} \times 3.2 \times 10^4 \ \text{mol Pa}^{-1} \ \text{m}^{-3} \times 2.41 \times 10^{-3} \ \text{m}^3\right)$$

$$= 2.993 \times 10^{-2} \ \text{mol Pa}^{-1} \ \text{day}^{-1}$$

The D-value for food uptake from diet D_D is given by

$$D_D = k_D Z_D V_F = \left(1.32_{42} \times 10^{-2} \ \text{day}^{-1} \times 1.00 \times 10^4 \ \text{mol Pa}^{-1} \ \text{m}^{-3} \times 2.41 \times 10^{-3} \ \text{m}^3\right)$$

$$= 0.3191 \ \text{mol Pa}^{-1} \ \text{day}^{-1}$$

The egestion D-values, D_E is

$$D_E = k_E Z_F V_F = \left(3.311 \times 10^{-3} \ \text{day}^{-1} \times 3.2 \times 10^4 \ \text{mol Pa}^{-1} \ \text{m}^{-3} \times 2.41 \times 10^{-3} \ \text{m}^3\right)$$

$$= 0.2553 \ \text{mol Pa}^{-1} \ \text{day}^{-1}$$

The metabolic D-value D_M is

$$D_M = k_M Z_F V_F = \left(6.930 \times 10^{-4} \ \text{day}^{-1} \times 3.20 \times 10^4 \ \text{mol Pa}^{-1} \ \text{m}^{-3} \times 2.41 \times 10^{-3} \ \text{m}^3\right)$$

$$= 5.344 \times 10^{-2} \ \text{mol Pa}^{-1} \ \text{day}^{-1}$$

For growth, the D-value is

$$D_G = k_G Z_F V_F = \left(1.115 \times 10^{-3} \ \text{day}^{-1} \times 3.20 \times 10^4 \ \text{mol Pa}^{-1} \ \text{m}^{-3} \times 2.41 \times 10^{-3} \ \text{m}^3\right)$$

$$= 8.602 \times 10^{-2} \ \text{mol Pa}^{-1} \ \text{day}^{-1}$$

Here the fish volume is increasing by a factor of 0.0011 per day, i.e., 0.1% per day. Finally, the total loss D-value is the sum of all parallel processes, i.e.,

$$D_T^{\text{LOSS}} = D_V + D_E + D_M + D_G$$

$$= \left(2.993 + 25.53 + 5.344 + 8.602\right) \times 10^{-2} \ \text{mol Pa}^{-1} \ \text{day}^{-1}$$

$$= 0.4247 \ \text{mol Pa}^{-1} \ \text{day}^{-1}$$

Note that, in each case, the fugacity capacity used is associated with the medium in which the transfer occurs, i.e., for food intake we use Z_D, for gill intake, Z_W. The same D_V value applies to uptake and loss by respiration.

vi. Determine the fugacity of hypothene-6 in the fish at steady state by mass balance:
   At steady state, the input and outputs of hypothene-6 from the fish must balance. This requirement allows us to write a mass-balance statement, from which the fugacity of hypothene-6 in the fish follows directly:

$$\sum_{i=1}^{\text{Inputs}} f_i D_i = \sum_{j=1}^{\text{Outputs}} f_F D_j$$

or

$$f_F = \frac{\sum\limits_{i=1}^{\text{Inputs}} f_i D_i}{\sum\limits_{j=1}^{\text{Outputs}} D_j}$$

For our scenario:

$$\sum_{i=1}^{\text{Inputs}} f_i D_i = f_W D_1 + f_D D_D$$

$$= 1.435 \times 10^{-8}\,\text{Pa} \times 2.993 \times 10^{-2}\,\text{mol Pa}^{-1}\,\text{day}^{-1} + 3.333 \times 10^{-7}\,\text{Pa} \times 0.3191\,\text{mol Pa}^{-1}\,\text{day}^{-1}$$

$$= 4.2950 \times 10^{-10}\,\text{mol day}^{-1} + 1.0637 \times 10^{-7}\,\text{mol day}^{-1}$$

$$= 1.068 \times 10^{-7}\,\text{mol day}^{-1}$$

$$f_F = \frac{f_W D_1 + f_D D_D}{D_T^{\text{Loss}}} = \frac{1.068 \times 10^{-7}\,\text{mol day}^{-1}}{0.4247\,\text{mol Pa}^{-1}\,\text{day}^{-1}} = 2.515 \times 10^{-7}\,\text{Pa}$$

Note that the fish fugacity is much higher than that of the water but somewhat lower than that of the diet.

vii. Determine the concentration of hypothene-6 in the fish at steady state:
The molar concentration of hypothene-6 in the fish is given by

$$C_F = Z_F f_f = 3.20 \times 10^4\,\text{mol Pa}^{-1}\,\text{m}^{-3} \times 2.515 \times 10^{-7}\,\text{Pa} = 8.048 \times 10^{-3}\,\text{mol m}^{-3}$$

It is common to express this concentration in terms of mass of chemical (in mg) per mass of fish (in kg):

$$C_F^{M/M} = \frac{C_F M^m}{\rho_F} = \left(\frac{8.048 \times 10^{-3}\,\text{mol m}^{-3} \times 300\,\text{g mol}^{-1}}{1000\,\text{kg m}^{-3}} \right) \times 10^3\,\text{mg g}^{-1} = 2.414\,\text{mg kg}^{-1}$$

The *FISH* model is available from the CEMC website in spreadsheet format. Users may define the dietary, biological, environmental, and physico-chemical parameters to calculate the steady-state uptake of a particular chemical of interest by adjusting the parameters associated with the physico-chemical, environmental, dietary, and organism properties.

For pedagogical purposes, it may be useful to estimate the rate constants approximately for the desired chemical and fish and then proceed through the calculation using these rate constants and equivalent *D*-values to obtain an estimate of the concentration in the selected fish as a function of the concentrations in the water and the diet.

Several features and limitations of models such as *FISH* are worthy of note:

1. Rate constant regressions are subject to error, and it is useful to test a variety of these regressions available in the literature.
2. The respiratory uptake rate constant, the food intake rate, and the growth rate should be bioenergetically consistent.

3. For persistent chemicals the egestion rate proves to be very important. To address this uncertainty, it may be preferable to treat the GIT as a separate compartment and to break down the dietary components into lipids, proteins, carbohydrates, and fibrous materials, each with its own uptake efficiency as expressed in the AQUAWEB model of Arnot and Gobas (2004).
4. Conditions in the water column including the content of dissolved organic matter and temperature can be variable spatially and temporarily.
5. The biotransformation rate constant is often very uncertain; thus, it may prove useful, at least for persistent chemicals to assume a worst case of k_M of zero.
6. As the fish grows and matures, its diet and dietary intake rate also change. The respiration rate also changes with size and temperature.
7. Rarely does the fish rely on a single dietary source. Multiple sources can be included if desired.

Most meaningful is a description of the entire food web from plankton to large fish. Mathematically, the preferred approach is to start at the base of the food web and calculate concentrations and fugacities sequentially from a series of prey to predators of increasing trophic levels as described by Gobas (1993). The trophic magnification factor (TMF) can be obtained from the slope of a plot of the log of fugacity or lipid normalized concentration versus trophic level.

7.2   AQUATIC FOOD WEBS

The theoretical approach taken in the *FISH* bioaccumulation model can be applied to an aquatic food web starting with water and then moving successively to phytoplankton, zooplankton, invertebrates, small fish, and to various levels of larger fish. Each level becomes food for the next higher level. If K_{OW} is relatively small, i.e., <10^5, the D_V terms dominate, and equifugacity is probable throughout the web; i.e., f_D, f_W, and f_F will be equal. For chemicals of larger K_{OW}, biomagnification is likely. For "super-hydrophobic" chemicals of $K_{OW} > 10^7$, the E_A term becomes small, uptake is slowed, and the growth and metabolism terms become critical. Association with suspended organic matter in the water column becomes important, i.e., "bioavailability' is reduced. A falloff in observed BCFs is (fortunately) observed for such chemicals; thus, there appears to be a "window" in K_{OW} of about 10^6–10^7 in which bioaccumulation is most significant and most troublesome. Chemicals such as DDT and PCBs lie in this "window." This issue has been discussed in detail and modeled by Thomann (1989).

Several features of food web biomagnification are worthy of note. Humans usually eat creatures close to the top of food webs, and strive to remain at the top of food webs, avoiding being eaten by other predators. Fish consumption is often the primary route of human exposure to hydrophobic chemicals. Creatures high in food webs are invaluable as bioindicators or biomonitors of contamination of lakes by hydrophobic chemicals. But to use them as such requires knowledge of the D-values, especially the D-value for metabolism. A convincing argument can be made that, if we live in an ecosystem in which wildlife at all trophic levels is thriving, we can be fairly optimistic that we humans are not being severely affected by environmental chemicals. This is a (selfish) social incentive for developing, testing, and validating better environmental fate models, especially those employing fugacity.

A food web model treating multiple species can be written by applying the general bioaccumulation equation to each species (with appropriate parameters). The final set of equations for n organisms has n unknown fugacities that can be solved sequentially, starting at the base of the food web and proceeding to other species, with smaller animals becoming food for larger animals.

An alternative and more elegant method is to set up the equations in matrix form as described by Campfens and Mackay (1997) and solve the equations by a routine such as Gaussian elimination. This permits complex food webs to be treated with no increase in mathematical difficulty.

A DOS-based BASIC model *FoodWeb* is available from the CEMC website that performs these calculations as described in detail by Campfens and Mackay (1997), and a spreadsheet version is planned.

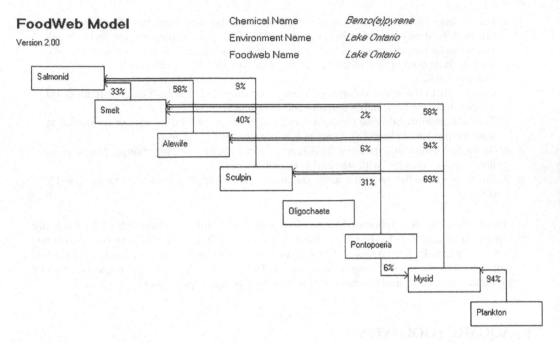

FIGURE 7.3 Diagrammatic output of the *FoodWeb* model, showing the percentage contributions of dietary sources in lower trophic levels to the load in a given organism in the food web.

It is essentially an expansion of the *FISH* model, and it treats any number of aquatic species that consume each other according to a dietary preference matrix. A steady-state condition is calculated using matrix algebra. All organism-to-organism fluxes (i.e., food consumption rates) are given. The model is also useful as a means of testing how concentrations in top predators respond to changes in food web structure. It is essentially a multibox model with one-way transfers from box to box.

A typical *FoodWeb* calculation results in the following diagrammatic output, in this case for benzo[a]pyrene in a model food web based on Lake Ontario, input data for both of which are included in the *FoodWeb* databases (Figure 7.3).

The diagram shows the percentage of chemical that comes to each organism from the dietary source in the trophic levels beneath it. For example, the salmonid organism is obtaining nearly half its load of benzo[a]pyrene from the alewife contributions to its diet, and the alewife obtains nearly all (94%) of its load from mysid consumption.

An obvious extension of such a food web model is to include non-aquatic species such as birds and mammals. This has been discussed by Clark et al. (1988), who showed that fish-eating birds could be included in an aquatic food web model. In the long term, it may be possible to build models containing all relevant biota, including fish, birds, mammals, insects, and vegetation. A framework for accomplishing this has been described by Sharpe and Mackay (2000) and deBruyn and Gobas (2006). The primary difficulties are in the development of species-specific mass-balance equations, determining appropriate parameters for the organisms and obtaining validation data. There is little doubt that comprehensive food web models will be developed and validated in the future, even models including humans.

7.3 BIOACCUMULATION IN AIR-BREATHING ORGANISMS

Homeotherms are air-breathing animals such as mammals and birds that regulate their body temperature independent of their environment. From the point of view of modeling, they are rather similar to fish, in that they take up oxygen from a fluid medium (here air instead of water) and nutrients from a (mostly) solid diet. As a result, modification of the *FISH* model to a homeotherm requires only that the parameters associated with various uptakes be modified to reflect the appropriate rates of transport, with air-based chemical intake by the lungs replacing water-based gill uptake and losses. Metabolic, egestion, growth dilution, and food uptake are all treated in the same manner.

The primary difference between a fish and any homeotherm is that the concentration of the chemical in water tends to be much greater than that in air, and as a result, transfer to and from the lungs may be very inefficient. This results in a much lower rate of loss for non-metabolized chemicals, and a much higher BMF.

7.4 MULTI-COMPARTMENT/PBPK MODELS

Physiologically based pharmacokinetic models (PBPK models) treat an animal as a collection of connected boxes or organs in which exchange between organs occurs primarily by blood flow, which circulates between all the boxes. The model can include uptake from air and food and possibly by dermal contact or injection. We then calculate the dynamics of the circulation of the chemical in venous and arterial blood, to and from various organs or tissue groups including adipose tissue, muscle, skin, brain, kidney, and liver. There may be losses by exhalation and metabolism, and in urine, feces, and sweat. In mammals, nursing mothers also lose chemical to their offspring in breast milk, and they lose tissue when giving birth. Analogous processes occur during egg laying in birds, amphibians, and reptiles. As in environmental models, partition ratios or Z-values can be deduced to quantify chemical equilibrium between air, blood, and various organs. Flows of blood to each organ can be expressed as D-values. Metabolic rates can be expressed using rate constants, usually invoking Michaelis–Menten kinetics, as described in Chapter 6, and translated into D-values. Mass-balance equations then can be assembled, describing the constant conditions that develop following exposure to long-term constant concentrations or the dynamic conditions that follow a pulse input. Experiments are done, often with rodents or fish, to follow the time course of chemical transport and transformation in the animal. The resulting data can be compared with model assertions to achieve a measure of validation. In some cases, the potential rate of metabolism can exceed the rate of flow to the liver, and a "flow limitation" may constrain the effective rate. A "first-pass" effect can also apply to ingested chemicals that are rapidly metabolized in the live and fail to enter the general circulation.

Much of the pharmacokinetic literature is devoted to assessment of the time course of the fate of therapeutic drugs within the human body. The aim is to supply a sufficient, but not too large and thus toxic, dose of drug to the target organ. Closer to environmental exposure conditions are PBPK models for occupational exposure to toxicants such as solvent or fuel vapors, which may be intermittent or continuous in nature. A pioneering example is the model of Ramsey and Andersen (1984), which was translated into fugacity terms by Paterson and Mackay (1986, 1987). Accounts of various aspects of pharmacokinetics and PBPK models and their contribution to environmental science are the works of Welling (1986), Parke (1982), Reitz and Gehring (1982), Tuey and Matthews (1980), Fiserova-Bergerova (1983), Menzel (1987), Nichols et al. (1996, 1998), and Wen et al. (1999). A fugacity model has been developed for whales by Hickie et al. (1999) and a rate constant model for birds by Clark et al. (1987), Drouillard et al. (2012), and Drouillard and Norstrom (2014).

Figure 7.4, which is adapted from Paterson and Mackay (1987), illustrates the fugacity approach to modeling the fate of a chemical in the human body. In principle, it is possible to calculate steady- and unsteady-state fugacities, concentrations, amounts, fluxes, and response times, thus linking external environmental concentrations to internal tissue concentrations. Ultimately, from a human health viewpoint, it is likely that it will be possible to undertake these calculations and compare levels of chemical contamination in vulnerable tissues with levels that are believed to cause adverse effects.

There is clearly a need to link environmental and pharmacokinetic modeling efforts to build up a comprehensive capability of assessing the journey of the chemical from source to environment to organism and ultimately to the target site.

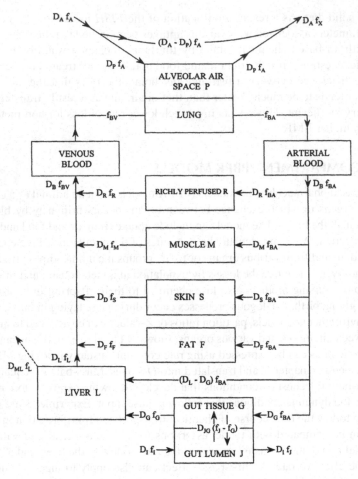

FIGURE 7.4 Transport and transformation in a multi-compartment pharmacokinetic model as applied to humans.

SYMBOLS AND DEFINITIONS INTRODUCED IN CHAPTER 7

Symbol	Common Units	Description
BCF	—	Bioaccumulation factor
BMF	—	Biomagnification factor
BSF	—	Bioavailable solute fraction
E_D	—	Dietary uptake efficiency parameter
F_D	kg day^{-1}	Food ingestion rate
k_E	day^{-1}	Fish egestion uptake rate constant
k_G	day^{-1}	Fish growth dilution rate constant
k_M	day^{-1}	Fish metabolic degradation rate constant
M_F	kg	Mass of fish
Q_D	—	Dietary loss to egestion parameter
Q_L	—	Lipid gill uptake transport parameter
Q_W	—	Hydrophilic gill uptake transport parameter
v_L^{f-D}	—	Volume fraction of lipid in diet

8 Human Health

8.1 COMPREHENSIVE MULTIMEDIA MODELS INCLUDING HUMAN EXPOSURE (e.g., THE *RAIDAR* MODEL)

Large numbers of new chemicals are produced each year, presenting a significant regulatory challenge. As a result, screening models that can be used to evaluate chemical risk in the environment are of great value and interest. An example of such a model is *RAIDAR* or the *Risk Assessment, IDentification, And Ranking* model, which combines chemical partitioning properties with those of reactivity, fate, transport, food web bioaccumulation, exposure, effect end-point, and emission rate in a coherent mass-balance framework. Other models of this type are listed in the online fugacity models tabulation from the CEMC website (www.trentu.ca/cemc/resources-and-models). The RAIDAR model was originally introduced in 2006 (Arnot et al., 2006), and subsequently updated in 2008 (Arnot and Mackay 2008) with further updates planned. More recently, an Indoor and Consumer Exposure variant of RAIDAR has been developed, called *RAIDAR-ICE* (Li et al., 2018). This version of RAIDAR is designed for high-throughput screening of human exposure to chemicals in indoor environments and due to products directly applied to the body.

The RAIDAR model employs the standard equilibrium criterion (EQC) environment, introduced by Mackay and co-workers (Mackay et al., 1996), but includes various aquatic, avian, and terrestrial biota as well as vegetation. Often the most uncertain input parameters in these models are chemical emission or discharge rates that may be subject to considerations of commercial confidentiality. This may be circumvented in part by selecting a reasonable concentration for the chemical in air, and deducing the input rate required to generate that concentration. This is termed "reverse modeling." The RAIDAR model can be employed using such an approach, by first determining the critical concentration for the most sensitive organism in the system, and then using that value to determine the emission rate that would be expected to give rise to this level of contamination. With this information in hand, the typically highly uncertain data for actual emissions can be reviewed and evaluated from the point of view of risk.

For any given chemical considered, RAIDAR generates three risk-hazard-exposure metrics, which may be used in screening of groups of chemicals. These are defined as follows:

Exposure Assessment Factor (EAF):

$$EAF = \frac{\text{Concentration in organism}}{\text{Exposure due to emission}} = \frac{C_U}{E_U}$$

Hazard Assessment Factor (HAF):

$$HAF = \frac{\text{Exposure Level}}{\text{Toxic Level}} = \frac{C_U}{C_T} = \frac{E_A}{E_U}$$

Risk Assessment Factor (RAF):

$$RAF = \frac{EAF}{HAF} = \frac{C_U/C_T}{E_A/E_U}$$

A high EAF indicates an organism that is vulnerable to contamination, but is not necessarily an organism that is at risk. Therefore, one must also consider exposure as well as chemical toxicity. The HAF takes exposure into account by essentially normalizing the exposure level with respect to the level at which toxicity is anticipated. If the HAF is low, the level of exposure is low and "at risk" organisms remain at sub-toxic exposure levels. The RAF combines these two considerations, and generates a metric that will only be high when the risk of exposure and organism sensitivity combined to give a high risk of toxicity due to exposure.

The RAIDAR model may be downloaded from the Arnot Research Consultants (ARC) website following a registration step (https://arnotresearch.com/raidar). Updates are expected in the future as the science underlying the model continues to progress.

8.2   THE PBT-LRT ATTRIBUTES AND THE STOCKHOLM CONVENTION

The multimedia environmental models described in Chapter 6 lead to estimates of fugacities and concentrations in air, water, soil, and sediments. These abiotic fugacities can be used to deduce fugacities and concentrations in fish, and possibly in other animals and plants. The primary weakness is probably that they do not yet adequately quantify partitioning into the variety of vegetable matter that is consumed by animals and humans. For some compounds, such as the dioxins, the air-grass-cow-milk-dairy product-human route of transfer is critical. In this chapter, we discuss briefly the principles by which these concentration data can be used to assess the impact of chemicals on humans and other organisms.

The first obvious use of these abiotic and biotic concentrations is to compare them with concentration levels that are believed to cause adverse effects. These levels are usually developed by regulatory agencies and published as guidelines, objectives, or effect concentrations of various types. Target or objective concentrations can be defined for most media. For example, from considerations of toxicity or esthetics, it may be possible to suggest that water concentrations should be maintained below 1 mg m^{-3}, air below 1 μg m^{-3}, and fish below 1 mg kg^{-1}.

These concentrations can be compared as a ratio or quotient to the estimated environmental concentrations. A hypothetical example is given in Table 8.1, illustrating the quotient method. In this example, the primary concern is with air inhalation and fish ingestion. The proximities of the estimated prevailing concentrations to the targets are expressed as quotients, which can be regarded as safety factors. A large quotient implies a large safety factor and low risk. The high-risk situations correspond to low quotients. This quotient is also called a *toxicity/exposure ratio* or *TER*. The concentration level in fish may not be directly toxic to fish but may pose a threat to humans if the fish is consumed on a regular basis. The reciprocal ratio is also used in the form of a PEC/PNEC ratio, i.e., predicted environmental concentration/predicted no effect concentration. In this case, a high value implies high risk. There is, however, a continuing controversy about whether or not it is possible to

TABLE 8.1

Comparison of Predicted or Measured Environmental Concentrations with Concentrations Producing a Specified Effect, or No Effect

	Concentrations		
Medium	Predicted Level	Effect Level	Quotient
Air (μg m^{-3})	3	60	20
Water (μg L^{-1})	10	3000	300
Fish (μg g^{-1})	2	10	5
Soil (μg g^{-1})	1	100	100

TABLE 8.2

Representative Exposure Rates for Four Human Age Classes Derived Principally from the US Environmental Protection Agency Exposure Factors Handbook

Route	\<1	1–5	6–19	20+	Units
			Age Classes		
Air inhalation	4.5	6.0	10.0	13.5	$m^3 day^{-1}$
Drinking water	300.0	694.0	904.0	1500.0	mL day^{-1}
Fruit	135.6	145.1	161.6	166.9	g day^{-1}
Vegetables	61.9	108.0	209.1	271.3	g day^{-1}
Grains	155.1	143.7	227.7	219.0	g day^{-1}
Meat	26.1	59.1	113.8	123.3	g day^{-1}
Fish	—	4.0	6.0	14.0	g day^{-1}
Dairy	570.9	339.6	438.7	249.6	g day^{-1}
Human milk	775.	0.0	0.0	0.0	g day^{-1}
Soil ingestion	—	0.11[a]	0.087	0.065	g day^{-1}
Dermal contact soil area	—	0.43	0.62	0.70	m^2
Duration	—	1440.0	600.0	320.0	h year^{-1}
Water area	—	0.65	1.25	1.81	m^2
Duration	—	82.4	82.4	82.4	h year^{-1}

[a] A pica child may ingest up to 10 g day^{-1}.

determine a "no-effect" concentration, since it is not amenable to direct measurement taking into account all organisms at all life stages and subject to other stresses.

Difficulties are encountered when suggesting target concentrations in soil and sediment, because these media are not normally consumed directly by organisms. Although simple lethality experiments can be designed using air, water, or food as vehicles for chemical exposure, it is not always clear how concentrations in the solid matrices of soils and sediments relate to exposure or intake of chemical by organisms. It is difficult to design meaningful bioassays involving interactions between organisms, soils, and sediments. One approach is to decree that whatever target fugacity is developed for water be applied to sediment. This effectively links the target concentrations by equilibrium partition ratios.

A second method is to use concentrations to estimate exposure or dosage in units such as mg day^{-1} of chemical to an organism, which for reasons of self-interest is usually a human. Individual and total dosages can be estimated to reveal the more important routes. This calculation of dose is enlightening, because it reveals which medium or route of exposure is of most importance. Presumably, steps can then be taken to reduce this route by, for example, restricting fish consumption. Table 8.2 lists representative exposure quantities for several human age classes.

An average human inhales some 14 m^3 of air per day. If the concentration in air is known in mg m^{-3}, the amount of chemical inhaled in this air is readily calculated as the product with units of mg day^{-1}. Not all this chemical may be absorbed, but at least a maximum dosage can be deduced. The same human may consume 1.5 L day^{-1} of water containing dissolved chemical, enabling this dosage to be estimated again in mg day^{-1}. Food, the other vehicle, is more difficult to estimate. A typical diet may consist of 1 kg day^{-1} of solids broken down as shown in Table 8.2. Fish concentrations can be estimated directly from water concentrations, but meat, vegetable, and dairy product concentrations are still poorly understood functions of the concentrations of chemical in air, water, soil, and animal feeds, and of agrochemical usage. Techniques are emerging for calculating food-environment concentration ratios, but at present the best approach is to analyze a typical purchased "food basket." This issue is complicated by the fact that much food is grown at distant locations and imported. Beverages, food, and water may also be treated for chemical removal commercially or domestically by washing, peeling, or cooking.

Significant chemical exposure may also occur in occupational settings (e.g., factories), in institutional and commercial facilities (e.g., schools, stores, and cinemas), and at home by personal care products, but these exposures vary greatly from individual to individual and depend on lifestyle. Much of the toxic chemical exposure comes from diet that may include local as well as imported foods. Some imports may be from countries that have lax regulations on acceptable contaminant concentrations and pesticide residue levels.

There emerges a profile of relative exposures by various routes from which the dominant route(s) can be identified. If desired, appropriate measures can be taken to reduce the largest exposures. The advantage of this approach is that it places the spectrum of exposure routes in perspective. There is little merit in striving to reduce an already small exposure.

Exposure routes vary greatly in magnitude from chemical to chemical, depending on the substance's physical chemical properties such as K_{AW} and K_{OW}, and it is not usually obvious which routes are most important.

Data from the environmental fate models can provide a sound basis for estimating risk when used to assess quotients and to determine dominant exposure routes. If such information can be presented to the public, the individuals will be, at least in principle, able to choose or modify their lifestyles to minimize exposure and presumably risk. Individuals then have the freedom and information to judge and respond to acceptability of risk from exposure to this chemical compared with the other voluntary and involuntary risks to which they are subject.

A regulatory issue in which evaluative mass balance models are playing an increasingly important role is in assessing the *persistence, bioaccumulation, toxicity and long-range transport (PBT-LRT)* attributes of chemicals. If chemicals that display these undesirable attributes can be identified, they can be considered for regulation, as was done by the United Nations Environment Programme (UNEP) and the Stockholm convention for the "dirty dozen" high-priority substances discussed earlier. Monitoring data are usually too variable to enable them to be used directly in this priority setting task, and monitoring is impossible for chemicals not yet in use. Since there are many thousands of chemicals of commerce that require assessment, and (it is hoped) most are innocuous, there is an incentive to develop a tiered system in which there are minimal data demands initially and perhaps 90% of chemicals evaluated are rejected as of no concern. The remaining 10% of potential concern can be more fully evaluated in a second tier with a similar rejection ratio. A third tier may be needed to select the (perhaps) 100 top priority chemicals from a universe of 100,000 chemicals in a three-tier system. The challenge is to devise models or evaluation systems that will accomplish this task efficiently.

Webster et al. (1999) have suggested using a Level III model similar to EQC, but with advection shut off, to evaluate persistence. Gouin et al. (2000) have described an even simpler Level II approach. This has the advantage that no "mode-of-entry" information is required. Regardless of which model is used, it seems inevitable that models will play a key role in assessing persistence or residence time, since these quantities cannot be measured directly in the environment.

Bioaccumulation can be evaluated most simply by calculating the equilibrium bioconcentration factors as the product of lipid content and K_{OW}. For more detailed evaluation involving considerations of bioavailability, metabolism, and possible biomagnification from food uptake, the *Bioaccumulation* model or a variant of it can be used. In some cases, a food web model may be required to determine if significant biomagnification occurs. *Foodweb* can be used for this purpose. Mackay and Fraser (2000) have suggested such a three-tiered approach for assessing the bioaccumulation potential of chemicals. More advanced models and tools of this type are listed on the CEMC website.

The PBT approach is favored by many regulatory agencies because it is relatively simple, requiring only the basic physico-chemical properties, estimates of production and emission rates, some environmental monitoring data and some toxicity data. Detractors can point out specific cases in which the evaluation is flawed (McLachlan 2018). Multimedia models of the type suggested therein can contribute to more reliable evaluations.

Finally, we note that toxicity is not considered within the scope of this book, and is not considered further here.

It is expected that new models will be developed to assist in the evaluation of these attributes, especially in situations where there is no easy method of obtaining the required information from environmental monitoring data.

8.3 CONCLUDING REMARKS

Modern society now depends on a wide variety of chemicals for producing materials, as components of fuels, for maintaining food production, for ensuring sanitary conditions and reducing the incidence of disease, for use in domestic and personal care products, and for use in medical and veterinary therapeutic drugs. We enjoy enormous benefits from these chemicals. Our industrial, municipal, and domestic activities also generate chemicals inadvertently by processes such as incineration and waste treatment. It is unlikely that multimedia models of environmental fate and effect can be fully "validated," but this need not discourage their application to regulatory actions. Perhaps "validity" is a bar too high and we should be content with models that are scientifically credible and have been shown to be broadly consistent with available environmental monitoring data. The challenge is to use chemicals wisely and prudently by reducing emissions or discharges to a level at which there is assurance that there are no adverse effects on the quality of life from chemicals, singly or in combination. It is hoped that the tools developed in these chapters can contribute to this process.

Perhaps the task addressed by this book is best summarized by Figure 8.1, which depicts many of the environmental processes to which chemical contaminants are subject. The aim has been to develop methods of calculating partitioning, transport, and transformation in the wide range of media that constitute our environment. Ultimately of primary concern to the public, and thus to regulators, is the effect that these chemicals may have on human well-being. But there are sound practical and ethical reasons for protecting wildlife, and indeed all fellow organisms in our ecosystem.

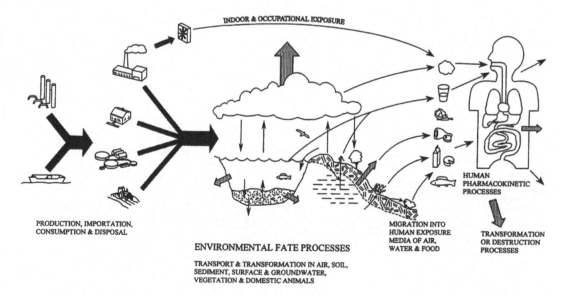

FIGURE 8.1 An illustration of a chemical's sources, environmental fate, human exposure, and human pharmacokinetics.

It is not yet clear how severe the effects of chemical contaminants are, nor is it likely that the full picture will become clear for some decades. Undoubtedly, there are chemical surprises or "time bombs" in store as analytical methods and toxicology improve and new chemicals of concern are identified. Of particular interest and concern is the issue of the combined impact of numerous chemicals, no one of which is present at sufficient concentrations to cause adverse effects, since the addition of exposures and toxicities presents considerable challenges.

Regardless of the incentive nurtured by public fear of "toxics," environmental science has a quite independent and noble objective of seeking, for its own sake, a fuller quantitative understanding of how the biotic and abiotic components of our multimedia ecosystem operate; how chemicals that enter this system are transported, transformed, and accumulate; and how they eventually reach organisms and affect their well-being.

Appendix A
Estimated Properties of Chemicals of Environmental Concern for use in Modeling Exercises

The following properties are reproduced from a series of Handbooks that provide data suitable for direct input to fugacity models described in this book. The Handbooks also give examples of these model applications. See Mackay, D., Shiu, W.Y., Ma, K.C. 2002. "Illustrated Handbook of Physical Chemical Properties and Environmental Fate for Organic Chemicals". Second edition containing 4 volumes, namely Vol. 1 Hydrocarbons, Vol.2 Halogenated Hydrocarbons, Vol.3 Oxygen Containing Compounds and Vol. 4 Nitrogen and Sulfur Containing Compounds and Pesticide Chemicals. Lewis Publishers/CRC Press New York, N.Y. Also available as a CD-ROM, CRCnetBASE 2000, Chapman & Hall, CRC Press LC., Boca Raton FL.

These property values were selected from the refereed scientific literature prior to 2002. The aim is to provide the reader with a convenient source of the required data for implementation in fugacity models. More recent and more accurate data may be available from research publications and from other property data bases. A convenient source is the US EPA EPI Suite data base that also provides estimation programs, namely: The EPI (Estimation Programs Interface) Suite a Windows®-based suite of physical/chemical property and environmental fate estimation programs developed by EPA's and Syracuse Research Corp. (SRC). EPI Suite contains a number of estimation programs including a fugacity Level III program, a STP model and Bioaccumulation programs.

TABLE A.1
Physical Chemical Properties of Selected Organic Chemicals at 25°C Including Estimated Half-Lives (h) and Toxicity Expressed as Oral LD$_{50}$ to the Rat

Chemical Name	Molar Mass (g mol⁻¹)	Vapor Pressure (Pa)	Aqueous Solubility (g m⁻³)	Log K_{OW}	Melting Point (°C)	Air	Water	Soil	Sediment	Rat Oral LD$_{50}$ (mg kg⁻¹)
Benzene	78.11	12,700	1780	2.13	5.53	17	170	550	1700	930
1,2,4-Trimethylbenzene	120.2	270	57	3.6	−43.8	17	550	1700	5500	3550
Ethylbenzene	106.2	1270	152	3.13	−95	17	550	1700	5500	5460
n-Propylbenzene	120.2	450	52	3.69	−101.6	17	550	1700	5500	6040
Styrene	104.14	880	300	3.05	−30.6	5	170	550	1700	2650
Toluene	92.13	3800	515	2.69	−95	17	550	1700	5500	5000
Nitrobenzene	123.11	20	1900	1.85	5.6	5	1700	1700	5500	349
2-Nitrotoluene	137.14	17.9	651.42	2.3	−3.85	17	55	1700	5500	891
4-Nitrotoluene	137.14	0.653	254.4	2.37	51.7	17	55	1700	5500	1960
2,4-Dinitrotoluene	182.14	0.133	270	2.01	70	17	55	1700	5500	268
Chlorobenzene	112.6	1580	484	2.8	−45.6	170	1700	5500	17000	1110
1,4-Dichlorobenzene	147.01	130	83	3.4	53.1	550	1700	5500	17,000	500
1,2,3-Trichlorobenzene	181.45	28	21	4.1	53	550	1700	5500	17,000	756
1,2,3,4-Tetrachlorobenzene	215.9	4	7.8	4.5	47.5	1700	5500	5500	17,000	1470
Pentachlorobenzene	250.3	0.22	0.65	5	86	5500	17,000	17,000	17,000	11,000
Hexachlorobenzene	284.8	0.0023	0.005	5.5	230	7350	55,000	55,000	55,000	3500
Fluorobenzene	96.104	10,480	1430	2.27	−42.21	17	170	550	1700	4399
Bromobenzene	157.02	552	410	2.99	−30.8	170	1700	5500	17,000	2383
Iodobenzene	204.01	130	340	3.28	−31.35	170	1700	5500	17,000	1749
n-Pentane	72.15	68,400	38.5	3.45	−129.7	17	550	1700	5500	90,000
n-Hexane	86.17	20,200	9.5	4.11	−95	17	550	1700	5500	30,000
1,3-Butadiene	54.09	281,000	735	1.99	−108.9	5	170	550	1700	5480
1,4-Cyclohexadiene	80.14	9010	700	2.3	−49.2	5	170	550	1700	130
Dichloromethane	84.94	26,222	13,200	1.25	−95	1700	1700	5500	17,000	1600
Trichloromethane	`119.38	26,244	8200	1.97	−63.5	1700	1700	5500	17,000	1000
Carbon tetrachloride	153.82	15,250	800	2.64	−22.9	17,000	1700	5500	17,000	2350

(Continued)

TABLE A.1 (Continued)

Physical Chemical Properties of Selected Organic Chemicals at 25°C Including Estimated Half-Lives (h) and Toxicity Expressed as Oral LD$_{50}$ to the Rat

Chemical Name	Molar Mass (g mol⁻¹)	Vapor Pressure (Pa)	Aqueous Solubility (g m⁻³)	Log K_{OW}	Melting Point (°C)	Air	Water	Soil	Sediment	Rat Oral LD$_{50}$ (mg kg⁻¹)
Tribromomethane	252.75	727	3100	2.38	−8.3	1700	1700	5500	17,000	933
Bromochloromethane	129.384	19,600	14.778	1.41	−87.95	550	550	1700	5500	5000
Bromodichloromethane	163.8	6670	4500	2.1	−57.1	550	550	1700	5500	430
1,2-Dichloroethane	98.96	10,540	8606	1.48	−35.36	1700	1700	5500	17,000	750
1,1,2,2-Tetrachloroethane	167.85	793	2962	2.39	−36	17,000	1700	5500	17,000	200
Pentachloroethane	202.3	625	500	2.89	−29	17,000	1700	5500	17,000	920
Hexachloroethane	236.74	50	50	3.93	186.1	17,000	1700	5500	17,000	5000
1,2-Dichloropropane	112.99	6620	2800	2	−100.4	550	5500	5500	17,000	1947
1,2,3-Trichloropropane	147.43	492	1896	2.63	−14.7	550	5500	5500	17,000	505
Chloroethene (vinyl chloride)	62.5	354,600	2763	1.38	−153.8	55	550	1700	5500	500
Trichloroethylene	131.39	9900	1100	2.53	−73	170	5500	1700	5500	4920
Tetrachloroethylene	165.83	2415	150	2.88	−19	550	5500	1700	5500	2629
Methoxybenzene	108.15	472	2030	2.11	−37.5	17	550	550	1700	3700
Bis(2-chloroethyl)ether	143.02	206	10,200	1.12	−46.8	17	550	550	1700	75
Bis(2-chloroisopropyl)ether	171.07	104	1700	2.58	−97	17	550	550	1700	240
2-Chloroethyl vinyl ether	106.55	3566	15,000	1.28	−69.7	17	550	550	1700	210
Bis(2-chloroethoxy)methane	173.1	21.6	8100	1.26	0	17	550	550	1700	65
1-Pentanol	88.149	300	22,000	1.5	−78.2	55	55	55	170	3030
1-Hexanol	102.176	110	6000	2.03	−44.6	55	55	55	170	720
Benzyl alcohol	108.14	12	80	1.1	−15.3	55	55	55	170	1230
Cyclohexanol	100.16	85	38,000	1.23	25.15	55	55	55	170	1400
Benzaldehyde	106.12	174	3000	1.48	−55.6	5	55	55	170	1300
3-Pentanone	86.135	4700	34,000	0.82	−38.97	55	170	170	550	2410
2-Heptanone	114.18	500	4300	2.08	−35	55	170	170	550	1670
Cyclohexanone	98.144	620	23,000	0.81	−32.1	55	170	170	550	1540

(Continued)

TABLE A.1 (*Continued*)

Physical Chemical Properties of Selected Organic Chemicals at 25°C Including Estimated Half-Lives (h) and Toxicity Expressed as Oral LD$_{50}$ to the Rat

Chemical Name	Molar Mass (g mol⁻¹)	Vapor Pressure (Pa)	Aqueous Solubility (g m⁻³)	Log K_{OW}	Melting Point (°C)	Air	Water	Soil	Sediment	Rat Oral LD$_{50}$ (mg kg⁻¹)
Acetophenone	120.15	45	5500	1.63	19.62	550	170	170	550	815
Vinyl acetate	86.09	14,100	20,000	0.73	−92.8	55	55	170	550	2900
Propyl acetate	102.13	4500	21,000	1.24	−95	55	55	170	550	9370
Methyl methacrylate	100.12	5100	15,600	1.38	−42.8	17	55	55	170	7872
Diphenylamine	169.23	0.0612	300	3.45	52.8	5	55	170	550	2000
Aniline	93.12	65.19	36,070	0.9	−6.3	5	55	170	1700	250
Quinoline	129.16	1.21	6110	2.06	−14.85	55	170	550	1700	331
Thiophene	84.14	10,620	3015	1.81	−38	55	55	1700	5500	1400
Benzoic acid	122.13	0.11	3400	1.89	122.4	55	55	170	550	1700
Hexanoic acid	116.1	5	958	1.92	−3.44	55	55	170	550	6400
Phenylacetic acid	136.15	0.83	16600	1.41	77	55	55	170	550	2250
Salicylic acid	138.12	0.0208	2300	2.2	159	55	55	170	550	891
Anthracene	178.2	0.001	0.045	4.54	216.2	55	550	5500	17,000	8000
Benzo[a]pyrene	252.3	7×10^{-7}	0.0038	6.04	175	170	1700	17,000	55,000	n/a
Chrysene	228.3	5.7×10^{-7}	0.002	5.61	255	170	1700	17,000	55,000	n/a
Naphthalene	128.19	10.4	31	3.37	80.5	17	170	1700	5500	2400
Phenanthrene	178.2	0.02	1.1	4.57	101	55	550	5500	17,000	n/a
p-Xylene	106.2	1170	214.9488	3.18	13.2	17	550	1700	5500	4300
Pyrene	202.3	0.0006	0.132	5.18	156	170	1700	17,000	55,000	n/a
Benzo(b)thiophene	134.19	26.66	130	3.12	30.85	170	550	1700	5500	2200
1-Methylnaphthalene	142.2	8.84	28	3.87	−22	17	170	1700	5500	1840
Biphenyl	154.2	1.3	7	3.9	71	55	170	550	1700	3280
PCB-7	223.1	0.254	1.25	5	24.4	170	5500	17,000	17,000	n/a
PCB-15	223.1	0.0048	0.06	5.3	149	170	5500	17,000	17,000	n/a
PCB-29	257.5	0.0132	0.14	5.6	78	550	17,000	55,000	55,000	n/a
PCB-52	292	0.0049	0.03	6.1	87	1700	55,000	55,000	55,000	n/a

(*Continued*)

TABLE A.1 (*Continued*)

Physical Chemical Properties of Selected Organic Chemicals at 25°C Including Estimated Half-Lives (h) and Toxicity Expressed as Oral LD_{50} to the Rat

Chemical Name	Molar Mass (g mol⁻¹)	Vapor Pressure (Pa)	Aqueous Solubility (g m⁻³)	Log K_{OW}	Melting Point (°C)	Air	Water	Soil	Sediment	Rat Oral LD_{50} (mg kg⁻¹)
PCB-101	326.4	0.00109	0.01	6.4	76.5	1700	55,000	55,000	55,000	n/a
PCB-153	360.9	0.000119	0.001	6.9	103	5500	55,000	55,000	55,000	n/a
PCB-209	498.7	5.02×10^{-8}	10^{-6}	8.26	305.9	55,000	55,000	55,000	55,000	n/a
Total polychlorinated biphenyls	326	0.0009	0.024	6.6	0	5500	55,000	500,000	500,000	1900
Dibenzo-p-dioxin	184	0.055	0.865	4.3	123	55	55	1700	5500	1220
2,3,7,8-Tetrachlorodibenzo-p-dioxin	322	0.0000002	1.93×10^{-5}	6.8	305	170	550	17,000	5500	0.02
1,2,3,4,7,8-Hexatetrachlorodibenzo-p-dioxin	391	5.1×10^{-9}	4.42×10^{-6}	7.8	273	550	1700	55,000	55,000	0.8
1,2,3,4,6,7,8-Heptatetrachlorodibenzo-p-dioxin	425.2	7.5×10^{-10}	2.4×10^{-6}	8	265	550	1700	55,000	55,000	6.325
Octachlorodibenzodioxin	460	1.1×10^{-10}	7.4×10^{-8}	8.2	322	550	5500	55,000	55,000	1
Dibenzofuran	168.2	0.3	4.75	4.31	86.5	55	170	1700	5500	n/a
2,8-Dichlorodibenzofuran	237.1	0.00039	0.0145	5.44	184	170	550	5500	17,000	n/a
2,3,7,8-Tetrachlorodibenzofuran	306	2×10^{-6}	4.19×10^{-4}	6.1	227	170	550	17,000	55,000	n/a
Octachlorodibenzofuran	443.8	5×10^{-10}	1.16×10^{-6}	8	258	550	5500	55,000	55,000	n/a
4-Chlorophenol	128.56	20	27000	2.4	43	55	550	550	1700	500
2,4-Dichlorophenol	163	12	4500	3.2	44	55	550	550	1700	2830
2,3,4-Trichlorophenol	197.45	1	500	3.8	79	170	170	1700	5500	2800
2,4,6-Trichlorophenol	197.45	1.25	434	3.69	69.5	170	170	1700	5500	2800
2,3,4,6-Tetrachlorophenol	231.89	0.28	183	4.45	70	550	550	1700	5500	140
Pentachlorophenol	266.34	0.00415	14	5.05	190	550	550	1700	5500	210
2,4-Dimethylphenol	122.17	13.02	8795	2.35	26	17	550	1700	5500	2300
p-Cresol	108.13	13	20000	1.96	34.8	5	17	55	170	207
Dimethylphthalate (DMP)	194.2	0.22	4000	2.12	5	170	170	550	1700	2400
Diethylphthalate (DEP)	222.26	0.22	1080	2.47	−40.5	170	170	550	1700	8600
Dibutylphthalate (DBP)	278.34	0.00187	11.2	4.72	−35	55	170	550	1700	8000
Butyl benzyl phthalate	312.39	0.00115	2.69	4.68	−35	55	170	550	1700	13,500

(*Continued*)

TABLE A.1 (*Continued*)

Physical Chemical Properties of Selected Organic Chemicals at 25°C Including Estimated Half-Lives (h) and Toxicity Expressed as Oral LD$_{50}$ to the Rat

Chemical Name	Molar Mass (g mol⁻¹)	Vapor Pressure (Pa)	Aqueous Solubility (g m⁻³)	Log K$_{OW}$	Melting Point (°C)	Air	Water	Soil	Sediment	Rat Oral LD$_{50}$ (mg kg⁻¹)
Di-(2-ethylhexyl)-phthalate (DEHP)	390.54	1.33×10^{-5}	0.285	5.11	−50	55	170	550	1700	25,000
Aldicarb	190.25	0.004	6000	1.1	99	5	550	1700	17,000	0.5
Aldrin	364.93	0.005	0.02	6.5	104	55	5500	17,000	55,000	39
Carbaryl	201.22	0.0000267	120	2.36	142	55	170	550	1700	230
Carbofuran	221.3	0.00008	351	2.32	151	5	170	550	1700	5
Chloropyrifos	350.6	0.00227	0.73	4.92	41	17	170	170	1700	82
Cis-Chlordane	409.8	0.0004	0.056	6	103	55	17,000	17,000	55,000	500
p,p′-Dichlorodiphenyldichloroethylene	319	0.000866	0.04	5.7	88	170	55000	55,000	55,000	880
p,p′-Dichlorodiphenyltrichloroethane	354.5	0.00002	0.0055	6.19	108.5	170	5500	17,000	55,000	87
Dieldrin	380.93	0.0005	0.17	5.2	176	55	17,000	17,000	55,000	38.3
Diazinon	304.36	0.008	60	3.3	0	550	1700	1700	5500	66
γ-Hexachlorocyclohexane (lindane)	290.85	0.00374	7.3	3.7	112	1040	17,000	17,000	55,000	76
α-Hexachlorocyclohexane	290.85	0.003	1	3.81	157	1420	3364	1687	55,000	177
Heptachlor	373.4	0.053	0.056	5.27	95	55	550	1700	5500	40
Malathion	330.36	0.001	145	2.8	2.9	17	55	55	550	290
Methoxychlor	345.7	0.00013	0.045	5.08	86	17	170	1700	5500	1855
Mirex	545.59	0.0001	0.000065	6.9	485	170	170	55,000	55,000	235
Parathion	291.27	0.0006	12.4	3.8	6	17	550	550	1700	2
Methyl parathion	263.5	0.002	25	3	37	17	550	550	1700	6.01
Atrazine	215.68	0.00004	30	2.75	174	5	550	1700	1700	672
2-(2,4-Dichlorophenoxy) acetic acid	221.04	0.00008	400	2.81	140.5	17	55	550	1700	375
Dicamba	221.04	0.0045	4500	2.21	114	55	550	550	1700	1039
Mecoprop	214.6	0.00031	620	3.94	94	17	170	170	1700	650
Metolachlor	283.8	0.0042	430	3.13	0	170	1700	1700	5500	2200
Simazine	201.7	8.5×10^{-6}	5	2.18	225	55	550	1700	5500	971
Trifluralin	335.5	0.015	0.5	5.34	48.5	170	1700	1700	5500	1930
Thiram	240.4	0.00133	30	1.73	145	170	170	550	1700	560

These data have been selected from a number of sources, including Mackay (2000), RTECS (2000), and the Hazardous Substances Data Bank (2000).

Appendix B
Units

It is essential to define the system of units and dimensions that forms the foundation of all calculations. With few exceptions, we adopt the SI system, and some key aspects of the SI system are discussed below:

Length (metre, m): This base unit is defined as the specified number of wavelengths of a krypton light emission.

Area: (square metre, m^2): Occasionally, the hectare (ha) (an area $100 \times 100\,m$ or $10^4\,m^2$) or the square kilometer (km^2) is used. For example, pesticide dosages to soils are often given in kg ha^{-1}.

Volume (cubic metre, m^3): The liter (L) ($0.001\,m^3$) is also used because of its convenience in analysis, but it should be avoided in environmental calculations. In the United States, the spellings "meter" and "liter" are often used.

Mass (kilogram, kg): The base unit is the kilogram (kg), but it is often more convenient to use the gram (g), especially for concentrations. For large masses, the megagram (Mg) or the equivalent metric tonne (t) may be used.

Amount of matter (mole abbreviated to mol): This unit is the actual number of particles divided by Avogadro's number (6.022×10^{23}), which is defined as the number of atoms in 12 g of the carbon-12 isotope. When reactions occur, the amounts of substances reacting and forming are best expressed in moles rather than mass, since atoms or molecules combine in simple stoichiometric ratios. The need to involve atomic or molecular masses is thus avoided.

Molar mass (g mol^{-1}): This is the mass of 1 mole of matter and is sometimes (wrongly) referred to as molecular weight or molecular mass. Strictly, the correct unit is kg mol^{-1}, but it is often more convenient to use g mol^{-1}, which is obtained by adding the atomic masses. Benzene (C_6H_6) is thus approximately 78 g mol^{-1} or 0.078 kg mol^{-1}.

Time (second or hour, s or h): The standard unit of a second (s) is inconveniently short when considering environmental processes such as flows in large lakes when residence times may be many years. The use of hours, days, and years is thus acceptable. We generally use hours as a compromise.

Concentration: (mol m^{-3}): The preferred unit is the mole per cubic metre (mol m^{-3}) or the gram per cubic metre (g m^{-3}). Most analytical data are reported in amount or mass per liter, because a liter (L) is a convenient volume for the analytical chemist to handle and measure; however, it is not coherent with area or length, and therefore conversion to m^3 is necessary.

Density (kg m^{-3}): This has identical units to mass concentrations, but the use of kg m^{-3} is preferred, water having a density of 1000 kg m^{-3} and air a density of approximately 1.2 kg m^{-3}.

Force (Newton, N): The newton is the force that causes a mass of 1 kg to accelerate at 1 m s^{-2}. It is 10^5 dynes and is approximately the gravitational force operating on a mass of 102 g at the Earth's surface.

Pressure (Pascal, Pa): The pascal or newton per square metre (N m^{-2}) is inconveniently small, since it corresponds to only 102 g force over 1 sq. m, but it is the standard unit, and it is used here. The atmosphere (atm) is 101,325 Pa or 101.325 kPa. The torr or mm of mercury (mmHg) is 133 Pa and, although still widely used, should be regarded as obsolescent.

Energy (joule, J): The joule, which is one Nm or Pa m^3, is also a small quantity. It replaces the obsolete units of calorie (which is 4.184 J) and the British thermal unit (BTU) (1055 J).

Temperature (K): The kelvin is preferred, although environmental temperatures may be expressed in degrees Celsius, °C, and not centigrade, where 0°C is 273.15 K.

Frequency (Hertz, Hz): The hertz is one event per second (s^{-1}). It is used in descriptions of acoustic and electromagnetic waves, stirring, and in nuclear decay processes where the quantity of a radioactive material may be described in becquerels (Bq), where 1 Bq corresponds to the amount that has a disintegration rate of 1 Hz.

Gas constant (R): This constant, which derives from the gas law, is 8.314 J mol^{-1} K^{-1} or Pa m^3 mol^{-1} K^{-1}.

References

Abernethy, S., Bobra, A.M., Shiu, W.Y., Wells, P.G., and Mackay, D. (1986) Acute lethal toxicity of hydrocarbons and chlorinated hydrocarbons to two planktonic crustaceans: The key role of organism-water partitioning. *Aquat. Toxicol.* 8:163–174.

Abernethy, S., Mackay, D., and McCarty, L.S. (1988) Volume fraction correlation for narcosis in aquatic organisms: The key role of partitioning. *Environ. Toxicol. Chem.* 7:469–481.

Abraham, M.H., Grellier, P.L., Hamerton, I., McGill, R.A., Prior, D.V., and Whiting, G.S. (1988) Solvation of gaseous non-electrolytes *Faraday discuss. Chem. Soc.* 85:107–115.

Abraham, M.H. (1993) Scales of solute hydrogen-bonding: Their construction and application to physicochemical and biochemical processes. *Chem. Soc. Rev.* 22:73.

Abraham, M.H., Ibrahim, A., and Zissimos, A.M. (2004) Determination of sets of solute descriptors from chromatographic measurements. *J Chromatogr. A.* 1037:29–47.

Armitage, J.M., Toose, L., Embry, M., Foster, K.L., Hughes, L., and Arnot, J.A. (2018) *The Bioaccumulation Assessment Tool (BAT) Version 1.0.* ARC Arnot Research and Consulting Inc., Toronto.

Arnot, J.A. and Gobas, F.A. (2004) A food web bioaccumulation model for organic chemicals in aquatic ecosystems. *Environ. Toxicol. Chem.* 23: 2343–2355.

Arnot, J.A., Mackay, D., Webster, E., and Southwood, J.M. (2006) Screening level risk assessment model for chemical fate and effects in the environment. *Environ. Sci. Technol.* 40:2316–2313.

Arnot, J.A., and Mackay, D. (2008) Policies for chemical hazard and risk priority setting: Can persistence, bioaccumulation, toxicity, and quantity information be combined? *Environ. Sci. Technol.* 13:4648–4654.

Arp, H.P.H., Schwarzenbach, R.P., and Goss, K.-U. (2008) Ambient gas/particle partitioning. 2: The influence of particle source and temperature on sorption to dry terrestrial aerosols. *Environ. Sci. Technol.* 42:5951–5957.

Burns, L.A., Cline, D.M., and Lassiter, R.P. (1982) Exposure Analysis Modeling System (EXAMS): user manual and system documentation. EPA-600/3-82-023, U.S. EPA.

Baum, E.J. (1997) *Chemical Property Estimation: Theory and Application.* Lewis Publication, Boca Raton, FL.

Bennett, D.H., McKone, T.E., Matthies, M., and Kastenberg, W.E. (1998) General formulation of characteristic travel distance for semivolatile organic chemicals in a multimedia environment. *Environ. Sci. Technol.* 32:4023–4050.

Bennett, D.H. and Furtaw, E.J. (2004) Fugacity-based indoor residential pesticide fate model. *Environ. Sci. Technol.* 38:2142–2152.

Beyer, A., Mackay, D., Matthies, M., Wania, F., and Webster, E. (2000) Assessing long-range transport potential of persistent organic pollutants. *Environ. Sci. Tech.* 34:699–703.

Bidleman, T.F. (1988) Atmospheric processes. *Environ. Sci. Technol.* 22:361–367 and errata 726.

Bidleman, T.F. and Harner, T. (2000) Sorption to aerosols, pp. 233–260. In: *Handbook of Property Estimation Methods for Chemicals: Environmental and Health Sciences.* Boethling, R.S. and Mackay, D., Eds., Lewis Publishers, Boca Raton, FL.

Boethling, R.S. and Mackay, D. (2000) *Handbook of Property Estimation Methods for Chemicals: Environmental and Health Sciences.* Lewis Publishers, London.

Buser A., MacLeod M., Scheringer, M., Mackay D., Bonnell, M., Russell M., DePinto, J., and Hungerbuhler K. (2012) Good modeling practice guidelines for applying multimedia models in chemical assessments. *Integr. Environ. Assess. Manage.* 8:703–708.

Campfens, J. and Mackay, D. (1997) Fugacity-based model of PCB bioaccumulation in complex aquatic food webs. *Environ. Sci. Technol.* 31:577–583.

Clark, K.E., Gobas, F.A.P.C., and Mackay, D. (1990) Model of organic chemical uptake and clearance by fish from food and water. *Environ. Sci. Technol.* 24:1203–1213.

Clark, T., Clark, K., Paterson, S., Nostrom, R., and Mackay, D. (1988) Wildlife monitoring, modelling and fugacity. *Environ. Sci. Technol.* 22:120–127.

Clark, T.P., Norstrom, R.J., Fox, G.A., and Won, H.T. (1987) Dynamics of organochlorine compounds in herring gulls (*Larus argentatus*): II. A two compartment model and data for ten compounds. *Environ. Toxicol. Chem.* 6:547–559.

Cole, J.G. and Mackay, D. (2000) Correlation of environmental partitioning properties of organic compounds: The three solubilities approach. *Environ. Toxicol. Chem.* 19:265–270.

Connell, D.W. (1990) *Bioaccumulation of Xenobiotic Compounds*. CRC Press, Boca Raton, FL.

Connolly, J.P. and Pedersen, C.J. (1988) A thermodynamically based evaluation of organic chemical accumulation in aquatic organisms. *Environ. Sci. Technol.* 22:99–103.

Coulon, F., Whelan, M.J., Paton, G.I., Semple, K.T, Villa, R., and Pollard, S.J.T. (2010) Multimedia fate of petroleum hydrocarbons in the soil: Oil matrix of constructed biopiles. *Chemosphere* 81:1454–1462.

Cousins, I.T., Gevao, B., and Jones, K.C. (1999a) Measuring and modelling the vertical distribution of semi-volatile organic compounds in soils. 1: PCB and PAH soil core data. *Chemosphere* 39:2519–2534.

Cousins, I.T., Beck, A.J., and Jones, K.C. (1999b) A review of the processes involved in the exchange of semi-volatile organic compounds (SVOC) across the air/soil interface. *Sci. Tot. Environ.* 228:5–24.

Csiszar, S.A., Diamond, M.L., and Thibodeaux, L.J. (2012) Modeling urban films using a dynamic multimedia fugacity model. *Chemosphere* 87:1024–1031.

deBruyn, A.M.H. and Gobas, F.A.P.C. (2006) A bioenergetic biomagnification model for the animal kingdom. *Environ. Sci. Technol.* 40:1581–1587.

Di Guardo, A., Calamari, D., Zanin, G., Consalter, A., and Mackay, D. (1994a) A fugacity model of pesticide runoff to surface water: Development and validation. *Chemosphere* 28(3):511–531.

Di Guardo, A., Williams, R.J., Matthiessen, P., Brooke, D.N., and Calamari, D. (1994b) Simulation of pesticide runoff at Rosemaud Farm (UK) using the SoilFug model. *Environ. Sci. Pollut. Res.* 1(3):151–160.

Diamond, M., Mackay, D., Cornett, R.J., and Chant, L.A. (1990) A model of the exchange of inorganic chemicals between water and sediments. *Environ. Sci. Technol.* 24:713–722.

Diamond, M., Mackay, D., Poulton, D., and Stride, F. (1994) Development of a mass balance model of the fate of 17 chemicals in the Bay of Quinte. *J. Great Lakes Res.* 20(4):643–666.

DiToro, D.M. (1985) A particle interaction model of reversible organic chemical sorption. *Chemosphere* 14:1503–1538.

DiToro, D.M. (2001) *Sediment Flux Modeling*. Wiley-Interscience, New York, NY.

Dominguez-Morueco, N., Diamond, M.L., Sierra, J., Schuhmacher, M., Domingo, J.L., and Nadal, M. (2016) Application of the multimedia urban model to estimate the emissions and environmental fate of PAHs in Tarragona County, Catalonia, Spain. *Sci. Tot.Environ.* 573:1622–1629.

Doucette, W.J. (2000) Soil and sediment sorption coefficient, pp. 141–188. In: *Handbook of Property Estimation Methods for Chemicals: Environmental and Health Sciences*. Boethling, R.S. and Mackay, D., Eds., Lewis Publishers, Boca Raton, FL.

Drouillard, K.G. and Norstrom, R.J. (2014) Use of a vial equilibration technique to measure the change in fugacity capacity of avian food and feces samples for 1,2,3,4-tetrachlorobenzene. *Bull. Environ. Contam. Tox.* 93:561–566.

Drouillard, K.G., Paterson, G., Liu, J., and Haffner, G.D. (2012) Calibration of the gastrointestinal magnification model to predict maximum biomagnification potentials of polychlorinated biphenyls in a bird and fish. *Environ. Sci. and Technol.* 46:10279–10286.

Eisenreich, S.L. (1987) The chemical limnology of non-polar organic contaminants. PCBs in Lake Superior, pp. 393–470. In: *Sources and Fates of Aquatic Pollutants*. Hites, R.A. and Eisenreich, S.J., Eds., American Chemical Society, Advances in Chemistry Series 216, Washington, DC, Ch. 13.

Endo, S. and Goss, K.-U. (2014) Applications of polyparameter linear free energy relationships in environmental chemistry. *Environ. Sci. Technol.* 48:12477–12491.

Finizio, A., Mackay, D., Bidleman, T., and Harner T. (1997) Octanol-air partition coefficient as a predictor of partitioning of semi-volatile organic chemicals to aerosols. *Atmos. Environ.* 31:2289–2296.

Fiserova-Bergerova, V. (1983) *Modeling of Inhalation Exposure to Vapors*, Vols. I and II. CRC Press, Boca Raton, FL.

Formica, S.J., Baron, J.A., Thibodeaux, L.J., and Valsaraj, T. (1988) PCB transport into lake sediments. Conceptual model and laboratory simulation. *Environ. Sci. Technol.* 22:1435–1440.

Gawlik, B.M., Sotiriou, N., Feicht, E.A., Schulte-Hostede, S., and Kettrup, A. (1997) Alternatives for the determination of the soil adsorption coefficient, K_{OC}, of non-ionic organic compounds—A review. *Chemosphere* 34:2525–2551.

Gobas, F.A.P.C. and Mackay, D. (1987) Dynamics of hydrophobic organic chemical bioconcentration in fish. *Environ. Toxicol. Chem.* 6:495–504.

Gobas, F.A.P.C., Clark, K.E., Shiu, W.Y., and Mackay, D. (1989) Bioconcentration of polybrominated benzenes and biphenyls and related super-hydrophobic chemicals in fish: Role of bioavailability and elimination into the feces. *Environ. Toxicol. Chem.* 8:231–245.

Gobas, F.A.P.C. (1993) A model for predicting the bioaccumulation of hydrophobic organic chemicals in aquatic food-webs: Application to Lake Ontario. *Ecol. Model.* 69:1–17.

Gobas, F.A.P.C. and Morrison, H.A. (2000) Bioconcentration and biomagnification in the aquatic environment, pp. 189–232. In: *Handbook of Property Estimation Methods for Chemicals: Environmental and Health Sciences.* Boethling, R.S. and Mackay, D., Eds., Lewis Publishers, Boca Raton, FL.

Gossett, J.M. (1987) Measurement of Henry's law constants for C_1 and C_2 chlorinated hydrocarbons. *Environ. Sci. Technol.* 21:202–208.

Gouin T., Mackay D., Webster E., and Wania F. (2000) Screening chemicals for persistence in the environment. *Environ. Sci. Technol.* 34:881–884.

Green, J.C. (1988) Classification and properties of soils (Appendix C), pp. C1–C28. In: *Environmental Inorganic Chemistry: Properties, Processes and Estimation Methods.* Lyman, W.J., Reehl, W.F., and Rosenblatt, D.H., Eds., Pergamon Press, New York, NY.

Hamelink, J.L., Landrum, P.F., Bergman, H.L., and Benson, W.H., Eds. (1994) *Bioavailability: Physical, Chemical, and Biological Interactions.* SETAC Special Publication. CRC Press, Boca Raton, FL.

Hansch, C., Muir, R. M., Fujita, T., Maloney, P. P., Geiger, F., and Streich, M. (1963) The correlation of biological activity of plant growth regulators and chloromycetin derivatives with Hammett constants and partition coefficients. *J. Am. Chem. Soc.* 85:2817–2824.

Hansch, C., and Fujita, T. (1964) p-σ-π Analysis. A method for the correlation of biological activity and chemical structure. *J. Am. Chem. Soc.* 86:1616–1626.

Harner, T., Green, N.J.L., and Jones, K.C. (2000) Measurements of octanol-air partition coefficients for PCDD/Fs: A tool in assessing air-soil equilibrium status. *Environ. Sci. Technol.* 34:3109–3114.

Hazardous Substances Data Bank (2000) National Library of Medicine Toxicology Data Network (TOXNET). Available online at: http://toxnet.nlm.nih.gov/

Hiatt, M.H. (1999) Leaves as an indicator of exposure to airborne volatile organic compounds. *Environ. Sci. Technol.* 33:4126–4133.

Hickie, B.E., Mackay, D., and de Koning, J. (1999) A lifetime pharmacokinetic model for hydrophobic contaminants in marine mammals. *Environ. Toxicol. Chem.* 18:2622–2633.

Holysh, M., Paterson, S., Mackay, D., and Bandurraga, M.M. (1985) Assessment of the environmental fate of linear alkylbenzenesulphonates. *Chemosphere* 15:3–20.

Hughes, L., Webster, E., and Mackay, D. (2008) An evaluative screening level model of the fate of organic chemicals in sludge-amended soils including organic matter degradation. *Soil Sediment Contam.* 17: 564–585.

Hughes, L. and Mackay, D. (2011) Model of the fate of chemicals in sludge-amended soils with uptake in vegetation and soil-dwelling organisms. *Soil Sed. Contam. Int. J.* 20:938–960.

Jain, N. and Yalkowsky, S.H. (2001) Estimation of the aqueous solubility I: Application to organic nonelectrolytes. *J. Pharmceut. Sci.* 90:234-252.

Jantunen, L.M. and Bidleman, T. (1996) Air-water gas exchange of hexachlorocyclohexanes (HCHs) and the enantiomers of a-HCH in arctic regions. *J. Geophys. Res.* 101:28837–28246. (See correction in 102:19279–19282.)

John, L. Monteith, J.L., and Unsworth, M.H. (2013) *Principles of Environmental Physics: Plants, Animals, and the Atmosphere,* 4th ed., p. 111. Elsevier, Oxford.

Jones-Otazo, H.A., Clarke, J.P., Diamond. M.L., Archbold, J.A., Ferguson, G., Harner, T., Richardson, G. M., Ryan, J.J., and Wilford, B. (2005) Is house dust the missing exposure pathway for PBDEs? An analysis of the urban fate and human exposure to PBDEs? *Environ. Sci. Technol.* 39:5121–5130.

Jury, W.A., Spencer, W.F., and Farmer, W.J. (1983) Behavior assessment model for trace organics in soil: I Model description. *J. Environ. Qual.* 12:558.

Jury, W.A., Farmer, W.J., and Spencer, W.F. (1984a) Behavior assessment model for trace organics in soil: II Chemical classification and parameter sensitivity. *J. Environ. Qual.* 13:567.

Jury, W.A., Farmer, W.J., and Spencer, W.F. (1984b) Behavior assessment model for trace organics in soil: III Application of screening model. *J. Environ. Qual.* 13:573.

Jury, W.A., Spencer, W.F., and Farmer, W.J. (1984c) Behavior assessment model for trace organics in soil: IV Review of experimental evidence. *J. Environ. Qual.* 13:580.

Kaiser, K.L.E. (1984) *QSAR in Environmental Toxicology.* D. Reidel Publishing Co., Dordrecht.

Kaiser, K.L.E. (1987) *QSAR in Environmental Toxicology II.* D. Reidel Publishing Co., Dordrecht.

Karcher, W. and Devillers, J. (1990) *Practical Applications of Quantitative Structure-Activity Relationships (QSAR) in Environmental Chemistry and Toxicology.* Karcher, W. and Devillers, J., Eds., Kluwer Academic Publisher, Dordrecht.

Karickhoff, S.W. (1981) Semiempirical estimation of sorption of hydrophobic pollutants on natural sediments and soils. *Chemosphere* 10:833–849.

Klamt, A. and Schüürmann, G. (1993) COSMO: A new approach to dielectric screening in solvents with explicit expressions for the screening energy and its gradient. *J. Chem. Soc. Perkin Trans.* 2:799–805.

Klamt, A. (1995) Conductor-like screening model for real solvents: A new approach to the quantitative calculation of solvation phenomena. *J. Phys. Chem.* 99:2224–2235.

Klamt, A., Jonas, V., Bürger, T., and Lohrenz, J.C. (1998) Refinement and parametrization of COSMO-RS. *J. Phys. Chem. A.* 102:5074.

Klamt, A. (2005) *COSMO-RS: From Quantum Chemistry to Fluid Phase Thermodynamics and Drug Design.* Elsevier, Amsterdam.

Klamt, A. (2016) COSMO-RS for aqueous solvation and interfaces. *Fluid Phase Equilibr.* 407:152–158.

Klamt, A. (2018) The COSMO and COSMO-RS solvation models. *WIREs Comput. Mol. Sci.* 8:1–11.

Konemann, H. (1981) Quantitative structure-activity relationships in fish toxicity studies. Part I. Relationship for 50 industrial pollutants. *Toxicology* 19:229–238.

Leo, A. (2000) Octanol/water partition coefficient, pp. 89–114. In: *Handbook of Property Estimation Methods for Chemicals: Environmental and Health Sciences.* Boethling, R.S. and Mackay, D., Eds., Lewis Publishers, Boca Raton, FL.

Li, L., Westgate, J.N., Hughes, L., Zhang, X., Givehchi, B., Toose, L., Armitage, J. M., Wania, F., Egeghy, P., and Arnot, J.A. (2018) A model for risk-based screening and prioritization of human exposure to chemicals from near-field sources. *Environ. Sci. Tech.* 52:14235–14244.

Ling, H., Diamond, M., and Mackay, D. (1993) Application of the QWASI fugacity/aquivalence model to assessing sources and fate of contaminants in Hamilton Harbour. *J. Great Lakes Res.* 19:82–602.

Liss, P.S. and Slater, P.G. (1974) Flux gases across the air-sea interface. *Nature* 247:181–184.

Lun, R., Lee, K., De Marco, L., Nalewajko, C., and Mackay, D. (1998) A model of the fate of polycyclic aromatic hydrocarbons in the Saguenay Fjord. *Environ. Toxicol. Chem.* 17:333–341.

Lyman, W.J., Reehl, W.F., and Rosenblatt, D.H. (1982) *Handbook of Chemical Property Estimation Methods.* McGraw-Hill, New York, NY.

Mackay, D. and Leinonen, P.J. (1975) The rate of evaporation of low solubility contaminants from water bodies. *Environ. Sci. Technol.* 9: 1178–1180.

Mackay, D. (1979) Finding fugacity feasible. *Environ. Sci. Technol.* 13:1218.

Mackay, D., Shiu, W.Y., and Sutherland, R.P. (1979) Determination of air-water Henry's law constants for hydrophobic pollutants. *Environ. Sci. Technol.* 13:333–337.

Mackay, D. and Shiu, W.Y. (1981) A critical review of Henry's law constants for chemicals of environmental interest. *J. Phys. Chem. Ref. Data* 10:1175–1199.

Mackay, D. and Paterson, S. (1981) Calculating fugacity. *Environ. Sci. Technol.* 15(9):1006–1014.

Mackay, D. and Paterson, S. (1982) Fugacity revisited. *Environ. Sci. Technol.* 16:654A–660A.

Mackay, D. (1982a) Correlation of bioconcentration factors. *Environ. Sci. Technol.* 16:274–278.

Mackay, D. (1982b) Effect of surface films on air-water exchange rates. *J. Great Lakes Res.* 8:299–306.

Mackay, D. and Paterson, S. (1983) Indoor exposure to volatile chemicals. *Chemosphere* 12:143–154.

Mackay, D. and Yuen, A.T.K. (1983) Mass transfer coefficient correlations for volatilization of organic solutes from water. *Environ. Sci. Technol.* 17:211–216.

Mackay, D., Joy, M., and Paterson, S. (1983a) A quantitative water, air, sediment interaction (QWASI) fugacity model for describing the fate of chemicals in lakes. *Chemosphere* 12:981–997.

Mackay, D., Paterson, S., and Joy, M. (1983b) A quantitative water, air, sediment interaction (QWASI) fugacity model for describing the fate of chemicals in rivers. *Chemosphere* 12:1193–1208.

Mackay, D. and Hughes, A.I. (1984) Three-parameter equation describing the uptake of organic compounds by fish. *Environ. Sci. Technol.* 18:439–444.

Mackay, D., Paterson, S., Cheung, B., and Neely, W.B. (1985) Evaluating the environmental behaviour of chemicals with a level III fugacity model. *Chemosphere* 14:335–374.

Mackay, D., Paterson, S., and Schroeder, W.H. (1986) Model describing the rates of transfer processes of organic chemicals between atmosphere and water. *Environ. Sci. Technol.* 20:810–816.

Mackay, D. (1989) Modelling the long term behaviour of an organic contaminant in a large lake: Application to PCBs in Lake Ontario. *J. Great Lakes Res.* 15:283–297.

Mackay, D. and Diamond, M. (1989) Application of the QWASI (quantitative water air sediment interaction) fugacity model to the dynamics of organic and inorganic chemicals in lakes. *Chemosphere* 18:1343–1365.

Mackay, D. and Stiver, W.H. (1989) The linear additivity principle in environmental modelling: Application to chemical behavior in soil. *Chemosphere* 19:1187–1198.

Mackay, D. and Southwood, J.M. (1992) Modelling the Fate of Organochlorine Chemicals in Pulp Mill Effluents. *Water Poll. Res. J. Canad.* 27:509–537.

Mackay, D.;Shiu, W.Y.; Ma, K.C. (2002) Illustrated Handbook of Physical Chemical Properties and Environmental Fate for Organic Chemicals. 2nd ed. in 4 volumes: Vol 1. Hydrocarbons, Vol 2. Halogenated Hydrocarbons, Vol 3. Oxygen Containing Compounds and Vol 4. Nitrogen and Sulfur Containing Compounds and Pesticide Chemicals. Lweis Publishers/CRC PRess New York N.Y.

Mackay, D., Paterson, S., Di Guardo, A., and Cowan, C.E. (1996) Evaluating the environmental fate of a variety of types of chemicals using the EQC model. *Environ. Toxicol. Chem.* 15(9):1627–1637.

Mackay, D. (2000) Solubility in water, pp. 125–140. In: *Handbook of Property Estimation Methods for Chemicals: Environmental and Health Sciences.* Boethling, R.S. and Mackay, D., Eds., Lewis Publishers, Boca Raton, FL.

Mackay, D. and Boethling, R.S. (2000) *Handbook of Property Estimation Methods for Chemicals: Environmental and Health Sciences* CRC Press, Boca Raton, FL.

Mackay, D. and Fraser, A. (2000) Bioaccumulation of Persistent Organic Chemicals: Mechanisms and Models. *Environ. Pollut.* 110:375–391.

Mackay, D., Shiu, W.Y., and Ma, K.C. (2000) Physical-Chemical Properties and Environmental Fate and Degradation Handbook. CRCnetBASE 2000, Chapman & Hall CRCnetBASE, CRC Press LLC, Boca Raton, FL (CD-ROM).

Mackay, D., Hughes, L., Powell, D.E., and Kim, J. (2014) An updated quantitative water air sediment interaction (QWASI) model for evaluating chemical fate and input parameter sensitivities in aquatic systems: Application to D5 (decamethylcyclopentasiloxane) and PCB-180 in two lakes. *Chemosphere* 111:359–365.

Mackay, D., Celsie, A.K.D., and Parnis, J.M. (2019) Kinetic delay in partitioning and parallel particle pathways: Underappreciated aspects of environmental transport. *Environ. Sci. Technol.* 53(1):234–241.

Matoba, Y., Ohnishi, J., and Matsuo, M. (1995) Indoor simulation of insecticides in broadcast spraying. *Chemosphere* 30:345–365.

Matoba, Y., Takimoto, Y., and Kato, T. (1998a) Indoor behavior and risk assessment following space spraying of *d*-tetramethrin and *d*-resmethrin. *Am. Indust. Hygiene Assoc. J.* 59:181–190.

Matoba, Y., Takimoto, Y., and Kato, T. (1998b) Indoor behavior and risk assessment following space spraying of *d*-phenothrin and *d*-tetramethrin. *Am. Indust. Hygiene Assoc. J.* 59:191–199.

McLachlan, M. S. (2018) Can the Stockholm convention address the spectrum of chemicals currently under regulatory scrutiny? Advocating a more prominent role for modeling in POP screening assessment Environ. Sci: Processes Impacts 20:32-37

Menzel, D.B. (1987) Physiological pharmacokinetic modeling. *Environ. Sci. Technol.* 21:944–950.

Nazaroff, W.W. and Cass, G.R. (1986) Mathematical modeling of chemically reactive pollutants in indoor air. *Environ. Sci. Technol.* 20:924–934.

Nazaroff, W.W. and Cass, G.R. (1989) Mathematical modeling of indoor aerosol dynamics. *Environ. Sci. Technol.* 23:157–166.

Neely, W.B., Branson, D.R., and Blau, G.E. (1974) Partition coefficient to measure bioconcentration potential of organic chemicals in fish. *Environ. Sci. Technol.* 8(13):1113–1115.

Neely, W.B. and Mackay, D. (1982) Evaluative model for estimating environmental fate, p. 127. In: *Modeling the Fate of Chemicals in the Aquatic Environment.* Dickson, K.L., Maki, A.W., and Cairns, J., Eds., Ann Arbor Science Publishers, Ann Arbor, MI.

Nichols, J.W., McKim, J.M., Lien, G.J., Hoffman, A.D., Bertelsen, S.L., and Elonen, C.M. (1996) A physiologically based toxicokinetic model for dermal absorption of organic chemicals by fish. *Fund. Appl. Toxicol.* 31:229–242.

Nichols, J.W., Jensen, K.M., Tietge, J.E., and Johnson, R.D. (1998) Physiologically based toxicokinetic model for maternal transfer of 2,3,7,8-tetrachlorodibenzo-*p*-dioxin in brook trout (*Salvelinus fontinalis*). *Environ. Toxicol. Chem.* 17(12): 2422–2434.

Nguyen, T.H., Goss, K.-U., and Ball, W.P. (2005) Polyparameter linear free energy relationships for estimating the equilibrium partition of organic compounds between water and the natural organic matter in soils and sediments. *Environ. Sci. Technol.* 39:913–924.

Oreskes, N., Shrader-Frechette, K., and Belitz, K. (1994) Verification, validation, and confirmation of numerical models in the Earth sciences. *Science* 263:641–646.

Pankow, J.F. (1998) Further discussion of the octanol/air partition coefficient K_{OA} as a correlating parameter for gas/particle partitioning coefficients. *Atmos. Environ.* 32:1493–1497.

Parke, D.V. (1982) The disposition and metabolism of environmental chemicals by mammalia, pp. 141–178. In: *Handbook of Environmental Chemistry*, Vol. 2/Part B, Hutzinger, O., Ed., Springer-Verlag, Heidelberg.

Parnis, J.M., Mackay, D., and Harner, T. (2015) Temperature dependence of Henry's law constants and K_{OA} for simple and heteroatom-substituted PAHs by COSMO-RS. *Atmos. Environ.* 110:27–35.

Paterson, S. and Mackay, D. (1986) A pharmacokinetic model of styrene inhalation using the fugacity approach. *Toxicol. Appl. Pharmacol.* 82:444–453.

Paterson, S. and Mackay, D. (1987) A steady-state fugacity-based pharmacokinetic model with simultaneous multiple exposure routes. *Environ. Toxicol. Chem.* 6:395–408.

Paustenbach, D.J. (2000) The practice of exposure assessment: A state-of-the-art review. *J. Toxicol. Environ. Health Part B* 3:179–291.

Perrin, D.D., Dempsey, B., and Serjeant, E.P. (1981) *pK$_a$ Prediction for Organic Acids and Bases*. Chapman and Hall, New York, NY.

Pinsuwon, S., Li, A., and Yalkowsky, S.H. (1995) Correlation of octanol/water solubility and partition coefficients. *J. Chem. Eng. Data* 40:623–626.

Pollard, S.J.T., Hough, R.L., Kim, K.-H., Bellarby, J., Paton, G., Semple, K.T., and Coulon, F. (2008) Fugacity modelling to predict the distribution of organic contaminants in the soil: Oil matrix of constructed biopiles. *Chemosphere* 71:1432–1439.

Prausnitz, J.M., Lichtenthaler, R.N., and de Azevedo, E.G. (1986) *Molecular Thermodynamics of Fluid Phase Equilibria*, 2nd ed., Prentice Hall, Englewood Cliffs, NJ.

Ramsey, J.C. and Anderson, M.E. (1984) A physiologically based description of the inhalation pharmacokinetics of styrene in rats and humans. *Toxicol. Appl. Pharmacol.* 73:159–175.

Registry of Toxic Effects of Chemical Substances (RTECS) (2000) U.S. National Institute for Occupational Safety and Health (NIOSH). Available online at: http://ccinfoweb.ccohs.ca/rtecs/search.html

Reitz, R.H. and Gehring, P.J. (1982) Pharmacokinetic models, pp. 179–195. In: *The Handbook of Environmental Chemistry*, Vol. 2/Part B, Hutzinger, O., Ed., Springer-Verlag, Heidelberg.

Reuber, B., Mackay, D., Paterson, S., and Stokes, P. (1987) A discussion of chemical equilibria and transport at the sediment-water interface. *Environ. Toxicol. Chem.* 6:731–739.

Sanghvi, T., Jain, N., Yang, G., and Yalkowsky, S.H. (2003) Estimation of aqueous solubility by the general solubility equation (GSE) the easy way. *SAR Comb. Sci.* 22:258–262.

Satterfield, C.N. (1970) *Mass Transfer in Heterogeneous Catalysis*. MIT Press, Cambridge.

Schwarzenbach, R.P., Gschwend, P.M., and Imboden, D.M. (2nd ed. 2002) *Environmental Organic Chemistry*. John Wiley & Sons, New York, NY.

Seth, R., Mackay, D., and Muncke, J. (1999) Estimating of organic carbon partition coefficient and its variability for hydrophobic chemicals. *Environ. Sci. Technol.* 33:2390–2394.

Sharpe, S. and Mackay, D. (2000) A framework for evaluating bioaccumulation in food webs. *Environ. Sci. Technol.* 34:2373–2379

Southwood, J.M., Harris, R.C., and Mackay, D. (1989) Modeling the fate of chemicals in an aquatic environment: The use of computer spreadsheet and graphics software. *Environ. Toxicol. Chem.* 8:987–996.

Southwood, J., Muir, D.C.G., and Mackay, D. (1999) Modelling agrochemical dissipation in surface microlayers following aerial deposition. *Chemosphere* 38:121–141.

Spacie, A. and Hamelink, J.L. (1982) Alternative models for describing the bioconcentration of organics in fish. *Environ. Toxicol. Chem.* 1:309–320.

Sposito, G. (1989) *The Chemistry of Soils*. Oxford University Press, New York, NY.

Thibodeaux, L.J. (1996) *Environmental Chemodynamics: Movement of Chemicals in Air, Water and Soil*, 2nd ed., Wiley-Interscience Publication, New York, NY.

Thibodeaux, L.J. and Mackay, D. (2011) *Handbook of Chemical Mass Transport in the Environment*. CRC Press, Taylor & Francis, New York, NY.

Thomann, R.V. (1989) Bioaccumulation model of organic chemical distribution in aquatic food chains. *Environ. Sci. Technol.* 23:699–707.

Thompson, H.C., Jr., Kendall, D.C., Korfmacher, W.A., Rowland, K.L., Rushing, L.G., Chen, J.J., Kominsky, J.R., Smith, L.M., and Stalling, D.L. (1986) Assessment of the contamination of a multibuilding facility by polychlorinated biphenyls, polychlorinated dibenzo*p*-dioxins, and polychlorinated dibenzofurans. *Environ. Sci. Technol.* 20:597–603.

Tuey, D.B. and Matthews, H.B. (1980) Distribution and excretion of 2,2′, 4,4′, 5,5′-hexabromobiphenyl in rats and man: Pharmacokinetic model predictions. *Toxicol. Appl. Pharmacol.* 53:420–431.

Ulrich, N., Endo, S., Brown, T.N., Watanabe, N., Bronner, G., Abraham, M.H., and Goss, K.-U. (2017) UFZ-LSER database v 3.2.1 [Internet], Leipzig, Germany, Helmholtz Centre for Environmental Research-UFZ. Available from http://www.ufz.de/lserd.

Valsaraj, K.T. (1995) *Elements of Environmental Engineering: Thermodynamics and Kinetics.* CRC Press, Boca Raton, FL.

Van Pul, W.A.J., de Leeuw, F.A.A.M., van Jaarsveld, J.A., van der Gagg, M.A., and Sliggers, C. J. (1998) The potential for long-range transboundary atmospheric transport. *Chemosphere* 37:113–141.

Veith, G.D., Defoe, D.L., and Bergstedt, B.V. (1979) Measuring and estimating the bioconcentration factor of chemicals in fish. *J. Fish. Res. Board Can.* 6:1040–1048.

Veith, G.D., Call, D.J., and Brooke, L.T. (1983) Structure-toxicity relationships for the fathead minnow, *Pimephales promelas*: Narcotic industrial chemicals. *Can. J. Fish. Aquat. Sci.* 40:743.

Wania, F. and Mackay, D. (1993) Global fractionation and cold condensation of low volatile organochlorine compounds in polar regions. *Ambio* 22:10–18.

Wania, F. and Mackay, D. (1995). A global distribution model for persistent organic chemicals. *Sci. Tot. Environ.* 160/161:211–232.

Wania, F. and Mackay, D. (1996). Tracking the distribution of persistent organic pollutants. *Environ. Sci. Technol.* 30:390A–396A.

Wania, F., Hoff, J.T., Jia, C.Q., and Mackay, D. (1998) The effects of snow and ice on the environmental behaviour of hydrophobic organic chemicals. *Environ. Pollut.* 102:25–41.

Wania, F. and Mackay, D. (1999a). Global chemical fate of a-hexachlorocyclohexane. 2. Use of a global distribution model for mass balancing, source apportionment, and trend prediction. Environ. Toxicol. Chem. 18:1400–1407.

Wania, F. and Mackay, D. (1999b) The evolution of mass balance models of persistent organic pollutant fate in the environment. *Environ. Pollut.* 100:223–240.

Wania, F., Mackay, D., Li, Y.F., Bidleman, T.F., and Strand, A. (1999) Global chemical fate of a-hexachlorocyclohexane. 1. Modification and evaluation of a global distribution model. *Environ. Toxicol. Chem.* 18:1390–1399.

Webster, E., Mackay, D., and Qiang, K. (1999). Equilibrium lipid partitioning concentrations as a multimedia synoptic indicator of contaminant levels and trends in aquatic ecosystems. *J. Great Lakes Res.* 25:318–329.

Webster update of QWASI

Webster, E., Hubbarde, J., Mackay, D., Swanston, L., and Hodge, A. (2003) Development of Tools to improve Exposure Estimation for Use in Ecological Risk Assessment. Report to Environment Canada K2251-2-004 Report 1: The TaPL3 Upgrade.

Wegmann, F., Cavin, L., MacLeod, M., Scheringer, M., and Hungerbühler, K. (2009) The OECD software tool for screening chemicals for persistence and long-range transport potential. *Environ. Model. Softw.* 24, 228–237.

Welling, P.G. (1986) *Pharmacokinetics: Processes and Mathematics.* ACS Monograph 185. American Chemical Society, Washington, DC.

Wen, Y.H., Kalff, J., and Peters, R.H. (1999) Pharmokinetic modeling in toxicology: A critical perspective. *Environ. Rev.* 7:1–18.

Whelan, M.J., Coulon, F., Hince, G., Rayner, J., McWatters, R., Spedding, T., and Snape, I. (2015) Fate and transport of petroleum hydrocarbons in engineered biopiles in polar regions. *Chemosphere* 131:232–240.

Whitman, W.G. (1923) The two-film theory of gas absorption. *Chem. Metal Eng.* 29:146–150.

Woodfine, D.G., Seth R., Mackay D., and Havas, M. (2000) Simulating the response of metal contaminated lakes to reductions in atmospheric loading using a modified QWASI model. *Chemosphere* 41:1377–1388.

Xie, W.H., Shiu, W.Y., and Mackay, D. (1997) A review of the effect of salts on the solubility of organic compounds in seawater. *Marine Environ. Res.* 44:429–444.

Yalkowsky, S.H. and Banerjee, S. (1992) *Aqueous Solubility: Methods of Estimation for Organic Compounds.* Marcel Dekker, Inc., New York, NY.

Index

Printed in the United States
by Baker & Taylor Publisher Services